THE N

Professor Fritz Vahrenholt is

politician and industrialist. \

Vahrenholt has researched at the Max Planck Institute for Carbon Research at Mühleim. A former Senator and Deputy Environmental Minister for Hamburg, he has served on the Sustainable Advisory Board successively for Chancellors Gerhard Schroeder and Angela Merkel.

Dr habil. Sebastian Lüning holds a doctorate in Geology and Palaeontology. For twenty years he has worked on the refiguring of natural ecological changes in geology. In 2005–2006 he was a visiting professor at the University of Vienna. Since 2007 he has worked as an Africa expert in the oil and gas industry. A reviewer for major geoscience journals, Lüning has been a member of the American Geophysical Union (AGU) since 1991.

Other titles in the INDEPENDENT MINDS series:

THE HOCKEY STICK ILLUSION
A.W. MONTFORD

CLIMATE: THE COUNTER CONSENSUS
PROFESSOR ROBERT CARTER

THE WIND FARM SCAM
DR JOHN ETHERINGTON

CLIMATE: THE GREAT DELUSION
CHRISTIAN GERONDEAU

WHEN WILL THE LIGHTS GO OUT?
DEREK BIRKETT

CHASING RAINBOWS
TIM WORSTALL

CLIMATE CHANGE: NATURAL OR MANMADE?
JOE FONE

The Neglected Sun

The Neglected Sun

Die kalte Sonne

STACEY INTERNATIONAL

128 Kensington Church Street

London W8 4BH

Tel: +44 (0)20 7221 7166; Fax: +44 (0)20 7792 9288

Email: info@stacey-international.co.uk

www.stacey-international.co.uk

ISBN: 978-1-909022-24-9

This edition published in 2013 by Stacey International

Editor: Ruth Willats

Printed in Turkey

CIP Data: A catalogue record for this book is available from the British Library

The Neglected Sun

Why the Sun Precludes Climate Catastrophe

Fritz Vahrenholt

and

Sebastian Lüning

Translated from the German, Die kalte Sonne, *by Pierre and Doris Gosselin*

STACEY
INTERNATIONAL

The greatest challenge facing mankind is the challenge of distinguishing reality from fantasy, truth from propaganda. ... We must daily decide whether the threats we face are real, whether the solutions we are offered will do any good, whether the problems we're told exist are in fact real problems, or non-problems.

Michael Crichton, 2003

Contents

Acknowledgements

We thank Pierre Gosselin for his fruitful input. He not only translated the book into English but provided invaluable content suggestions. Furthermore, on his blog notrickszone.com Pierre regularly features key articles from our German-language blog kaltesonne.de in English, for which we are very grateful.

Preface

The day the book *Die kalte Sonne* was launched in Germany happened to be the coldest of 2012. And 6 February was remarkable for another reason too: on that day Germany's power grid teetered on the brink of collapse. Having decommissioned eight of its older nuclear reactors, the country was no longer able to guarantee its own power supply. Electricity from an old, mothballed, oil-fired power plant in neighbouring Austria and from Czech nuclear power plants had to be fed in to prevent Germany's power supply failing.

In 2011 Chancellor Angela Merkel announced that Germany would implement the *Energiewende* (energy turnaround) in an attempt to replace nuclear and fossil fuel power plants with renewable sources. At the same time, she promised that Germany would no longer need to import electricity and that electricity prices would not go up. Within less than a year this grand declaration proved to be little more than wishful thinking.[1] Today electricity prices in Germany are soaring out of control due to unlimited subsidies given to renewable power, and the German power supply can be secured only through emergency decrees. Power companies also have to keep unprofitable power plants on standby and large power consumers may find their supply cut off in the event of unexpected supply bottlenecks. Within less than a year, Germany has gone from having a power supply that was one of the world's most stable to one that is on the brink of collapse.

---

[1] June 2011, Chancellor Merkel in the Bundestag: 'The energy feed-in cost must not rise beyond that of today. Today it is at 3.5 cents per kilowatt-hour.' In November 2012 the feed-in cost had increased almost 50 per cent.

How did Germany reach this point?

Germany is implementing an energy policy driven by fear. After a catastrophic tsunami on the other side of the globe struck Japan in 2011, causing the Fukushima reactor accident, fear gripped Germany. While other leaders such as Britain's David Cameron and France's Nicolas Sarkozy soberly acknowledged that a tsunami could not be expected in their respective countries and that their reactors were deemed safe, Merkel lost her nerve and promptly shut down eight of Germany's nineteen reactors even though they had been rated as among the safest in the world.

At the same time, the German government made generating 80 per cent of the country's electricity from renewables – wind and solar energy – by 2050 a national priority. Gas, coal and oil would not play a role in the future because Germany's energy policy was being driven by fear of a climate catastrophe.

This fear was being fanned by climate scientists such as Professor Hans-Joachim Schellnhuber of the Potsdam Institute for Climate Impact Research. He had been promoted to the position of chief climate adviser to the German Chancellor. Schellnhuber and his group succeeded in apportioning all the blame for past and future climate change on CO_2 alone. He is on record as saying, 'We ... can show that there is an extremely simple, quasi-linear relation between the global mean temperature and the total amount of CO_2 that will be emitted into the atmosphere over the next four or five decades. The climate system's entire complexity can be boiled down to this simple linear relation' [1].

Politicians simply accept this as true and base energy and social policy on this. The climate scientists who shape public opinion and the IPCC postulate that an uncurbed rise of atmospheric CO_2 concentration will lead to a dramatic temperature increase of 2–6° C.

Spreading fear is poor policy

It is this fear-driven energy policy that has led to millions of tonnes of wheat being converted into biofuel, with some even being imported for that purpose. Our fear-driven energy policy has led to wind parks being erected in the middle of forests, thus destroying the function of the forest. Fear is why half of the world's photovoltaic capacity is installed in mostly overcast Germany, a country with no more sunshine than Alaska. This German energy policy mantra is what we question in our book. The reaction to our book from politicians and media was predictable: indignation, ostracism and marginalization of the issues and the authors.

Just what outrageous facts did we bring to light? We were able to cite hundreds of scientific studies showing that the changes in the sun's activity and oceanic decadal oscillations are responsible for at least half of the recent warming, which means that the contribution of CO_2 is at most half.

Yes, some warming can be traced to anthropogenic and natural sources, but the impact of CO_2 has been wildly exaggerated. A warming of 2–6° C is not to be expected by the end of the twenty-first century; a warming of about 1° C is more likely. Worse still: the sun and ocean decadal oscillations indicate that we are entering a period of modest cooling that will last decades.

Consequently, the earth-burning climate catastrophe, which has long been a creed for many in politics and the media, should be abandoned. That would mean cancelling the annual circuses of 20,000 participants in exotic venues like Doha, Cancun and Durban. The high priests of climate fear would no longer be welcome; political advisers and their huge research budgets would shrink. The much yearned for transformation envisaged by green ideologues, where a centrally controlled energy economy would be put in place and hollow out the nation's industrial base, would disintegrate. The alarming headlines of the globe burning up, which no doubt boost circulation and ratings, would quickly become a thing of the past. Indeed, there was a real threat to alarmism when Germany's most widely circulated daily *Bild* changed tack and ran a series titled 'The CO_2 Lie' just after our book was released.

Everything is fine – just don't voice any criticism!

There were plenty of reasons for the media, scientists and politicians to avoid spreading such realism. Especially active in this respect were the leftist-liberal weekly *Die Zeit* ('Vahrenholt as the front man of a new eco-reactionary movement') [2] and the conservative *Frankfurter Allgemeine Zeitung* ('Obsolete climate claims', 'ridiculous'). On the other hand, the weekly news magazine *Der Spiegel* and other dailies such as *Die Welt* gave the issues examined in our book broad coverage.

The reaction of some climate scientists was particularly harsh, among them Professor Jochem Marotzke, director of the Max Planck Institute for Meteorology, Hamburg. He kept it as simple as possible by claiming that we were not real climate scientists: 'If Vahrenholt studied the IPCC report, then he read a lot but understood little' [3]. Professor Mojib Latif of the Helmholtz Centre for Ocean Sciences, Kiel took it a level higher, claiming that our arguments 'belong in a well-deserved place in the graveyard of Absurdistan' [4]. Universities, academies and other institutes came under pressure to cancel scheduled speaking engagements by the authors.

Changing winds

However, the arguments raised in our book were welcome overseas. Invitations came from the University of Oslo and the London Royal Society. Speaking engagements in Chicago, Vienna and Berne were met with positive resonance. We were even able to present our arguments before the European Parliament in Strasbourg, thanks to an invitation from the European People's Party.

Over the course of the year we were able to gain support even in Germany. The longer that *Die kalte Sonne* stayed on the bestseller list, the more obvious it became that our arguments were very well supported by a growing number of scientific publications, and as a result the more support we got from politicians and the media. Former Chancellor Helmut Schmidt found the reaction of some in the media ill-advised. He invited Fritz Vahrenholt to an hours-long meeting, and even allowed Vahrenholt and Lüning to quote him as follows: 'I find your line of argument plausible.'

Science always progresses

Hardly a week went by that new scientific publications underpinning the fact that CO_2 had been exaggerated did not appear. Contrary to the supposed IPCC consensus that natural climate variability does not play a major role in today's or the past climate, many scientists continue to work on this important subject. New papers documenting the great importance of the inconvenient natural climate drivers are published in international, peer-reviewed journals almost every week. Many disciplines contributing to the climate puzzle are still in the early stages of research and many fundamental questions remain unanswered. Controversial scientific debates are an essential part of science and are taking place today despite all the claims coming from the IPCC that the 'science is settled'. Indeed, nothing could be further from the truth.

Scientists find themselves in a quandary. How are they to deal with the politically incorrect scientific results? Two American researchers, Jackson Davis and Peter Taylor, recently came across something astonishing. While studying an Antarctic ice core covering the past 12,000 years, they identified a total of forty-six strong natural warming events throughout the pre-industrial era. The mean warming rate of these events was approximately 1.2° C per century, more than the 0.7° C warming we have seen since 1900. While the material, methods and analysis of the study were sound and unchallenged, not a single major scientific journal was interested enough to publish these important results. Was it the study's powerful, yet inconvenient conclusion that deterred the journals? Contrary to what is always claimed by IPCC-affiliated scientists, the warming that occurred over the past one and a half centuries is not unprecedented after all. Desperate to share their results with fellow scientists and the public, Davis and Taylor eventually posted their paper on the largest climate discussion blog (www.wattsupwiththat.com) for maximum distribution [5]. Similar pre-industrial warming events over the past 2000 years were also reported from China [6]. While papers from well-connected climate alarmists are routinely published within a few weeks, papers refuting the IPCC often struggle to get into print. Similar problems occur with the media coverage of new scientific climate studies. Interestingly, while new results inconvenient to the IPCC are often ignored, the latest climate scare stories – some paid for by insurance companies with a vested interest – are widely carried by the mainstream media. For example, how many of us have heard that winter temperatures at the Antarctic Ross Sea have significantly cooled over the past 30 years? [7]. Has anyone read that current temperatures on the Antarctic peninsula were at their present level for 7000 out of the last 10,000 years? [8]. One might think that a study documenting that temperatures in southern Italy during the Roman Warm Period were slightly higher than they are today would be a worth reporting [9]. But such reports are largely ignored. Why do so many journalists shy away from spreading good news? The US National Oceanographic Data Center, for example, recently found in a new study that the ocean is not warming up as aggressively as the IPCC had predicted [10]. And the Gulf Stream is remarkably more stable than predicted earlier by the IPCC-affiliated climate scientist Stefan Rahmstorf. In western Europe the supposed record summer heatwave of 2003 was recently downgraded to second place because it turns out a heatwave in 1540 was markedly warmer [11]. If the media are truly after an explosive news story, then they could begin by investigating the dubious temperature 'corrections' that are now being made to the measured data before they are input into official databases. Is it really justifiable that temperatures from the 1930s warm period are routinely corrected downwards while modern values are often inflated?

Here comes the sun

The past is the key to the present and to the future. Data provide us with a picture of pre-industrial, natural climate patterns. They reveal that when the sun was active, temperatures were high; and when the sun was quiet, temperatures were low. This was always the relation during pre-industrial times. That is one of the key findings of this book and is thoroughly documented in Chapter 3. Reconstructions based on ice cores, dripstones, tree rings and ocean or lake sediment cores reveal that temperature history was characterized by significant temperature changes of more than 1° C. Warm and cold phases alternated according to thousand-year cycles. Examples include the Minoan Warm Period three thousand years ago and the Roman Warm Period two thousand years. During the Medieval Warm Period, around a thousand years ago, Greenland was colonized and grapes suitable for winemaking were cultivated in England. Cold periods prevailed between the warm phases, among them the Little Ice Age which lasted from the fifteenth to the nineteenth centuries. All these temperature fluctuations occurred at a time when atmospheric CO_2 concentration was essentially stable, which means that only natural processes could have been responsible for the historical climate variations. Is it really credible to think that these natural variations came to a halt about 150 years ago?

Let us consider for a moment what the climate since 1850 would have looked like had the natural pattern simply continued. 1850 marks the end of the Little Ice Age, a natural cold period associated with low solar activity. Based solely on the natural pattern, we see that solar activity has increased since 1850, more or less in parallel with an increase in temperature. When we compare this with real-world climate data for the past 160 years, we are surprised to learn that this is exactly what happened. Both the timing and the 1° C warming fit nicely into the natural scheme. The solar magnetic field has more than doubled over the past century. According to the solar physicist Sami Solanki, the past decades have been among the most active in terms of solar activity in the last ten thousand years [12].

Does it really make sense to assume that the sun has almost nothing to do with modern climate warming, as the IPCC claims? Hard-core IPCC supporters such as Rahmstorf deny that solar-driven millennial-scale climate cycles exist and insist it's a cul-de-sac for climate science. But many researchers disagree. Since the first edition of our book appeared in German in early 2012, many studies have been published confirming the great importance of natural climate cycles in the past, and therefore they also must apply to the present and the future [13]. We find that solar-driven millennial-scale cycles have controlled wet and drought phases in the Mediterranean region during Roman times [14]. Along the French Mediterranean coast, storms occurred in millennial cycles in line with solar activity [15]. In Germany too, the sun has driven the climate over the past 10,000 years [16]. Likewise, the temperatures of the Swiss Alpine lakes fluctuated according to the same rhythm [17]. Millennial-scale solar cycles were also found to be responsible for Alpine glacier movements [18]. Similar cycles were found in Finnish Lapland. Interestingly, each successive warm phase over the last 2500 years was colder than the one that preceded it [19], marking a long-term cooling, which is not compatible with the climate catastrophe now being proposed by the IPCC.

Solar-driven millennial-scale climate cycles are also reported in North America by a number of new studies. For example, temperatures along the coast of Cape Hatteras pulsated according to the rhythm of the thousand-year solar cycle [20]. Florida was drier when the sun was weak and wetter when the sun was strong [21]. The climate of British Columbia has been driven by solar activity over the past 11,000 years [22] and in South America the sun regulated the distribution and intensity of the monsoon rains [23].

In China's Taklamakan desert, oases blossomed according to solar millennial-scale cycles [24]. Likewise, temperatures on the Tibetan plateau followed the sun's pattern [25]. The East Asian monsoon too was controlled by solar activity [26]. The currents of the East China Sea varied according to the sun's activity [27]. Even the climate of Lake Baikal fluctuated in accordance to the solar rhythm [28]. Natural climate cycles led to the collapse of the mighty Indus civilization [29]. Finally, the rains in south-east Australia followed the solar pattern [30]. Could all this be a coincidence? All these studies affirm the need to include the sun as a key climate driver. And any models used to project future climate trends need to be tested rigorously by using the climate of the pre-industrial 10,000 years. Only models capable of reproducing the known climate past can be approved for use in future modelling. Unfortunately, not a single climate model used today by the IPCC is able to reproduce the climate cycles of the past.

We find the sun everywhere

Besides long-term millennial-scale solar cycles, researchers have also found evidence that changes in solar activity strongly contribute to climate development on human timescales, that is to say in years and decades. For example, Norwegian studies have revealed that a significant part of the warming in their country has been caused by the sun [31–34]. In Sweden too, climate and solar activity are tightly linked [35]. In neighbouring Finland, solar cycles have been discovered in tree rings [36]. The extent of Baltic Sea ice is now known to be influenced by solar activity [37], as is the ice on the Rhine in central Europe [38]. A massive cold period in central Europe 2800 years ago appears to have been triggered by a weak sun. [39]. The north Atlantic deep water formation was found to be modulated by the sun [40]. The notorious rains in Northern Ireland are affected by changes in solar activity [41]. Winds in Portugal were particularly strong when the sun was weak [42]. Solar activity fluctuations and the North Atlantic Oscillation (NAO) have contributed to Italy's climate over the past 10,000 years [43]. A solar influence can even be detected in Italy's salt marshes [44].

In Asia, monsoon rains have waxed and waned according to the rhythm of the sun over the past 150 years [45]. Rains on the Tibetan plateau ceased whenever the sun weakened [46]. Coral reefs in Japan died during cold phases triggered by low solar activity [47]. A marked solar influence on Japan's climate was also found in other recent studies [48–49]. Wet phases in the Aral Sea were associated with solar high activity phases [50]. The rains in Maine over the past 7000 years have been controlled by the sun [51]. A solar influence on precipitation has now been found for Brazil [52–53]. Solar cycles have even been detected in the water masses of the deep sea [54]. The field of research in solar–climate interaction is more active than ever [55–58]. Unfortunately, the IPCC has chosen to marginalize and underrate this important subject. Therefore, books like this one provide thousands of active researchers in this field with a much-deserved public platform and recognition for their painstaking and fascinating work.

The illusive CO_2 fingerprint in the middle atmosphere

In the past, IPCC-friendly scientists always argued that the enormous climate potency of CO_2 could easily be demonstrated. In the middle atmosphere, namely the stratosphere, temperatures had been cooling, they said. And the reason for this could only be the CO_2 greenhouse effect because warming in the lower atmosphere would always be associated with cooling in the middle atmosphere. Activist scientists like Mojib Latif have used this logic in numerous public lectures. People hear this and have no choice but to believe it because they don't have knowledge or literature to verify the claim.

However, when we take a closer look at this proposed CO_2 'proof', the story quickly falls apart. First, while the temperature in the stratosphere did indeed decline between 1980 and 1995, since then it has been fairly stable. Contrary to Latif's claim, the stratosphere has not cooled at all over the last 15 years. This is not a good start for the alleged CO_2 warming 'proof'. Unfortunately, there is more: the cooling of 1980–95 coincided with the thinning of the ozone layer. Since the mid 1990s, however, the ozone layer has been recovering due to the reduction of chlorofluorocarbons (CFCs) and other substances addressed by the Montreal Protocol. This is precisely when stratospheric cooling stopped. Could temperatures in the middle atmosphere possibly be linked to the ozone concentration rather than to the CO_2 greenhouse effect as Latif claims? Research conducted at Columbia University and the Leibniz Institute of Atmospheric Physics, Kühlungsborn, appears to indicate precisely that. It is indeed mostly ozone that drives temperatures at those atmospheric levels, and not CO_2 [59–60]. And what really drives the ozone concentration in the stratosphere and mesosphere besides the CFCs? A series of papers published in 2010–12 provide the answer [61–65]: it's the sun, stupid! The CO_2 fingerprint in the middle atmosphere has disappeared, but hardly anyone has acknowledged it, especially not the old climate guard of the IPCC.

Extreme views on extreme weather

Major new developments have also occurred in the field of extreme weather since the German edition of our book came out. In March 2012, the IPCC published a special report on extreme weather [66], which stated that there will be no detectable influence on the earth's weather systems by mankind for at least 30 years, and possibly not until the end of the century. If and when mankind's influence becomes apparent, then it may just as likely reduce the number of extreme weather events as increase them [67]. New studies from central Europe confirm that our weather is still well within the range of natural variability [68]. In the Alps, weather extremes have even declined [69]. As discussed in Chapter 5, there is currently no scientific evidence that storms have become more extreme in recent decades.

When a severe drought struck the United States in 2012, many pundits viewed it as a portent of the coming climate catastrophe. While this event was certainly a catastrophe for the areas affected, an individual event like this has little relevance for the long-term climatic drought trend. A study carried out by researchers at Princeton University and the Australian National University, Canberra was published in the science journal Nature in late 2012 [70]. The results are unequivocal: droughts have not increased in frequency over the past 60 years. Another recent study of the Mediterranean found that rainfall today remains within the range of natural variability [71].

Other studies have revealed that the most severe droughts in Sweden and Spain occurred during the seventeenth to nineteenth centuries, during the Little Ice Age [72–74]. What has long been ignored is that marked natural drought-wet cycles operating over timescales of decades, centuries and millennia do exist. Many of these cycles are driven at least in part by changes in solar activity [75]. Studies have documented such cycles all over the world – Norway [76], the Mediterranean [14], the north-eastern United States [51], Mexico [77–78], South America [23, 79–80], the Sahel, [81], Lake Malawi [82], China and East Asia [24, 26, 46, 83–85], the Aral Sea [50] and south-east Australia [30]. A team from the US National Oceanic and Atmospheric Administration (NOAA) found that current climate models are still not able to reproduce regional trends in precipitation [86]. Most notably, the models significantly underestimate natural variability, according to these authors.

River flooding is still within the range of natural variability

Research has also moved forward on the question of whether river floods have already spiralled out of control and beyond the range of natural variability during the current Modern Warm Period, as some IPCC-affiliated players have claimed. The first surprising news was that global precipitation has become less extreme over the past 70 years [87]. Yet studies in the United States and Africa could not detect any statistically significant increase in flooding events [88–89]. Greater damage has more to do with ever more people settling in areas vulnerable to flooding and higher property values. Evidence for a link to anthropogenic global warming has not been found [90]. Prior to the floods of 2011 and 2012 in Australia, the IPCC suggested that droughts would be the greatest environmental threat to the country. Abruptly, the floods were re-interpreted and explained by alarmist activists as ominous signs of an imminent manmade climate catastrophe. However, a subsequent and in-depth scientific analysis revealed that the Australian floods had a natural cause – the La Niña phenomenon, enhanced by the negative phase of the Pacific Decadal

Oscillation (PDO) [91]. Another interesting result comes from the central European Alps, where research has shown that floods were more frequent there during cold rather than warm periods [92–93].

The climate sciences are still in an early and turbulent phase, where new research often exposes previously held scientific beliefs to be misconceptions. It is therefore essential to keep asking critical questions whenever sensational climate claims are made. All too often such concepts have collapsed when subjected to rigorous testing. This book aims to investigate the fundamental facts relevant to the climate catastrophe claims proposed by the IPCC and industries with vested interests, such as the insurance sector. Prepare yourself for an eye-opening journey through a climate science Wild West. You will be surprised to read about scientific distortions that you never would have thought possible in the supposedly enlightened twenty-first century.

References

1. Schellnhuber, H. J. (2009) Bundespressekonferenz Berlin, 23 November, www.youtube.com/watch?v=QECr4kskNZO.
2. Drieschner, F., C. Grefe and C. Tenbrock (2012) Störenfritz des Klimafriedens. *Zeit Online*, 10 February. http://www.zeit.de/2012/07/Klimawandel-Vahrenholt.
3. Marotzke, J. (2012) *Hamburger Abendblatt*, 7 February.
4. Latif, M. (2012) *Financial Times Deutschland*, 15 February; pers. comm., 5 March.
5. Davis, W. J. and P. Taylor (2012) Does the current global warming signal reflect a recurrent natural cycle? *WUWT*. http://wattsupwiththat.com/2012/09/05/is-the-current-global-warming-a-natural-cycle.
6. GE, Q., X. Zhang, Z. Hao and J. Zheng (2011) Rates of temperature change in China during the past 2000 years. *Science China Earth Sciences* 54 (11), 1627–34.
7. Sinclair, K. E., N. A. N. Bertler and T. D. van Ommen (2012) Twentieth-century surface temperature trends in the Western Ross Sea, Antarctica: evidence from a high-resolution ice core. *Journal of Climate* 25 (10), 3629–36.
8. Mulvaney, R., N. J. Abram, R. C. A. Hindmarsh, C. Arrowsmith, L. Fleet, J. Triest, L. C. Sime, O. Alemany and S. Foord (2012) Recent Antarctic Peninsula warming relative to Holocene climate and ice-shelf history. *Nature* 489, 141–4.
9. Chen, L., K. A. F. Zonneveld and G. J. M. Versteegh (2011) Short term climate variability during the 'Roman Classical Period' in the eastern Mediterranean. *Quaternary Science Reviews* 30 (27–8), 3880–91.
10. Levitus, S., J. I. Antonov, T. P. Boyer, O. K. Baranova, H. E. Garcia, R. A. Locarnini, A. V. Mishonov, J. R. Reagan, D. Seidov, E. S. Yarosh and M. M. Zweng (2012) World ocean heat content and thermosteric sea level change (0–2000 m), 1955–2010. *Geophys. Res. Lett.* 39 (10), L10603.

11. Wetter, O. and C. Pfister (2012): An underestimated record breaking event: why summer 1540 was very likely warmer than 2003. *Climate of the Past Discussion* 8, 2695–730.
12. Solanki, S. K., I. G. Usoskin, B. Kromer, M. Schüssler and J. Beer (2004) Unusual activity of the sun during recent decades compared to the previous 11,000 years. *Nature* 431, 1084–7.
13. Steinhilber, F., J. A. Abreu, J. Beer, I. Brunner, M. Christl, H. Fischer, U. Heikkilä, P. W. Kubik, M. Mann, K. G. McCracken, H. Miller, H. Miyahara, H. Oerter and F. Wilhelms (2012) 9,400 years of cosmic radiation and solar activity from ice cores and tree rings. *Proceedings of the National Academy of Sciences* 109 (16), 5967–71.
14. Dermody, B. J., H. J. de Boer, M. F. P. Bierkens, S. L. Weber, M. J. Wassen and S. C. Dekker (2012) A seesaw in Mediterranean precipitation during the Roman Period linked to millennial-scale changes in the North Atlantic. *Climate of the Past* 8, 637–51.
15. Sabatier, P., L. Dezileau, C. Colin, L. Briqueu, F. Bouchette, P. Martinez, G. Siani, O. Raynal and U. Von Grafenstein (2012) 7000 years of paleostorm activity in the NW Mediterranean Sea in response to Holocene climate events. *Quaternary Research* 77 (1), 1–11.
16. Fohlmeister, J., A. Schröder-Ritzrau, D. Scholz, C. Spötl, D. F. C. Riechelmann, M. Mudelsee, A. Wackerbarth, A. Gerdes, S. Riechelmann, A. Immenhauser, D. K. Richter and A. Mangini (2012) Bunker Cave stalagmites: an archive for Central European Holocene climate variability. *Climate of the Past* 8, 1751–64.
17. Niemann, H., A. Stadnitskaia, S. B. Wirth, A. Gilli, F. S. Anselmetti, J. S. Sinninghe Damsté, S. Schouten, E. C. Hopmans and M. F. Lehmann (2012) Bacterial GDGTs in Holocene sediments and catchment soils of a high-alpine lake: application of the MBT/CBT-paleothermometer. *Clim. Past* 8, 889–906.
18. Nussbaumer, S. U., F. Steinhilber, M. Trachsel, P. Breitenmoser, J. Beer, A. Blass, M. Grosjean, A. Hafner, H. Holzhauser, H. Wanner and H. J. Zumbühl (2011) Alpine climate during the Holocene: a comparison between records of glaciers, lake sediments and solar activity. *Journal of Quaternary Science* 26 (7), 703–13.
19. Esper, J., D. C. Frank, M. Timonen, E. Zorita, R. J. S. Wilson, J. Luterbacher, S. Holzkämpe, N. Fischer, S. Wagner, D. Nievergelt, A. Verstege and U. Büntgen (2012) Orbital forcing of tree-ring data. *Nature Climate Change* 2, 862–6.
20. Cléroux, C., M. Debret, E. Cortijo, J.-C. Duplessy, F. Dewilde, J. Reijmer and N. Massei (2012) High-resolution sea surface reconstructions off Cape Hatteras over the last 10 ka. *Paleoceanography* 27 (1), PA1205.
21. Schmidt, M. W., W. A. Weinlein, F. Marcantonio and J. Lynch-Stieglitz (2012) Solar forcing of Florida Straits surface salinity during the early Holocene. *Paleoceanography* 27 (3), PA3204.
22. Gavin, D. G., A. C. G. Henderson, K. S. Westover, S. C. Fritz, I. R. Walker, M. J. Leng and F. S. Hu (2011) Abrupt Holocene climate change and potential response to solar forcing in western Canada. *Quaternary Science Reviews* 30 (9–10), 1243–55.
23. Vuille, M., S. J. Burns, B. L. Taylor, F. W. Cruz, B. W. Bird, M. B. Abbott, L. C. Kanner, H. Cheng and V. F. Novello (2012) A review of the South American monsoon history as recorded in stable isotopic proxies over the past two millennia. *Climate of the Past* 8, 1309–21.

24. Zhao, K., X. Li, J. Dodson, P. Atahan, X. Zhou and F. Bertuch (2012) Climatic variations over the last 4000 yr BP in the western margin of the Tarim Basin, Xinjiang, reconstructed from pollen data. *Palaeogeography, Palaeoclimatology, Palaeoecology* 321–322, 16–23.
25. LIU Yu, C. Q., SONG HuiMing, AN ZhiSheng, Hans W. Linderholm (2011) Amplitudes, rates, periodicities and causes of temperature variations in the past 2485 years and future trends over the central-eastern Tibetan Plateau. *Chinese Science Bulletin* 56 (28–9), 2986–94.
26. Yu, F., Y. Zong, J. M. Lloyd, M. J. Leng, A. D. Switzer, W. W.-S. Yim and G. Huang (2012) Mid Holocene variability of the East Asian monsoon based on bulk organic δ13C and C/N records from the Pearl River estuary, southern China. *The Holocene* 22 (6), 705–15.
27. Wu, W., W. Tan, L. Zhou, H. Yang and Y. Xu (2012) Sea surface temperature variability in southern Okinawa Trough during last 2700 years. *Geophys. Res. Lett.* 39 (14), L14705.
28. Murakami, T., T. Takamatsu, N. Katsuta, M. Takano, K. Yamamoto, Y. Takahashi, T. Nakamura and T. Kawai (2012) Centennial- to millennial-scale climate shifts in continental interior Asia repeated between warm–dry and cool–wet conditions during the last three interglacial states: evidence from uranium and biogenic silica in the sediment of Lake Baikal, southeast Siberia. *Quaternary Science Reviews* 52, 49–59.
29. Giosan, L., P. D. Clift, M. G. Macklin, D. Q. Fuller, S. Constantinescu, J. A. Durcan, T. Stevens, G. A. T. Duller, A. R. Tabrez, K. Gangal, R. Adhikari, A. Alizai, F. Filip, S. VanLaningham and J. P. M. Syvitski (2012) Fluvial landscapes of the Harappan civilization. *Proceedings of the National Academy of Sciences.*
30. Kemp, J., L. C. Radke, J. Olley, S. Juggins and P. De Deckker (2012) Holocene lake salinity changes in the Wimmera, southeastern Australia, provide evidence for millennial-scale climate variability. *Quaternary Research* 77 (1), 65–76.
31. Solheim, J.-E., K. Stordahl and O. Humlum (2011) Solar activity and Svalbard temperatures. *Advances in Meteorology*, 8.
32. Solheim, J.-E., K. Stordahl and O. Humlum (2012) The long sunspot cycle 23 predicts a significant temperature decrease in cycle 24. *Journal of Atmospheric and Solar–Terrestrial Physics* 80, 267–84.
33. Humlum, O., J.-E. Solheim and K. Stordahl (2011) Identifying natural contributions to late Holocene climate change. *Global and Planetary Change* 79 (1–2), 145–56.
34. Vorren, K.-D., C. E. Jensen and E. Nilssen (2012) Climate changes during the last c. 7500 years as recorded by the degree of peat humification in the Lofoten region, Norway. *Boreas* 41 (1), 13–30.
35. Kokfelt, U. and R. Muscheler (2012) Solar forcing of climate during the last millennium recorded in lake sediments from northern Sweden. *The Holocene.*
36. Ogurtsov, M., E. Sonninen, E. Hilasvuori, I. Koudriavtsev, V. Dergachev and H. Jungner (2011) Variations in tree ring stable isotope records from northern Finland and their possible connection to solar activity. *Journal of Atmospheric and Solar-Terrestrial Physics* 73 (2–3), 383–7.
37. Leal-Silva, M. C. and V. M. Velasco Herrera (2012) Solar forcing on the ice winter severity index in the western Baltic region. *Journal of Atmospheric and Solar–Terrestrial Physics* 89, 98–109.

38. Sirocko, F., H. Brunck and S. Pfahl (2012) Solar influence on winter severity in Central Europe. *Geophys. Res. Lett.* 39 (16), L16704.
39. Martin-Puertas, C., K. Matthes, A. Brauer, R. Muscheler, F. Hansen, C. Petrick, A. Aldahan, G. Possnert and B. v. Geel (2012) Regional atmospheric circulation shifts induced by a grand solar minimum. *Nature Geoscience* doi:10.1038/ngeo1460.
40. Morley, A., M. Schulz, Y. Rosenthal, S. Mulitza, A. Paul and C. Rühlemann (2011) Solar modulation of North Atlantic central water formation at multidecadal timescales during the late Holocene. *Earth and Planetary Science Letters* 308 (1–2), 161–71.
41. Swindles, G. T., R. T. Patterson, H. M. Roe and J. M. Galloway (2012) Evaluating periodicities in peat-based climate proxy records. *Quaternary Science Reviews* 41, 94–103.
42. Costas, S., S. Jerez, R. M. Trigo, R. Goble and L. Rebêlo (2012) Sand invasion along the Portuguese coast forced by westerly shifts during cold climate events. *Quaternary Science Reviews* 42, 15–28.
43. Scholz, D., S. Frisia, A. Borsato, C. Spötl, J. Fohlmeister, M. Mudelsee, R. Miorandi and A. Mangini (2012) Holocene climate variability in north-eastern Italy: potential influence of the NAO and solar activity recorded by speleothem data. *Climate of the Past* 8, 1367–83.
44. Di Rita, F. A possible solar pacemaker for Holocene fluctuations of a saltmarsh in southern Italy. *Quaternary International.*
45. van Loon, H. and G. A. Meehl (2012) The Indian summer monsoon during peaks in the 11-year sunspot cycle. *Geophys. Res. Lett.* 39 (13), L13701.
46. Sun, J. and Y. Liu (2012) Tree ring based precipitation reconstruction in the south slope of the middle Qilian Mountains, northeastern Tibetan plateau, over the last millennium. *J. Geophys. Res.* 117 (D8), D08108.
47. Hamanaka, N., H. Kan, Y. Yokoyama, T. Okamoto, Y. Nakashima and T. Kawana (2012) Disturbances with hiatuses in high-latitude coral reef growth during the Holocene: correlation with millennial-scale global climate change. *Global and Planetary Change* 80–1, 21–35.
48. Yamaguchi, Y. T., Y. Yokoyama, H. Miyahara, K. Sho and T. Nakatsuka (2010) Synchronized Northern Hemisphere climate change and solar magnetic cycles during the Maunder Minimum. *Proceedings of the National Academy of Sciences* 107 (48), 20697–702.
49. Muraki, Y., K. Masuda, K. Nagaya, K. Wada and H. Miyahara (2011) Solar variability and width of tree ring. *Astrophys. Space Sci. Trans.* 7, 395–401.
50. Huang, X., H. Oberhänsli, H. von Suchodoletz and P. Sorrel (2011) Dust deposition in the Aral Sea: implications for changes in atmospheric circulation in Central Asia during the past 2000 years. *Quaternary Science Reviews* 30 (25–6), 3661–74.
51. Nichols, J. E. and Y. Huang (2012) Hydroclimate of the northeastern United States is highly sensitive to solar forcing. *Geophys. Res. Lett.* 39 (4), L04707.
52. Gusev, A. A. and I. M. Martin (2012) Possible evidence of the resonant influence of solar forcing on the climate system. *Journal of Atmospheric and Solar–Terrestrial Physics* 80, 173–8.
53. Rampelotto, P. H., N. R. Rigozo, M. B. da Rosa, A. Prestes, E. Frigo, M. P. Souza Echer and D. J. R. Nordemann (2012) Variability of rainfall and temperature (1912–2008) parameters measured from Santa Maria (29°41′S, 53°48′W) and

their connections with ENSO and solar activity. *Journal of Atmospheric and Solar-Terrestrial Physics* 77, 152–60.

54. Seidenglanz, A., M. Prange, V. Varma and M. Schulz (2012) Ocean temperature response to idealized Gleissberg and de Vries solar cycles in a comprehensive climate model. *Geophys. Res. Lett.* 39 (22), L22602.
55. Dudok de Wit, T. and J. Watermann (2010) Solar forcing of the terrestrial atmosphere. *Comptes Rendus Geoscience* 342 (4–5), 259–72.
56. Stauning, P. (2011) Solar activity–climate relations: A different approach. *Journal of Atmospheric and Solar-Terrestrial Physics* 73 (13), 1999–2012.
57. Raspopov, O. M., V. A. Dergachev, M. G. Ogurtsov, T. Kolström, H. Jungner and P. B. Dmitriev (2011) Variations in climate parameters at time intervals from hundreds to tens of millions of years in the past and its relation to solar activity. *Journal of Atmospheric and Solar-Terrestrial Physics* 73 (2–3), 388–99.
58. Kern, A. K., M. Harzhauser, W. E. Piller, O. Mandic and A. Soliman (2012) Strong evidence for the influence of solar cycles on a late Miocene lake system revealed by biotic and abiotic proxies. *Palaeogeography, Palaeoclimatology, Palaeoecology* 329–30, 124–36.
59. Polvani, L. M. and S. Solomon (2012) The signature of ozone depletion on tropical temperature trends, as revealed by their seasonal cycle in model integrations with single forcings. *J. Geophys. Res.* 117 (D17), D17102.
60. Berger, U. and F. J. Lübken (2011) Mesospheric temperature trends at mid latitudes in summer. *Geophys. Res. Lett.* 38 (22), L22804.
61. Keckhut, P., A. Hauchecorne, T. Kerzenmacher and G. Angot (2012) Modes of variability of the vertical temperature profile of the middle atmosphere at mid latitude: Similarities with solar forcing. *Journal of Atmospheric and Solar-Terrestrial Physics* 75–76, 92–7.
62. Fadnavis, S., G. Beig and T. Chakraborti (2012) Decadal solar signal in ozone and temperature through the mesosphere of Northern tropics. *Journal of Atmospheric and Solar-Terrestrial Physics* 78–9, 2–7.
63. Dall'Amico, M., L. J. Gray, K. H. Rosenlof, A. A. Scaife, K. P. Shine and P. A. Stott (2010) Stratospheric temperature trends: impact of ozone variability and the QBO. *Climate Dynamics* 34 (2–3), 381–98.
64. Oberländer, S., U. Langematz, K. Matthes, M. Kunze, A. Kubin, J. Harder, N. A. Krivova, S. K. Solanki, J. Pagaran and M. Weber (2012) The influence of spectral solar irradiance data on stratospheric heating rates during the 11-year solar cycle. *Geophys. Res. Lett.* 39 (1), L01801.
65. Smyshlyaev, S. P., V. Y. Galin, E. M. Atlaskin and P. A. Blakitnaya (2010) Simulation of the indirect impact that the 11-year solar cycle has on the gas composition of the atmosphere. *Izvestiya, Atmospheric and Oceanic Physics* 46 (5), 623–34.
66. IPCC (2012) Managing the risks of extreme events and disasters to advance climate change and adaptation. *Special Report of the Intergovernmental Panel on Climate Change.*
67. GWPF (2012) Natural Variability to Dominate Weather Events Over Coming 20–30 Years. http://wattsupwiththat.com/2011/11/18/the-gwpf-responds-to-new-ipcc-report.
68. Büntgen, U., R. Brázdil, K.-U. Heussner, J. Hofmann, R. Kontic, T. Kyncl, C. Pfister, K. Chromá and W. Tegel (2011) Combined dendro-documentary evidence of Central European hydroclimatic springtime extremes over the last millennium. *Quaternary Science Reviews* 30 (27–8), 3947–59.

69. Böhm, R. (2012) Changes of regional climate variability in Central Europe during the past 250 years. In F. Prodi and A. Sutera (eds.), Focus Point on Earth's Climate as a Problem in Physics. *The European Physical Journal Plus.*
70. Sheffield, J., E. F. Wood and M. L. Roderick (2012) Little change in global drought over the past 60 years. *Nature* 491, 435–8.
71. Camuffo, D., C. Bertolin, N. Diodato, C. Cocheo, M. Barriendos, F. Dominguez-Castro, E. Garnier, M. J. Alcoforado and M. F. Nunes (2012) Western Mediterranean precipitation over the last 300 years from instrumental observations. *Climate Change doi*: 10.1007/s10584-012-0539-9.
72. Seftigen, K., H. W. Linderholm, I. Drobyshev and M. Niklasson (2012): Reconstructed drought variability in southeastern Sweden since the 1650s. *International Journal of Climatology.*
73. Martin-Puertas, C., I. Dorado-Linan, A. Brauer, E. Zorita, B. L. Valero-Garcés and E. Gutierrez (2011) Hydrological evidence for a North Atlantic oscillation during the Little Ice Age outside its range observed since 1850. *Clim. Past Discuss.* 7, 4149–71.
74. Dominguez-Castro, F., P. Ribera, R. Garcia-Herrera, J. M. Vaquero, M. Barriendos, J. M. Cuadrat and J. M. Moreno (2012): Assessing extreme droughts in Spain during 1750–1850 from rogation ceremonies. *Climate of the Past* 8, 705–22.
75. Hajian, S. and M. S. Movahed (2010) Multifractal detrended cross-correlation analysis of sunspot numbers and river flow fluctuations. *Physica A: Statistical Mechanics and its Applications* 389 (21), 4942–57.
76. Paetzel, M. and T. Dale (2010) Climate proxies for recent fjord sediments in the inner Sognefjord region, western Norway. *Geological Society, London, Special Publications* 344 (1), 271–88.
77. Metcalfe, S. E., M. D. Jones, S. J. Davies, A. Noren and A. MacKenzie (2010) Climate variability over the last two millennia in the North American Monsoon region, recorded in laminated lake sediments from Laguna de Juanacatlán, Mexico. *The Holocene* 20 (8), 1195–206.
78. Lachniet, M. S., J. P. Bernal, Y. Asmerom, V. Polyak and D. Piperno (2012) A 2400 year Mesoamerican rainfall reconstruction links climate and cultural change. *Geology* 40 (3), 259–62.
79. Morales, M. S., D. A. Christie, R. Villalba, J. Argollo, J. Pacajes, J. S. Silva, C. A. Alvarez, J. C. Llancabure and C. C. Soliz Gamboa (2012) Precipitation changes in the South American Altiplano since AD 1300 reconstructed by tree-rings. *Climate of the Past* 8, 653–66.
80. Strikis, N. M., F. W. Cruz, H. Cheng, I. Karmann, R. L. Edwards, M. Vuille, X. Wang, M. S. de Paula, V. F. Novello and A. S. Auler (2011) Abrupt variations in South American monsoon rainfall during the Holocene based on a speleothem record from central-eastern Brazil. *Geology* 39 (11), 1075–8.
81. Seaquist, J. W., T. Hickler, L. Eklundh, J. Ardö and B. W. Heumann (2009) Disentangling the effects of climate and people on Sahel vegetation dynamics. *Biogeosciences* 6, 469–77.
82. Morgan, A. and M. Kalk (1970): Seasonal changes in the waters of Lake Chilwa (Malawi) in a drying phase, 1966–68. *Hydrobiologia* 36 (1), 81–103.
83. Zhao, J., Y.-B. Han and Z.-A. Li (2004) The effect of solar activity on the annual precipitation in the Beijing area. *Chin. J. Astron. Astrophys.* 4 (2), 189–97.

84. Zhou, Y. and G. Ren (2011) Change in extreme temperature event frequency over mainland China, 1961–2008. *Climate Research* 50 (2–3), 125–39.
85. Yamamoto, N., A. Kitamura, T. Irino, T. Kase and S.-i. Ohashi (2010) Climatic and hydrologic variability in the East China Sea during the last 7000 years based on oxygen isotope records of the submarine cavernicolous micro-bivalve Carditella iejimensis. *Global and Planetary Change* 72 (3), 131–40.
86. Hoerling, M., J. Eischeid and J. Perlwitz (2009) Regional precipitation trends: distinguishing natural variability from anthropogenic forcing. *Journal of Climate* 23 (8), 2131–45.
87. Sun, F., M. L. Roderick and G. D. Farquhar (2012) Changes in the variability of global land precipitation. *Geophys. Res. Lett.* 39 (19), L19402.
88. Hirsch, R. M. and K. R. Ryberg (2011) Has the magnitude of floods across the USA changed with global CO_2 levels? *Hydrological Sciences Journal* 57 (1), 1–9.
89. Di Baldassarre, G., A. Montanari, H. Lins, D. Koutsoyiannis, L. Brandimarte and G. Blöschl (2010) Flood fatalities in Africa: from diagnosis to mitigation. *Geophysical Research Letters* 37, 1–5.
90. Villarini, G. and J. A. Smith (2010) Flood peak distributions for the eastern United States. *Water Resources Research* 46, 1–17.
91. Cai, W. and P. van Rensch (2012) The 2011 southeast Queensland extreme summer rainfall: a confirmation of a negative Pacific Decadal Oscillation phase? *Geophys. Res. Lett.* 39 (8), L08702.
92. Swierczynski, T., A. Brauer, S. Lauterbach, C. Martín-Puertas, P. Dulski, U. von Grafenstein and C. Rohr (2012) A 1600 yr seasonally resolved record of decadal-scale flood variability from the Austrian Pre-Alps. *Geology* doi: 10.1130/G33493.1.
93. Giguet-Covex, C., F. Arnaud, D. Enters, J. Poulenard, L. Millet, P. Francus, F. David, P.-J. Rey, B. Wilhelm and J.-J. Delannoy (2012) Frequency and intensity of high-altitude floods over the last 3.5 ka in northwestern French Alps (Lake Anterne). *Quaternary Research* 77 (1), 12–22.

1. It's the sun, stupid!

For many of us the term 'climate change' stirs strong emotions. The string of catastrophe reports coming from the media seems endless. Every month we hear or see news of another disturbing climate record. Carbon dioxide produced by mankind is dramatically changing the climate, we are told. Unprecedented temperature extremes, storms, floods, widespread death and a slew of other horrors are said to be imminent. If we fail to apply the emergency brake now, and hard, then the climate will be irreparably damaged and there will be little hope of averting the approaching cataclysm. In just a few more years it may be too late, we are constantly warned. We are also told that the measures proposed for averting disaster are costly, very costly, but that the anticipated damage from climate change will be even more expensive, so there is no alternative but to act quickly and decisively. Politicians have come under pressure. Based on a chorus of dire warnings, laws for changing the direction society is taking have been enacted and billion-dollar decisions have been made.

There's no question about the general thrust. Eventually, we are going to have to utilize energy more efficiently and we shall need new technologies to replace much of our finite supplies of oil, gas and coal. Renewables will become one of the main pillars of our energy supply. However, the crucial question remains: How much time do we really have to carry out a comprehensive transformation of society?

The key to answering this question can be found in the climate sciences. It boils down to solving the problem of determining what share of the observed climate change has been truly caused by human activity and how much is due to natural factors. Meanwhile, the world seems to have split into two camps: those who are convinced mankind is significantly changing the climate through emissions of industrial CO_2 and those who see natural fluctuations at work. In the heat of the debate, the fact that nature is rarely a black-and-white picture is often lost. In reality, there are many indications that show our sun plays a more important role than CO_2 in the climate and that at times they may be mutually enhancing, and at times can offset each other.

Very different reasons compelled us to examine the sun and other natural events as the possible triggers for climate fluctuations. Sebastian Lüning is a geoscientist who has spent almost 20 years of his working life studying the climate and the earth's history. Time and again he has asked himself: Why is it that natural forces were able to dominate climate events in the past, but today they are believed to have become practically impotent? Is this a realistic assumption?

In December 2009 Fritz Vahrenholt, an energy expert, was asked by the UN Intergovernmental Panel on Climate Change (IPCC) to review the draft of a renewable energy report. I (Vahrenholt) found 293 errors and deficiencies in the 1000-page report, and then found at an IPCC meeting of experts held in Washington on 1 February 2010 that my remarks met no objection. So I asked myself: Could it be that a similar superficial and flawed approach had been taken to the topic of climate change? I am not a climate scientist, but I do have a comprehensive, in-depth knowledge of the renewable energy sector. Up to that point I had trusted the IPCC's pronouncements on climate protection. Taking such an unscientific approach to the main issue of climate as had been taken to the report on renewable energies would have been absolutely unthinkable to me. Up to that point, I had trusted all IPCC reports and never doubted the recommendations based on them.

But signs of deficits and deficiencies with the consensus finding process of the IPCC climate report started piling up. First, the warning that Himalayan glaciers would melt completely by 2035 was an alarming statement in the 2007 IPCC report. However, that claim had never been confirmed by studies from the Indian Ministry of Environment. IPCC Chairman Rajendra Pachauri initially called the results 'voodoo science', before lamely admitting 2 years later that the 2035 glacier melting claim originated from a telephone interview with a scientist, Syed Hasnain, who said the statement was intended as pure speculation. This telephone interview was quoted by the World Wildlife Fund (WWF) and the claim found its way into the 2007 report. Pachauri finally expressed his regret over the blunder in January 2010.

My uncertainty only grew after the Climategate scandal, when thousands of emails were made public and gave the impression that a crucial temperature data series from the University of East Anglia Climate Research Unit (CRU) had been changed to depict a growing warming trend. The CRU data series had been used prominently by the IPCC. CRU's head, Phil Jones, denied every irregularity, but resigned during the ensuing investigation. Subsequent investigations confirmed that infractions of the scientific obligation to make data available to other scientists had been committed, but threw out the charges of manipulation. My suspicions were aroused by all of these incidents. This left me with no choice but to take a much deeper look into climate science itself, and especially examine alternative scientific views that had failed to find their way into the official reports.

The second factor that compelled me to take a closer look at the sun and other natural impacts on climate came from within my own company. In early 2008 I had been appointed managing director of RWE Innogy. For decades RWE had failed to invest in renewable energy. As Europe's fifth largest power provider, the lion's share of RWE's power had been produced by burning lignite and anthracite coal. With the European emissions trading certificates looming, it was clear that from 2013, unless it reduced its CO_2 emissions, RWE would have to cough up billions of euros over the long term. The new chairman, Jürgen Grossmann, decided to cut the company's CO_2 emissions by implementing an ambitious investment programme in renewable energies, in addition to replacing older power plants with newer, more efficient coal- and gas-fired plants, and extending the operating lifetimes of its nuclear plants. From that point on approximately 1.2 billion euros a year were invested in wind, biomass and hydroelectric power plants. A respectable portfolio resulted after just 3 years: over 2300 megawatts of renewable power capacity is now online.

However, much to our surprise, the outputs from the wind power plants ended up falling far short of our expectations because the wind during the winters simply failed to materialize. In 2009 winds were down 10 per cent, in 2010 they were down a whopping 20 per cent, and in 2011 about 10 per cent. We discovered that across all the northern European countries (Britain, the Netherlands and Poland), the wind had simply taken a break. For years I had been chairman of a wind power company, REpower Systems AG, and so I was familiar with the unpredictability of the wind. However, such a widespread, multi-year fluctuation was completely unprecedented. Was this the first sign of climate change, brought about by anthropogenic global warming? Were the wind patterns of Europe changing permanently? This had to be investigated because we intended to invest up to another 5 billion euros in onshore and offshore wind parks over the next 5 years. In fact, we were planning to become one of the largest offshore investors in the North Sea, and so we had to be certain that the turbine-driving winds would not peter out. This scenario was nowhere to be found in the official IPCC statements.

One day I happened by chance to come across a paper by Michael Lockwood on the connection between cold winters and solar radiation [1]. I was completely absorbed! We always knew that whenever we had a stubborn system of cold easterly winds in the winter, the wind park yields would drop dramatically. On the other hand, strong westerly winds blowing in from the Atlantic provided enough energy to power the wind parks to near full capacity.

In Europe, whether westerly winds or easterly winds prevail depends in large part on the atmospheric pressure difference between Greenland and the Azores. This is known as the North Atlantic Oscillation (NAO). A positive NAO means there is a large difference between both pressure systems (a powerful Icelandic low and a powerful Azores high), a negative NAO means there is a weak difference (a weak Icelandic low and a weak Azores high). With a negative NAO, the powerful westerly winds are driven to the south and the weaker Siberian-influenced easterly wind systems make their way across northern Europe more frequently, thus making winters there colder and less windy (Figure 1.1).

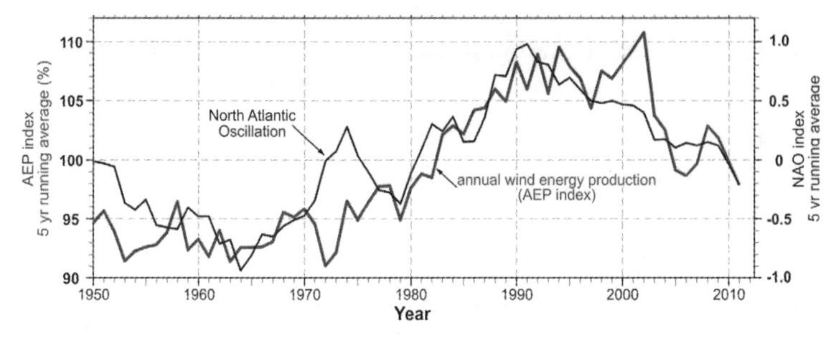

Figure 1.1 Annual energy production in Germany is closely coupled with the North Atlantic Oscillation (NAO). The NAO is a naturally occurring climate oscillation which appears to be influenced by solar activity.

But what is it that drives the NAO? Lockwood provided the decisive clue to solving this riddle. He linked the NAO and the British winter temperatures to solar activity. He determined that solar activity had strongly diminished at the end of the last decade and was able establish a statistically strong correlation between solar activity, the NAO and how cold British winters were [1]. I was stunned by his conclusion: despite global warming, Britain and Europe would have to reckon with cold winters in the near future [1–2]. The physical processes involved have meanwhile been successfully simulated in climate models [3–4].

In Lockwood's paper I read for the first time how the 11-year solar cycle had an impact on our weather and climate. I turned to sunspots next and came across more surprising relationships between solar activity and climatic change, both in the earth's history and mankind's recent history. I found relationships the IPCC had not reported – for example, that there is a 210-year (Suess/de Vries) cycle, an 87-year (Gleissberg) cycle and an 11-year (Schwabe) cycle with which solar activity oscillated. I was also surprised to learn that there is a scientific consensus on the fact that these cycles had an impact on climate development in the past – along with volcanic events and the 100,000-year Milankovitch cycles, which triggered the huge ices ages and warm interglacials.

The more I delved into the literature, the more obvious became the discrepancy between my knowledge and what I had shown as the 'hockey stick' in my presentations over the years. The hockey stick is a temperature reconstruction that depicts almost 900 years of relatively subdued temperature change from AD 1000 to 1900, followed by a sharp warming over the last century (Figure 1.2). During my time as an environmental senator, as a manager for Shell for renewable energies, as chairman of REpower Systems and as RWE Innogy CEO, I demonstrated the exceptional features of the warming since the middle of the twentieth century in hundreds of presentations, speeches and conferences.

I used Michael Mann's 'hockey stick' even though I should have known that relatively warm eleventh-century Greenland was not called 'green' for no reason and that the Little Ice Age depicted

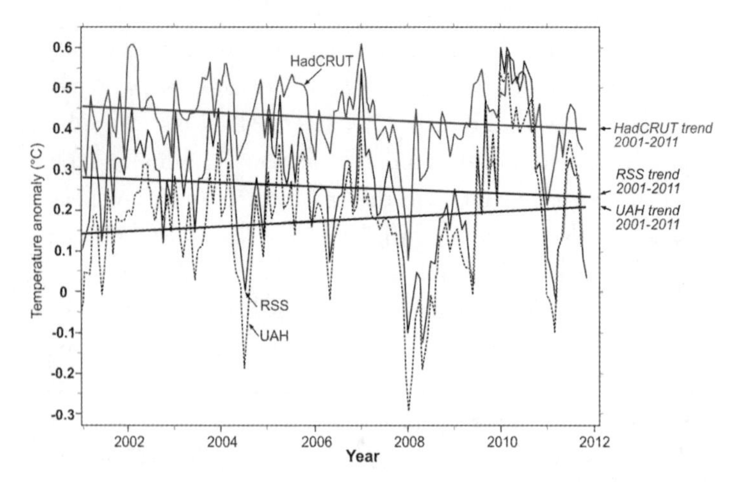

Figure 1.2 Temperature development and trends of the last ten years as illustrated by three established global temperature data sets (surface stations and satellite measurements). Global warming has essentially stopped since the year 2000. Temperatures have fluctuated about a stable plateau.

in Pieter Bruegel's paintings were familiar. In the meantime, a series of studies has been published that show the Medieval Warm Period around the year AD 1000 had a similar temperature level to today's and that the Little Ice Age of the sixteenth century was about 1° C cooler than today [5].

The geologist and, at the time, my RWE colleague Sebastian Lüning referred me to two books: *The Hockey Stick Illusion* by Andrew Montford [6] and *The Chilling Stars* by Henrik Svensmark and Nigel Calder [7], which I tore through over a matter of days. In Montford's book I read the following by David Deming, who received this in an email from an IPCC scientist: 'We have to get rid of the medieval warmth.' As I went through the ruses used to create the hockey stick, which feigned an unprecedented global temperature increase during recent decades, I was upset. I felt the wool had been pulled over my eyes.

To initiate a broad discussion on this obviously flawed procedure used by the IPCC, I wrote an essay that was published in the German daily *Die Welt*. It described the attempt by IPCC climate scientists to trivialize the natural variability of the climate in order to be able to make climate gas CO_2 the single determining effect on our future climate. This simplification was obviously necessary to give the political demand of radically changing the global energy supply badly needed momentum.

The reaction to my piece from traditional climate scientists was frightening. The paper, titled *The Cold Sun*, was characterized by Stefan Rahmstorf, lead scientist at the Potsdam Institute for Climate Impact Research and co-author of the IPCC Fourth Assessment Report, as 'an extraordinary example of twisting scientific facts'. I had pointed out that the warming had ceased in 1998, yet Rahmstorf called this a distortion, claiming that all established global climate data sets showed a rising trend [8]. Is that true? Let's put it to the test by looking at temperatures over the last 10 years. The result: two important data sets (HadCRUT and RSS) show a cooling trend (Figure 1.2). The rise in other data sets is minimal when compared to the strong warming phase of 1977–2000.

The confrontation served to initiate one thing: that we take a deeper look into natural climate change. Together with Lüning I became more intensely involved with the natural causes of climate fluctuations and tracked down the real extent of the CO_2 threat. Manfred Bissinger, a leading name in German journalism, encouraged us to go one step further and write a synthesis about it. This book is the result. It reports on the astonishing findings that stem from our research and our discussions with palaeontologists, astrophysicists, solar scientists, oceanographers and theoretical physicists. For more than a year we talked to many scientists of varying opinions from all over the world and exchanged or requested information about new research results.[2]

Our conclusions? At a sociopolitical level, they are highly explosive. There is no question that CO_2, methane and other climate gases have a limited warming effect on our climate. But there is also no doubt that a large part of the warming measured so far can be traced back to natural causes, with the sun having the most powerful impact on our climate.

Currently, the sun is in the process of switching to a longer-term phase of weak activity [9–10] and as a result we can expect a cooling period over the next decades. This is why we call it the 'cold sun'. Over the coming decades the cold sun will give us the time we need to put the energy supply system on a sustainable basis without putting our civilization's prosperity in jeopardy by implementing irrational, knee-jerk measures. Yes, of course it is necessary to free ourselves of our long-term dependency on fossil fuels for a number of reasons. We have to research new technologies for a sustainable energy supply, and we can develop renewable energies so that they

---

[2] Dr David Archibald, Dr Raimund Brunner, Professor Joachim Curtius, Professor Don Easterbrook, Dr Martin Enghoff, Professor Klaus Hasselmann, Professor Reinhard Hüttl, Dr Natalie A. Krivova, Dr Ben Laken, Professor Mojib Latif, Dr Rainer Link, Professor Horst-Joachim Lüdecke, Professor Jochem Marotzke, Professor Ullrich Müller, Professor Ron Prin, Dr John Reilly, Professor Nicola Scafetta, Professor Nir Shaviv, Professor Fred Singer, Professor Sami K. Solanki, Dr Leif Svalgaard, Professor Henrik Svensmark, Dr Ilya Usoskin, Professor Jan Veizer, Professor Hans von Storch, Professor Werner Weber, Dr Richard Willson.

become an economic alternative to traditional energy sources. This conversion process will take decades. But the cold sun will give us the time to do it in a measured way.

So why have the IPCC, many scientists and politicians been so successful in designating CO_2 as the sole cause of the warming during the second half of the twentieth century? As we shall show, there is a multitude of natural causes – the changing solar irradiative intensity in harmony with the large solar cycles, the solar magnetic field in connection with cosmic radiation and cloud formation, the oscillating Pacific and Atlantic warming and cooling processes, stratospheric ozone, atmospheric water vapour and also poorly understood additional anthropogenic causes such as soot and aerosols. In this book we shall show that the earth's climate depends in large part on multiple natural effects that are complexly interconnected, and are the main cause of the 1977–2000 warming.

Designating a single factor – namely, CO_2 – as the only meaningful climate driver has truly been the dubious crowning achievement of political and scientific communication. The conclusion that one derives is utterly misleading in that it suggests that we 'only' need to reduce manmade CO_2 emissions and then everything will be lovely. That was the simple message that the media and politicians readily accepted. It became the guiding narrative of every privately and publicly held discussion. Because of the simplicity of the equation for describing the climate system, it is easy for everyone to grasp. Unfortunately, it is false, as we shall demonstrate.

Physics and nature simply do not allow themselves to be influenced by such facile messages. It is becoming obvious that since the start of the millennium the CO_2 equation for explaining the climate really doesn't hold water. Despite continued increases in CO_2 emissions, the global temperature has not risen over the last 13 years. The sun has weakened and is now showing us its cold side. The Pacific Decadal Oscillation – which describes the alternation from warm and cold water areas in the northern Pacific Ocean – does the rest.

In the next chapter we summarize the most important facts and interrelations. Assessing the mass of scientific data has led us to conclude that warming in the twentieth century is due only a small extent to CO_2. More importantly, the 2-degree limit, which has been the mantra for all energy-political objectives in the political debate, will in all probability not be exceeded in the current century.

We hope that *The Neglected Sun* will make a contribution to initiating an urgent and much needed political discussion on realigning our climate policy, to opening up scientific research on the natural causes of climate change and to reorienting energy policy towards the politics of energy efficiency. The focus must return to ensuring that massive financial resources are rationally allocated to the truly urgent social, societal and ecological problems of the nine billion people on our planet.

References

1. Lockwood, M., R. G. Harrison, T. Woollings and S. K. Solanki (2010) Are cold winters in Europe associated with low solar activity? *Environ. Res. Lett.* 5, 1–7.
2. Lockwood, M., R. G. Harrison, M. J. Owens, L. Barnard, T. Woollings and F. Steinhilber (2011) The solar influence on the probability of relatively cold UK winters in the future. *Environ. Res. Lett.* 6, 1–11.
3. Ineson, S., A. A. Scaife, J. R. Knight, J. C. Manners, N. J. Dunstone, L. J. Gray and J. D. Haigh (2011) Solar forcing of winter climate variability in the Northern Hemisphere. *Nature Geoscience* 4, 753–7.
4. Matthes, K. (2011) Solar cycle and climate predictions. *Nature Geoscience.*
5. Ljungqvist, F. C. (2010) A new reconstruction of temperature variability in the extra-tropical Northern Hemisphere during the last two millennia. *Geografiska Annaler*: Series A 92 (3), 339–51.
6. Montford, A. W. (2010) *The Hockey Stick Illusion.* Stacey International, London.
7. Svensmark, H. and N. Calder (2007) *The Chilling Stars.* Icon Books, Cambridge.
8. Rahmstorf, S. (2011) Es winkt die RWE-Lobby. http://www.scilogs.de/wblogs/blog/klimalounge/medien-check/2011-02-23/klimawandel-vahrenholt-rwe.
9. Barnard, L., M. Lockwood, M. A. Hapgood, M. J. Owens, C. J. Davis and F. Steinhilber (2011) Predicting space climate change. *Geophysical Research Letters* 38, 1–6.
10. Clilverd, M. A., E. Clarke, T. Ulich, H. Rishbeth and M. J. Jarvis (2006) Predicting solar cycle 24 and beyond. *Space Weather* 4, 1–7.

2. Climate catastrophe deferred – a summary

No topic has dominated national and international politics for as long and as forcefully as concern over global warming. People in Europe and in many countries across the globe fear climate change is going to fundamentally and adversely change their lives and those of future generations. Global warming is making an impact at every political level. Today, hardly any local or state political decision is made or legislation enacted without first examining its impact on climate protection. Whether building a community cycle track, educating our children or passing laws on energy, transportation or social issues, everything first has to be discussed with respect to the fundamental climate question of how to prevent global warming. The main culprit has long been identified: mankind and the uncontrolled emissions of carbon dioxide, which are allegedly transforming the world. It is what we read day after day in our newspapers, what we hear from legislators, political parties and heads of institutions, and what we experience as laws that attempt to drastically curb the rise of CO_2 emissions. It is true that the earth has warmed at least 0.8° C over the last 150 years, with 0.5° C of this having occurred since 1977 alone, as some data sets show. It is also true that mankind has boosted atmospheric CO_2 concentrations over the last 150 years. In around 1750, atmospheric CO_2 concentrations were 0.028 per cent; since then concentrations have climbed to 0.039 per cent due primarily to the burning of fossil fuels. And because fossil fuels are still used in large quantities, global atmospheric CO_2 concentrations are increasing at about 0.0002 per cent a year (2 ppm). IPCC scientists insist that the climate system will be catastrophically damaged as a result, most likely irreversibly. Apparently, we are dealing with two variables that are increasing parallel to each other over time. Because both the temperature and CO_2 curves appear to have long-term similarities, it is tempting to assume that a causal relationship exists. This is precisely what the IPCC does, and based on that, this UN body has developed theoretical models that leave little room for the impact of other important climate factors.

However, it's just not that simple

The simplistic view that manmade gases are driving the temperature would be satisfactory and possibly adequate were it not for another major factor that has gained considerably in strength over the last 150 years, namely solar radiation. The sun's activities are cyclical. The most well known is the 11-year cycle. But equally important are the longer cycles of about 87, 210 and 1000 years. Just how much warming should be attributed to CO_2 and how much to the sun remains one of the most important open questions in the climate debate. But instead of applying themselves to this important question without prejudice, IPCC officials have opted to keep it as simple as possible. As a result the all-important sun has all but been deleted from the climate equation in theoretical calculations, thus disqualifying CO_2's meddlesome rival. According to the IPCC, the main cause of the past and expected future warming is and remains manmade CO_2 in conjunction with other anthropogenic greenhouse gases. Natural processes, such as solar activity, play hardly any role in today's climate, according to the IPCC.

The IPCC's main reason for ignoring the sun is its minimal (0.1 per cent) variation in total solar irradiance measured over the course of the 11-year solar cycles. With such a marginal change in irradiance, one simply cannot expect the global temperature to be significantly affected, the IPCC says. However, it overlooks a crucial detail. In the subdomain of UV irradiation, strong fluctuations do indeed occur, of up to 70 per cent. UV light is converted to heat in the ozone layer and ionosphere and, as a result, leads to a formidable temperature roller-coaster with magnitudes of change reaching several degrees Celsius – all synchronized with the 11-year solar cycle. An explanation for the causal relation between the massive stratospheric fluctuations and the tropospheric climate events below 15 km, however, has not been found so far.

The sun's magnetic field also fluctuates in sync with the 11-year solar cycles, and this has a profound impact of 10–20 per cent on cosmic rays. (Cosmic rays are showers of charged subatomic particles that strike the earth's atmosphere.) The number of particles reaching the earth depends on the strength of the sun's magnetic field. When the magnetic field is strong, it shields the earth. When it is weak, cosmic rays easily penetrate deep into the earth's atmosphere in increased quantities. These cosmic particles seed clouds in the lower atmosphere, which cool the earth. In other words, intense solar activity leads to less cloud cover, resulting in more warming, whereas weak solar activity leads to more cloud cover and so causes cooling. Recent studies have shown that low cloud cover over parts of the earth oscillates in line with solar activity. Low clouds form a huge umbrella which keeps a large part of the solar radiation energy from reaching the earth. And just a few per cent variation in cloud cover results in a change in the earth's irradiative energy budget equal to the projected amount of the warming the IPCC claims that anthropogenic CO_2 causes.

In 2009 a series of remarkable experiments were initiated at the European Organization for Nuclear Research (CERN) in Geneva, in an attempt to explain the interrelation between cloud formation and cosmic rays. A stream of particles from the CERN Proton Synchrotron imitated cosmic rays. The particles were streamed through a 3-metre diameter cylindrical chamber which contained different mixes of atmospheric gases. After the bombardment, scientists looked to see if suspended particles that could serve as seeds for clouds had formed. In the summer of 2011 researchers presented their first spectacular findings. The experiments showed that inside the chamber, up to ten times more suspended particles (aerosols) were present than what is found in a neutral chamber. Next, scientists want to find out whether larger particles can form from these small suspended particles and thus serve as condensation seeds for cloud formation.

There are strong indications that the solar contribution to climate is not only created by UV radiation, but also through the functional chain of solar magnetic field (cosmic rays) clouds. The two mechanisms may act independently, yet work in tandem. It seems that the observed small amount of variability in the total solar irradiation spectrum plays no major role in climatic events. The IPCC argument against the sun is solely based on these small changes and therefore does not hold water. Despite rapidly growing evidence of solar mechanisms having a profound impact on climate, the IPCC still refuses to take either of the two solar amplification processes into account in its climate model scenarios, even though the IPCC's fundamental policy requires every possibility to be examined in a statistically exhaustive manner. The IPCC bases its dismissal of the sun on the claim that the physical processes are simply not understood sufficiently.

Yet, that kind of reasoning has never kept these UN officials from warmly embracing another natural effect that is even less understood, namely water vapour feedback. The direct warming caused by atmospheric CO_2 is an unspectacular 1.1° C for each doubling of CO_2 concentration. This is generally accepted. But CO_2-induced atmospheric warming increases the air's ability to carry water vapour, which in turn may lead to increased net water vapour concentrations. It is well known that water vapour is an even more potent greenhouse gas than CO_2. Thus, according to the IPCC, CO_2's modest warming contribution of 1.1° C is amplified by the additional water vapour in the atmosphere. This is the questionable assumption on which the UN bases its alarming prediction of a temperature rise of up to 4.5° C for each doubling of CO_2.

The decision to deploy this amplification effect is truly astounding, especially when we consider that many experts are still vehemently debating just how great the amplification effect really is. More water vapour can also lead to more cloud cover, which would act to block out the sun and thus have the opposite effect – it would actually slow down CO_2 warming. Moreover, we have to keep in mind that the water vapour amplifier has to be applied to warming of every type independently of their origin, and not just to manmade greenhouse gases. Therefore, any warming due to increased solar activity would also have to be amplified by water vapour just as amplification for CO_2 is postulated by the IPCC. And, lo and behold, it turns out that the specific water content at an elevation of 10 km just happens to have pulsated in sync with the sun's activity since measurements first began 60 years ago.

There is evidence to show that the sun is responsible for at least half of the 0.8° C warming that the earth has experienced since 1850. But assigning a climate contribution to the sun would mean reducing CO_2's contribution and thus diminish the IPCC's threat of warming. If CO_2 is truly such a potent climate driver and the sun also turns out to be a much more powerful climate driver than previously thought, then the warming we have seen since the end of the Little Ice Age would have been much higher than it actually is. Consequently, it is not in the IPCC's interest to place more importance on the sun. Its climate models work only if the sun is excluded as a factor.

The fickle sun

What makes us so sure that the sun, which is dismissed by the IPCC, plays a central role in climate events? That is relatively easy to answer. Geological climate reconstructions exhaustively show that temperatures on earth have followed solar activity for thousands of years. That is not surprising when we consider that 99.98 per cent of the total energy of the world's climate comes from the sun. Would it not make sense to suspect that even small changes in solar energy could have huge impacts?

Fluctuations in solar activity are manifested over a wide range of cycles and have characteristic cycle lengths of between 11 and 2300 years. Especially important in today's context is the 1000-year cycle, which led to unusually high irradiation intensities during the second half of the twentieth century. Over recent decades, the sun has been in one of its most active phases of the last 10,000 years. Solar magnetic field activity more than doubled between 1901 and 1995. Similar irradiation maximums occurred 1000 years ago (the Medieval Warm Period) and 2000 years ago (the Roman Warm Period). In both periods, pronounced climate warming took place. The Roman Warm Period, the Medieval Warm Period and today's Modern Optimum (since 1850) are well documented. Between those warm periods the sun's activity decreased and this led to distinct cold phases – the Vandal Cold Period and the Little Ice Age.

Many studies investigating different oceans and several continents have shown that similar cycles shaped climate events throughout the entire 10,000-year post-ice age period. These cycles are visible in the lower, middle and upper latitudes and include all the climate zones from the Arctic to the Tropics. Temperature fluctuations over the last 10,000 years were at times up to several degrees Celsius, and so on a global average they had a similar or even larger range than the 0.8° C or more of warming that we have seen since the Little Ice Age (1400–1800). So just how plausible is it to ignore the link between the sun and climate?

The IPCC models run into trouble

When it comes to the time-point and magnitude, the temperature increase of today's Modern Warm Phase should come as no surprise. CO_2 has probably enhanced the current natural cycle, but it is obvious that the main climate driver is the sun. Only a small part of the 0.8° C temperature rise seen since 1850 can be attributed to CO_2. There is serious doubt over the IPCC's claimed impact of CO_2 on the climate, especially when it comes to CO_2's alleged water vapour feedback mechanism. There is also evidence indicating that the IPCC models are full of gross errors because they massively underestimate the role of the sun and other natural climate oscillations. Even the most powerful and most expensive computers are useless if they are input with fundamentally flawed assumptions. Therefore, conclusions drawn from the climate models need to be subjected to rigorous review.

Currently, less than half of the temperature increase postulated by the IPCC climate models can be found in real-life temperature curves. The explanation for this discrepancy may be the inflated value placed on CO_2's climate sensitivity and/or the cooling effect of aerosols. Not surprisingly, the IPCC favours the latter and assumes that aerosols have significantly offset CO_2's warming. Yet it warns that in the future emissions from aerosols will decline, at which point the CO_2 effect will be felt in full. Only by applying such assumptions is the IPCC able to come up with its dramatic projected temperature increases.

Fortunately, it is not going to turn out that way. That the temperature has stalled and not warmed over the last 13 years is a little-known fact. Global temperatures levelled off beginning in the year 2000, even though CO_2 emissions and atmospheric CO_2 concentrations continued to rise every year. Yet the IPCC still reported that the temperature would rise 0.2° C per decade. Had the IPCC scientists taken a closer look, they would have spared themselves this embarrassment for two reasons.

First, the earth is in the process of putting the Great Solar Maximum behind. That means a cold sun will leave its mark on the decades and centuries to come. Second, the IPCC should have noticed that the warming of the last 150 years took place over three distinct stages: 1860–80, 1910–40 and 1975–2000. The temperature increases for these three episodes were similar, at about 0.15° C per decade. Between these warming phases global temperatures cooled or stagnated.

What stands out is that the warming and the cooling phases were synchronized with the Pacific Decadal Oscillation (PDO), which is an oscillation within the climate system itself. One complete PDO cycle takes between 40 and 60 years. Every time the PDO enters a negative phase, global warming ceases. The PDO is superimposed over long-term solar activity and CO_2-triggered climate trends, and so raises or lowers the temperature by a few tenths of a degree depending on whether or not the PDO is in its warm or cold phase. Other oceanic oscillations such as the Atlantic Multidecadal Oscillation (AMO) and the North Atlantic Oscillation (NAO) contribute to this process. These well-established interrelationships certainly have to be valid for the future development of the earth's climate. However, the IPCC also refused to integrate this crucial factor in its calculations and, in doing so, displayed an unwarranted confidence in every climate model, which have all since failed. It is truly remarkable that not a single IPCC model predicted the halt to warming we have seen over the last decade. In October 2011 the BEST study, conducted at the University of California, Berkeley, was able to show that global temperatures are influenced by natural oceanic cycles such as the AMO: 'Since 1975, the AMO has shown a gradual but steady rise from –0.35 C to +0.2 C, a change of 0.55 C. During this same time, the land-average temperature has increased about 0.8 C ... Some of the long-term change in the AMO could be driven by natural variability ... In that case the human component of global warming may be somewhat overestimated' [1].

The IPCC climate experts tried to salvage what they could and rushed out the notion that increased cooling was due to sulphur emissions from coal burning in China. However, since 2005, China has equipped most of its coal-burning power plants with desulphurization systems. The same argument was also used to explain the cold phase of the 1970s. And because the last warming stage ended 10 years ago, the same dubious explanation has once again been retrieved from the IPCC's basement of climate tricks. Again it appears not to matter that these sulphur emissions occurred in the Northern Hemisphere and that the bulk of the cooling since 2000 has taken place in the Southern Hemisphere. By playing the sulphur joker (global dimming), the IPCC scientists once again have corrected their climate models downward as the need arose, and then claimed that their climate models represent the past and thus have the future well under control.

In addition to sulphur, there is a host of other airborne particles and droplets (aerosols) in the atmosphere that affect the climate. However, no one knows what impact they really have. This leads us to conclude that the aerosol effect is fraught with uncertainty. The IPCC admits this and even rated scientific knowledge of the aerosol effect as 'moderate to little'. This means that aerosols are the largest uncertainty in the IPCC 2007 report. Some researchers were very eager to play this potent climate modelling wild card. Consequently, the aerosol effect is scattered throughout the models, varying by a magnitude of ten, and hence it is readily employed whenever CO_2 and reality need to be reconciled. In the end it becomes child's play to generate whatever value you wish to apply to CO_2.

Soot is another factor that has given climate scientists a severe headache. Soot particles absorb sunlight and radiate the warmth back into the atmosphere. These dark particles also reduce the reflectivity of snow and ice for sunlight, which leads to more warming. In the IPCC's 2007 report, soot played practically no role as a climate driver. But the latest scientific results have significantly changed this. New findings show that soot should be assigned up to 55 per cent of the warming effect that CO_2 has, and this needs to be taken into account in climate models. But doing so would require CO_2's warming effect to be scaled back. This of course must be avoided at all costs, so a simple bookkeeping trick is deployed. The IPCC scientists exploit the large climatic uncertainties associated with airborne particles and simply increase the cooling effect of other aerosols by the amount that soot is shown to contribute to warming. Hey presto! The offsetting works perfectly: CO_2's impact is undiminished.

But there is good news: soot can be avoided relatively simply. And because it stays in the atmosphere for only a few days or at most weeks, one can rapidly get the suspected warming effect of soot under control. As soot emissions mainly originate from developing countries, technical support by developed countries within the framework of a worldwide UN anti-soot programme would be one of the most effective and financially sensible climate protection measures for limiting anthropogenic warming.

The multi-purpose aerosol joker is only one example of the many subjective degrees of freedom that the IPCC practises and shows how parameters are chosen freely and then built into climate models, which can be adjusted to get the curves you want. These adjustment factors, though difficult to quantify, are nothing more than fudge factors that serve as guarantors for the alleged agreement between models and reality. Consequently, the virtual climate world is beset with problems. For example, there are huge uncertainties when it comes to the interrelationship between clouds and water vapour, the interaction between oceans and the atmosphere, and the flow processes of the ice sheets. Also the cycle of trace gases like CO_2, methane, nitrous oxide and ozone, which cannot be calculated, is inadequately known. Hence, this has to be estimated in models. The natural ocean cycles such as the PDO are largely ignored in simulations. The spatial resolution of the models is, as a rule, limited to a few hundred kilometres. All processes that take place within this level of resolution, such as cloud formation or precipitation, simply cannot be expressly formulated and because these processes cannot be computed individually, the model must be rounded out with guesstimates.

An unexpected *déjà-vu*: melting ice and stormy episodes

If a large part of the warming is indeed due to natural causes, the consequences remain profound. With a temperature increase of about 1 ° C over the last 250 years, glaciers, ice caps and polar sea ice will begin to melt. The currently observed and media-sensationalized melting of the ice is not unusual, but is to be expected. The Greenland and Antarctic ice sheets have certainly taken a hit over the last decade. However, recently it has been determined that ice loss has been dramatically overstated. New studies have shown that the melting is significantly less than what was assumed just a few years ago, and thus the situation on the earth's large ice sheets is far less dramatic than what a few climate protagonists had alarmingly announced. Moreover, the central region of the east Antarctic ice refuses to take part in the ice-melt jamboree and is in fact slowly expanding.

The melting of Arctic sea ice should come as no surprise. In principle it is simply just a repeat of the Medieval Warm Period of the ninth to the fourteenth centuries, which occurred during the last solar activity maximum as part of the 1000-year solar cycle. Back then so much Arctic sea ice melted that the Vikings were able to undertake expeditions to Iceland and Greenland in the ninth century before settling there. Later, around 1420, the Chinese too sailed into the Arctic with a fleet of exploration ships and found hardly any ice.

Sea level development must be examined within the context of climate events over the last 250 years because as ice melts, water flowing into the oceans naturally causes sea levels to rise. During the Little Ice Age, a few hundred years ago, most glaciers and ice sheets gained in mass due to the colder temperatures. Sea level rises during this natural cold period came to a standstill. As the Little Ice Age ended naturally, the glaciers and ice sheets melted and the sea level rise slowly accelerated until the start of the twentieth century, at which point the rate of rise more or less stabilized. Acceleration in the rate of rise cannot currently be detected, as some scientists claim.

The often heard claim that the frequency of storm activity has increased in recent years has to be viewed over the long term. It is becoming clear that global storm activity is strongly dependent on ocean cycles. Rises and falls in long-term storm activity in the Tropics and the middle latitudes are synchronized in large part by natural climate cycles such as the PDO, AMO and NAO. The synchronicity between these oceanic cycles and storm frequency is well established. A periodic intensification of storm activity is therefore not unusual and, moreover, corresponds with the recurring pattern that is shown by historical reconstructions. Unfortunately, the IPCC has failed to address this phenomenon sufficiently.

Time and again glaring errors have been found in IPCC reports over the past couple of years. Of course, errors are human and even the IPCC is expected to make some, especially when it involves the mammoth, data-intensive works that are the IPCC reports. But isn't it odd that all these errors go in one direction only and serve to dramatize what is actually happening? The problems began with a forecast made in the second volume of the 2007 IPCC report, which claimed that 80 per cent of all Himalayan glaciers would have melted by 2035. It wasn't until 2 years after the original publication that the IPCC was forced to recant. Another major lapse occurred on the subject of flood risks. The IPCC wrote that 55 per cent of the Netherlands was below sea level when in fact the figure is 26 per cent. Other gaffes include the IPCC forecast that the anchovies fishery off West Africa would shrink by 50 per cent, that 40 per cent of the Amazon rainforest was threatened by climate change and that agricultural yields in Africa would drop by as much as half. All were exaggerations.

Strategy change in climate forecasting

Our examination of the IPCC argument shows that the IPCC did not want to take fundamental relationships into account, and so climate models from the start never had a chance of depicting reality in all its complexity. The IPCC's biggest error is its logic-defying assumption that natural climate fluctuations play hardly any role and that CO_2 and other anthropogenic greenhouse gases dominate climate events over the long term.

That is how the IPCC epically misjudged the 1977–2000 warming phase. In believing that this temperature increase was almost exclusively due to CO_2 and other anthropogenic greenhouse gases, the warming trend of that 24-year period was hastily declared to be normal and was simplistically extended to the year 2100. Here the fact that the positive warming flank of the 60-year PDO cycles and other natural cycles, including solar activity, were at play in causing the pronounced warming after the cool dip of the 1970s completely escaped the IPCC. They failed to recognize that the rapid warming was a cyclical special case and that the actual long-term warming is considerably less. When the turning point of the PDO was reached in the year 2000, the story ended. The temperature rise came to a halt.

We are by no means claiming that CO_2 has had no impact on today's climate events. However, we are able to show that at least half of the warming of the last 30 years can be attributed to the sun and oceanic oscillations. CO_2 could be responsible for the other half of the warming, but its share may be even less than that.

The era of simplistic IPCC climate prognoses is over. It would have been reasonable if the IPCC in the forthcoming 2013/14 report had abandoned its implausible linear upward projections and brought them in line with observed data. However, report drafts leaks in December 2012 reveal that this revision has not been made. It seems that the IPCC is sticking stubbornly to its original line. Better-quality prognoses would, however, have to incorporate refined temperature curves that depict the up-and-down trends that have their origins in overlapping natural cycles. Carbon dioxide and other climate gases are of course components of the climate equation. However, the alleged dominance of CO_2 will have to be relinquished and CO_2 be seen as an equal partner in a mix of climate-regulating factors that have all played an important role in climate events over millions of years. Man is influential and has changed the earth considerably over the course of time. However, we should not overrate ourselves and believe that we can turn off the natural forces and processes of the earth and solar system at a stroke.

The cold sun

The 11-year solar cycle minimum of 2005–10 was unusually weak. The current cycle which peaked in 2012–13 is even weaker and is probably the weakest of the last century. A similar development at the start of the nineteenth century led to a dramatic fall in solar irradiation, and is anticipated to occur now. This means the sun will be set at low-burn mode for the decades ahead. In 2010 the intensity of the solar magnetic field dropped to one of its lowest levels in the last 150 years and accordingly cosmic rays reached their highest level since records began 50 years ago. The magnetic flux density has been decreasing steadily since 1998 and the large plasma flows on the sun have slowed down over the last year. Currently we are in the middle of a very weak solar cycle 24.

American, Russian and British scientists see clear signs that the 'sun is shifting to a low-activity period ... the start of solar cycle 25 could be delayed, or not even take place at all. It may well be that this one [cycle 24] could be the last solar maximum we will see for a few decades' (Frank Hill, National Solar Observatory) [2]. The course of the other longer periodic solar activity cycles indicates the same when projected into the future. There's little doubt that the sun has put its warming plateau phase of the last decades behind it and has started on its frosty decline to a phase of low activity.

Taking both the important natural and anthropogenic climate factors into account, we can expect a slight cooling of about 0.2–0.3° C by the year 2035. Only cooling can be expected from the sun over the coming decades. The Gleissberg 87-year and Suess/de Vries 210-year cycles will reach their low points between 2020 and 2040, and so in 2035 solar activity will reach a level that is comparable with the Dalton minimum, which occurred between 1790 and 1830, when mankind faced harsh and trying living conditions. Back then the temperature was nearly 1° C lower than it is today and this was at least half due to the weak sun. The PDO, which superimposes itself on the climate, will remain in its cool phase until 2035 and this too will certainly contribute to cooling.

By the year 2035 the cooling that will have come about from the natural climate-regulating factors is expected to be about 0.4–0.6° C compared to today's temperature. This cooling, however, will be offset in part by the anthropogenic greenhouse effect. Using the IPCC-A1B emissions scenario, the CO_2 concentration in the atmosphere will then be close to 450 ppm. Using a realistic CO_2 climate sensitivity range of 1–1.5° C per doubling of CO_2, this corresponds to a positive temperature contribution of 0.2–0.3° C. This therefore will yield a net temperature reduction of 0.2–0.3° C. The dramatic drop in solar activity will thus deliver a cooling phase over the coming decades, one that CO_2 will not be able to offset in full.

Based on the established solar cycles, we can conclude that the sun will again become more active in the second half of the twenty-first century and that the earth will warm again as a consequence. However, the sun may not reach the high levels of the 1980s and 1990s. This moderate intermediate phase will then come to a halt towards the end of the century when the sun reaches another point resembling the Dalton minimum. This solar lull may contribute to a cooling of about 0.3–0.4° C compared to today. The PDO presumably will move to a low or moderate level.

By the year 2100, in accordance with the A1B emissions scenario, CO_2 atmospheric concentrations will approach 700 ppm. With a CO_2 climate sensitivity of 1–1.5° C per doubling of CO_2, this will yield a CO_2-related warming contribution of 0.8–1.3° C compared to today. Thus the sun, PDO and CO_2 together will produce a net temperature increase of 0.6–1.0° C, depending on CO_2's real climate sensitivity. This is very different from the IPCC's projected temperature increase of 2.8° C for the same emissions scenario. Based on our assumptions, the so-called 2-degree target by 2100 will, in any case, be easily met, and done so without a panicked and risky transformation of the industrial landscape over the next couple of decades.

By no means does this imply that the strategy of decarbonizing our power generation should be abandoned. First, an additional warming of up to 1° C by the end of the century would lead to significant changes to our climate. Second, we cannot be certain how the sun and its natural impacts will act in the second half of the century. However, it is certain that the sun, and the impact it has on the earth, will bestow colder times over the first half of the century. This means we gain valuable decades in which to convert our energy supply without incurring massive prosperity losses. Because today's climate politics determines the distribution of economic growth, it also determines future prosperity. Europe's current policy of transferring hundreds of billions of euros to developing countries with the aim of avoiding increased CO_2 emissions can be attributed to a single hypothesis: that the current century will see a warming of 1.8–4° C due to rising CO_2 emissions.

But if you conclude that the IPCC's projections are unfounded and that only a maximum of 1° C of warming is to be feared, then the setting of priorities for an energy agenda changes profoundly. This in turn will release resources to assure an adequate supply of food and water, and a rising standard of living globally. Factors concerning economic feasibility or social justice then again will equitably decide energy policy, along with climate protection. By understanding that natural climate factors will continue to play an important role, we gain time for rational decarbonizing. This can be achieved by implementing new, renewable technologies, through higher energy efficiency and better material consumption, and by improving the generation of conventional fossil energy in a rational, cost-effective, truly sustainable way.

References

1. Muller, R. A., J. Curry, D. Groom, R. Jacobsen, S. Perlmutter, R. Rohde, A. Rosenfeld, C. Wickham and J. Wurtele (submitted) Decadal Variations in the Global Atmospheric Land Temperatures. http://berkeleyearth.org/Resources/Berkeley_Earth_Decadal_Variations.
2. SpaceRef (2011) Major Drop in Solar Activity Predicted. http://www.spaceref.com/news/viewpr.html?pid=33826.

3. Our temperamental sun

Why do we think the sun's the culprit? Are there good reasons to suspect this? After all, the sun has never let us down before. Every morning it rises in the east and in the evening gracefully sets in the west, as reliably as a Swiss watch. It's powerful, provides warmth and its rays outshine everything. These are the qualities that have earned our mother star a prominent, semi-divine position in a number of religions. So wouldn't it make sense to think our seemingly infallible sun could have precocious moods when it really plays up, before shifting down a level or two and finally ceasing from all the activity? Has such solar volatility ever happened in the past? If so, when and for how long? Just how large were these fluctuations and what impact did our sun's mood swings have on the earth's climate?

As far as the IPCC is concerned, the book is closed when it comes to the sun's contribution. In its view there is little doubt that the sun has played only a minor role in climate change over the last 40 years. In the IPCC's Fourth Assessment Report of 2007 (AR4) the sun's influence on our climate was halved with respect to their 2001 report. The IPCC's message to the world's citizens and politicians was that manmade CO_2 is by far the most important regulator in the current climate schema and the impact of natural processes on climate development will be minuscule by comparison in the coming decades.

Because of the profound implications of this claim, and its critical importance to the world, we believe there is a pressing need to examine the facts. We have to ask ourselves whether AR4 really took into consideration all the scientific results from the sun-climate field available at the time. And in the years since the 4AR was published, have there been any developments that could change our views?

In this chapter we show how the sun was in fact much more involved in twentieth-century warming than first assumed. Even more important is the fact that the sun is a repeat offender, as we shall demonstrate. One thing will become very clear during the analysis of the scientific evidence: the earth's temperature has been rising and falling for thousands and millions of years in step with the sun. Just how probable is it that this time-honoured partnership abruptly came to an end a century ago?

How it all began

To better understand the variability of today's sun, we need to take a brief look at its birth and development. It was not born when the universe first began to form. Rather, our mother star took its time before it made its debut on the galactic stage. The Big Bang had occurred eight billion years before the first birth contractions commenced inside an interstellar cloud of hydrogen gas. A violent pressure wave, triggered by the explosion of a nearby supernova, swept through this cloud and caused the gas to compress so massively that hydrogen atoms started fusing into helium atoms at temperatures of several million degrees Celsius. This in turn triggered the atomic fusion motor for hundreds of new stars, among them our sun. Under the force of gravity, the freshly formed stars steadily compressed, which in turn further fuelled the fusion process. The new stellar power plants released enormous amounts of energy, emitted as light and heat. The increasing irradiative pressure countered the force of gravity until a point of equilibrium was reached, thus preventing further contraction. The stars and their illumination stabilized.

As lighter hydrogen atoms continued their transformation into heavier helium atoms, the stars continuously compressed, thus further firing up the fusion reactor. It is therefore assumed that in the early phases of its first four and half billion years, our sun had only 70 per cent of today's illuminative power and has since increased steadily until reaching today's value. The sun currently is about halfway through its life. Based on the development we have seen so far, it is projected that the sun's irradiative power will continue to grow linearly at about 1 per cent every 110 million years into the future [1]. That means a brightness increase of 10 per cent over a period of one billion years. Eventually, the earth will become so hot that it will practically be uninhabitable [1–2]. At that point, the sun will still have another six billion years of life ahead of it before its hydrogen supply is exhausted. When that occurs the irradiative pressure will quickly begin to diminish and gravity will once again regain the upper hand and cause the sun to contract. In the final stage the sun will briefly expand multiple times into a giant red star, expelling its gas shell, which in turn will incinerate the inner planets. In the end only a white dwarf will remain – the melancholy terminal stage of the life of the relatively low-mass, unspectacular star that is our sun.

The sun warms the earth

Although by itself it is relatively meaningless in space, the sun is absolutely essential for life on our planet. And here a huge amount of luck comes into play. The earth is situated at an ideal distance from the sun. The inner planets (Mercury and Venus) are too hot for human life and the outer planets (Mars, Jupiter, Saturn, etc.) are too cold. Let us briefly conduct a small thought experiment. What would happen if the sun suddenly were to go on strike and stopped shining?

First, we would expect to see the same effects as a normal sunset and a light night-time cooling would set in. There would still be huge amounts of heat stored in the oceans, land and atmosphere, but over a few days this would steadily dissipate and it would continuously get colder. The lakes and oceans would freeze over after just a few weeks. Because of the absence light, all plant life would die and thus the food chain would be interrupted. The earth would rapidly be transformed into an uninhabitable ball of ice. Luckily, from what we know today, an abrupt extinction of the sun is extremely improbable.

The sun is undoubtedly the source of life on earth. Of the total energy contribution to the earth's climate, 99.98 per cent originates from the sun. The nugatory remainder is supplied from the earth's warmth, which originates partly from the heat left over from the earth's creation over four billion years ago and is partly a product of the radioactive decay of the earth's interior. Due to the sun's paramount importance to the earth's energy budget, it is highly plausible that fluctuations in solar irradiation will have an impact on the earth's climate. The equilibrium on earth is based on the balance between irradiated solar energy and the share radiated back into space. Any alteration to this balance, for example changed solar irradiation, has the potential to bring about significant climate change [3].

The solar power plant

Let's take a closer look at the path sunlight takes from its place of production to its destination on earth. The sun, our power plant, is made up of 73.5 per cent hydrogen and 25 per cent helium. The remaining 1.5 per cent is heavy elements, foremost among them oxygen and carbon. The fusion zone is located in the sun's core, which occupies the inner quarter of the sun's radius. Here at 15 million degrees and under high pressure, hydrogen atoms fuse into helium atoms over several interim steps. Why is energy released here at all? The mass of each helium atom created during fusion is slightly less than the sum of the mass of the original four hydrogen atoms needed to form it. The difference in mass is released in the form of energy. This energy is precisely equal to the mass difference times the square of the speed of light, according to Albert Einstein's mass-energy equivalence equation, $E = mc^2$.

The transport of this energy from the fusion zone to the outside takes place in the form of radiation. In the upper levels of the sun's body, huge rolls of fire, so-called convection cells, take over the transport of energy. The energy emerges at the surface of the sun in the form of radiation and is emitted into space. The radiation first passes through the atmosphere of the sun, the corona, which is the halo one can see during a total eclipse of the sun. Once the outer areas of the sun have been traversed, the radiation enters the vast expanses of space.

Composition of solar radiation

What is the sun's radiation made of? The solar core fusion reactor produces gamma rays which are converted into a broad range of electromagnetic waves extending from radio waves to visible light and beyond to x-rays. The radiation maximum occurs as visible light in the yellow to green colour spectrum, whereby the irradiative intensity on both sides of the maximum diminishes steadily to short and long wavelengths (Figure 3.1). Indeed, evolution on earth had to respect the composition of the sun's spectrum. This is how the eyesight of man and many other animals developed, according to the spectral range of the irradiative maximum.

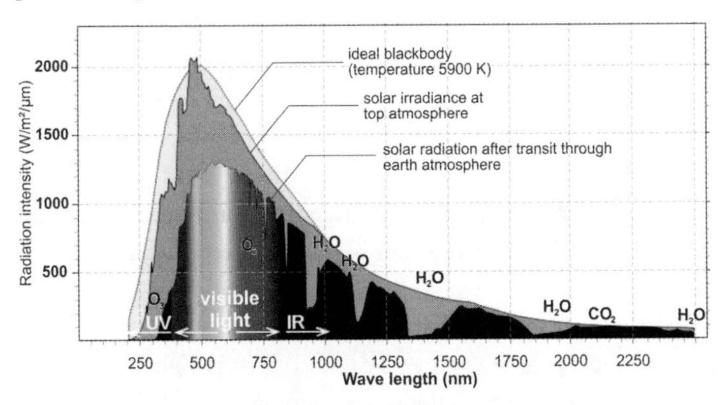

Figure 3.1 The radiation spectrum of the sun at its interior (upper, smooth curve), after passing the radiation-filtering solar atmosphere (middle curve) and the earth's atmosphere (lower ragged curve). Water vapour (H_2O), ozone (O_3) and carbon dioxide (CO_2) in the earth's atmosphere filter out certain wavelengths from solar radiation with precisely defined energy and thus are missing from the sun's radiation spectrum at the earth's surface. UV = ultraviolet radiation; IR= infrared [4].

However, solar radiation's bell-shaped distribution over the various wavelengths is not quite as perfect as one might assume.

A number of interruptions occur and appear as dark lines in the spectrum – so-called Fraunhofer lines. These radiation dropouts occur when sunlight passes through the gas layer of the outer solar layer (chromosphere) and the earth's atmosphere (Figure 3.1).

The chemical elements of the filtering gases screen out certain wavelengths from the solar radiation with a precisely defined energy, and thus are missing from the sun's radiation spectrum.[3] The precise energy amounts vary among individual chemical elements and molecules. In the outer solar layers, predominantly hydrogen, helium, iron, calcium, magnesium and sodium take a bite out of the radiation spectrum. In the earth's atmosphere predominantly water vapour, CO_2 and ozone are at work. Ozone mainly absorbs over a broad range of wavelengths and protects the earth from much of the UV radiation, which is hazardous to life.

In addition to electromagnetic radiation, the sun ejects solid matter from its outer core throughout its vicinity at the rate of approximately one million tons of mass a second. The high-energy particle stream jets at 400–900 km per second and consists mainly of protons, electrons and helium nuclei. Because the charged particles have different speeds, they also form wandering magnetic fields, which in turn can influence the path curves of other charged particles. In addition, the rolls of fire of the sun's outer layer create powerful magnetic fields that can reach the interplanetary levels.

Passage through space and arrival on earth

The sun's flux of electromagnetic energy travels through space at the speed of light. The 150 million km journey to earth takes a mere eight minutes. When we look at the sun (and that of course only with the necessary safety precautions) what we actually see is the past.

The composition of solar radiation remains intact during its passage through material-free space. Only its strength diminishes

---

3 Photons of these wavelengths carry the exact amount of energy that is necessary for catapulting an electron in the gas atom to a higher energy level.

little by little with growing distance from the sun because its energy is distributed over an ever-increasing area. When solar radiation enters the outer layers of the earth's atmosphere, it embarks on a formidable obstacle course. An entire cast of characters stands ready to block solar radiation.

The problems begin 300 km above the earth's surface. The most energy-rich component of solar radiation, the hard UV radiation and x-rays, smash the electrons out of the upper atmosphere gas molecules and consequently weaken. This produces a huge quantity of ions and free electrons which characterize the ionosphere, which can reach a level down to 80 km above the earth's surface (Figure 3.2). The ionosphere makes up the basis for global shortwave communication. Radio waves sent from the earth's surface are reflected multiple times by the electrically charged ionosphere. This is how radio waves can bounce back and forth between the earth's surface and the ionosphere, and under good transmission conditions reach almost every part of the globe.

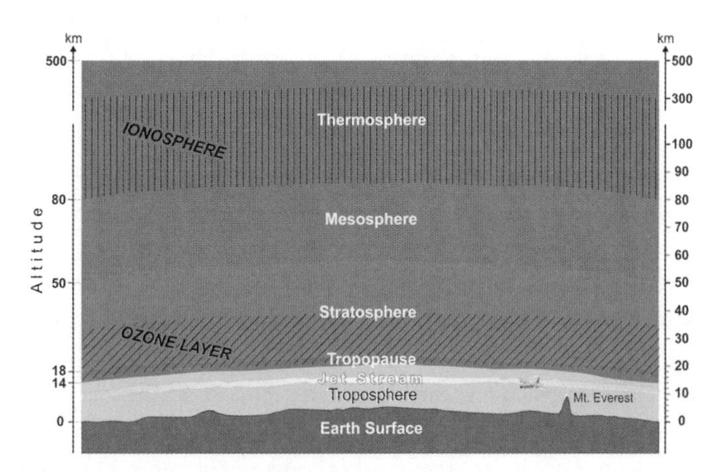

Figure 3.2 The structure of the earth's atmosphere.

At altitudes of 50–15 km, solar radiation takes another battering. Solar UV radiation within this zone splits the oxygen (O_2)

molecules into two oxygen radicals (O), which then combine with other oxygen molecules to form ozone (O3). The resulting ozone layer acts as the earth's protective shield against UV rays, which are almost completely filtered out and converted into heat (Figure 3.2). The somewhat less dangerous UV-B radiation also weakens.

Below 15 km infrared radiation is next in the firing line. In this atmospheric tier we find about 90 per cent of earth's entire air and practically all its water vapour. Water vapour, and to a lesser extent CO_2, has a great interest in certain infrared wavelength segments – they filter them out. The rest of the infrared radiation emitted by the sun can, however, pass through this atmospheric zone scot free.

The range of visible light wavelengths is hardly affected by the chemical atmospheric layers it passes through. This spectral range, visible to the human eye, is largely spared the absorption effect of the earth's atmosphere. However, not all visible light gets past the clouds. Twenty per cent of the sun's total incoming radiation fails to penetrate the white plumes of water droplets and other airborne particles on its way to the surface of the earth, but is reflected back into space. Once each atmospheric layer has finished filtering out its part of the solar radiation spectrum and part of it is reflected back into space, only about half of the original solar radiation reaches the earth's surface.

Electromagnetic solar radiation thus takes a real battering during its eight-minute journey to earth. But what happens to the solid material particles that are sent towards earth? Do they encounter fewer problems on their journey? Quite the opposite, very few of these particles reach the earth's surface. Almost all the electrically charged particles from solar winds and other cosmic sources are caught by the earth's magnetic field thousands of kilometres up on their path to earth.

In the Van-Allen radiation belt these particles are first forced to oscillate back and forth for some time between the earth's magnetic poles. The only ways of breaking out of this magnetic prison are found at the magnetic poles where the magnetic field lines are perpendicular to the earth's surface. If a particularly violent solar wind blows, the particles escaping from the earth's magnetic prison put on quite a show in the polar regions – the Northern Lights and Southern Lights spectacle.

During one particularly violent solar storm in March 1989, a huge amount of charged solar particles reaching the earth's magnetic field not only put on a colourful polar light show, but were accompanied by a number of hostile effects [5–6]. The avalanche of particles pulsated in the earth's magnetic field so violently that it changed rhythmically and induced gigantic electrical currents. The victims included power transmission lines, which greedily absorbed the energy like antennae. A massive electrical surge spread through the power grid at great speed. In Quebec a number of substations were unable to handle the loads and immediately failed. As a result the entire grid in the province was destabilized and crashed. The inhabitants of Canada's second largest city, Montreal, were plunged into darkness. It took nine hours to restore power.

This rare solar storm caused problems elsewhere too. Pipelines in the polar regions were damaged because the powerful, induced electrical current caused metal pipes to corrode. Satellites were thrown out of their orbits because the density and atmospheric friction were changed. The occupants of the space station *Mir* exceeded their safe annual dosage of radiation in a single hit. In California garage doors opened and closed during the storm, as if operated by a phantom.

Sunspots

Sunsets have fascinated man since the beginning of time. Just before the sun slips below the horizon, its brightness diminishes immensely. And if there happens to be a haze, then under the right conditions one can look at the sun briefly without being blinded. (Warning: Don't try this! Under unfavourable conditions you risk burning your retinas and going blind.)

Two thousand years ago Chinese astronomers took advantage of such conditions and were able to observe unusual dark spots on the surface of the sun. The western world stubbornly ignored this discovery because the 'spotty sun' did not fit its worldview of a perfect God in Heaven. This did not change until the start of the seventeenth century when the telescope was invented and the existence of sunspots could no longer be denied. Galileo Galilee was among the sunspot pioneers who at the time, in connection with other 'errors', became the subject of an inquisition.

Sunspots can be observed using amateur equipment. To do this a telescope, or a (non-plastic) field glass mounted on a tripod, is aligned with the sun. The sun's image is then projected onto a shaded screen positioned behind. (Warning: Never look through the ocular of the telescope or the viewfinder of the camera. There's a real danger of being blinded!)

Exactly what are sunspots and how are they generated? Sunspots are relatively cold and thus are areas that appear to be dark on the sun's surface and emit less visible light than the rest of the surface (Figure 3.3).

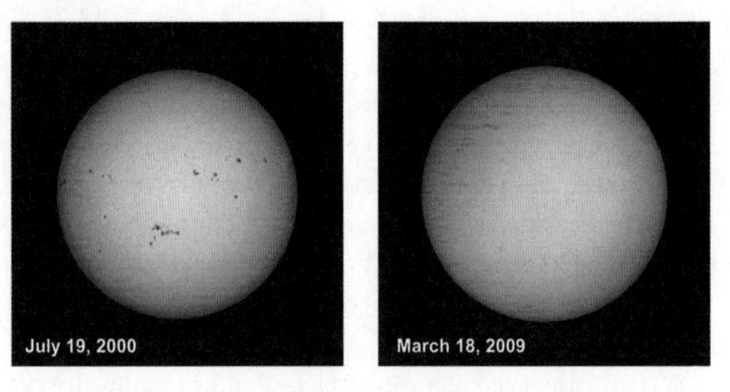

Figure 3.3 Sunspots on an active sun (left) compared with a spot-free, inactive sun (right) [7].

These localized cool spots are caused by powerful magnetic fields that prevent convection and through which less energy reaches the sun's surface at the centre of the sunspot. Close to the sunspots, however, are areas that are considerably hotter than the rest of the sun's surface, the so-called solar faculae. Their energy more than compensates for the energy missing from the sunspots. Overall, more sunspots lead to an increase in the sun's brightness. Counting the number of spots and estimating their size are simple ways of determining solar activity. This traditional method is still important, as it is fully independent of the effects of the earth's atmosphere from solar radiation.

Nowadays, sunspot measurement is supplemented by two modern processes. One is the measurement of the solar magnetic field strength, which we know increases as solar activity increases. The other process is measuring the solar irradiance strength at the top of the atmosphere directly with satellites.[4] This has been done since 1979. But here there are calibration problems with the instruments, thus making the analysis of long-term trends complicated [8–9].

The fickle sun: the discovery of the 11-year solar cycle

It was an amateur astronomer who discovered the cyclical nature of sunspots. At the age of forty, Heinrich Schwabe of Dessau had had enough of his pharmacy and sold it in 1829 so that he could devote his time to his real passions: astronomy and botany. Why he began with the systematic observation of sunspots and tenaciously stuck to it is unknown. The project appeared anything but promising because experts of the previous decades were convinced that there was no regularity in the occurrence of sunspots [10].

Schwabe didn't allow himself to be deterred and carried out his series of observations. After 17 years he had gathered enough data to recognize an 11-year cycle. He continued his sunspot observations almost until his eightieth birthday, and so was able to document the precise course of several cycles. After initial scepticism, professional circles finally accepted the discovery without reservation and even admitted Schwabe as an external member of the Royal Astronomical

[4] NASA's ACRIM satellite family and Nimbus 7.

Society in London. The 11-year sunspot cycle today is named the 'Schwabe cycle' in honour of its discoverer.

The Swiss astronomer Rudolf Wolf recognized the value of Schwabe's discovery early on and developed a practical measurement index, the *sunspot relative number*, with which observations can be recorded quantitatively and independent of the recorder. This parameter is still in use today. Wolf also analysed the early records from Galileo's time and thus was able to reconstruct the pattern of the cycles back to 1745.

Today, many sunspot cycles later, we know much more (but still not enough) about the 11-year cycle. In the meantime we have been able to document the course of the Schwabe cycle using telescopes and satellites. During the peak of the 11-year cycle, the so-called solar maximum, the sun puts on a heavenly display of fireworks. Gigantic eruptions and solar storms rage. The surface of the sun is covered with sunspots, which now have reached their greatest abundance. Also 'radio communication weather' for global shortwave communication greatly depends on the 11-year cycle and is coupled with the changing electrical charge of the ionosphere. During the cycle's maximum, a single watt of transmission power is enough to reach the opposite side of the earth. On the other hand, during the solar minimum, even the most powerful transmitters do little to help and the transmission signal diminishes after a few thousand kilometres.

Today, in addition to the 11-year cycle, we know of other solar activity cycles, particularly the 22-year Hale cycle [11], the 87-year Gleissberg cycle [12–18] and the 210-year DeVries/Suess cycle [19–25]. All these cycles are superimposed over each other, at times amplifying and at times weakening the net effect. They form the basic repertoire of solar activity fluctuations (Figure 3.4).

What could be causing cyclical flare-ups and quiet periods in solar output? The magnetic field of a quiet sun is about the same as that of a dipole. Every 11 years a reversal in field polarity takes place, which means that after 22 years the original alignment is reached once again [26]. The triggers for these magnetic changes are presumably oscillations in the solar dynamo. A relationship between the magnetic field change with the 11-year Schwabe and 22-year Hale sunspot cycles is obvious because of the period length similarity [27].

Cycle	Average period (years)	Fluctuation range (years)
Schwabe	11	9–14
Hale	22	18–26
Gleissberg	87	60–120
Suess/de Vries	210	180–220
Eddy	1000	900–1100
Hallstatt	2300	2200–2400

Figure 3.4 Solar activity cycles.

What fluctuates and by how much?

Let's take a look at how the amount of irradiation changes during the course of an 11-year cycle. Here we shall focus on which types of radiation carry the bulk of variability and which wavelength spectra are hardly influenced at all but remain more or less stable.

Let's go back to the upper edge of the earth's atmosphere in order to exclude all atmospheric effects on solar irradiation. Satellite measurements over recent decades have shown that the difference between the maximum and minimum of an 11-year solar cycle is about 0.1 per cent at the top of the atmosphere if the entire spectrum of solar radiation is taken into consideration (Figure 3.5). That doesn't sound like much. However, there is a big surprise in the UV part of the radiation spectrum, where variability during the 11-year cycle is more than ten times more pronounced than in the rest of the spectrum. The intensity of UV radiation fluctuates by a few percentage points [28–31] and in some wavelengths up to 70 per cent [32–33] (Figure 3.5). With radiation changes of these magnitudes, it is worth taking a closer look to see whether there are any impacts on the earth's climate system.

Let us recall briefly the processes that UV radiation initiates at altitudes of 15–50 km, that is to say within the ozone layer (Figure 3.2). An increase in UV radiation intensity here would certainly convert a greater number of oxygen molecules into ozone [34]. And a higher ozone concentration would in turn intercept more UV rays and convert their energy into heat, so that the ozone layer would heat up.

Well, that's the theory. Luckily since 2003 we have the NASA's SORCE satellite which has recorded the flux of individual solar irradiation types coming towards the earth. Its Spectral Irradiance Monitor (SIM) measuring instrument took readings during the transition from solar maximum to minimum between 2004 and 2007. The amount of change in UV radiation turned out to be five times greater [30, 35–36] than had previously been considered possible [37]. NASA satellite measurements also showed that ozone concentration during the solar maximum was higher than during the subsequent minimum [35]. Furthermore, the temperature of the ozone layer during the maximum was almost 2° C higher than it was during the minimum [28, 30].

Clearly, the fluctuating UV radiation intensity of the 11-year solar cycle can cause significant atmospheric impacts at altitudes of 15–50 km. And if a process that connects the stratospheric fluctuations to the tropospheric climate events below the 15 km level exists, then there's a possible solar amplification process that has not yet been taken into account by the IPCC climate models. There are numerous indications showing such solar amplification processes do exist (see Chapter 6).

Amazing observations incidentally were also made in the overlying ionosphere during the last solar minimum (2007–9). When solar UV radiation plummeted, the ionosphere cooled significantly. First, the density and expansion of the ionosphere dropped considerably and reached its lowest value in the 43 years

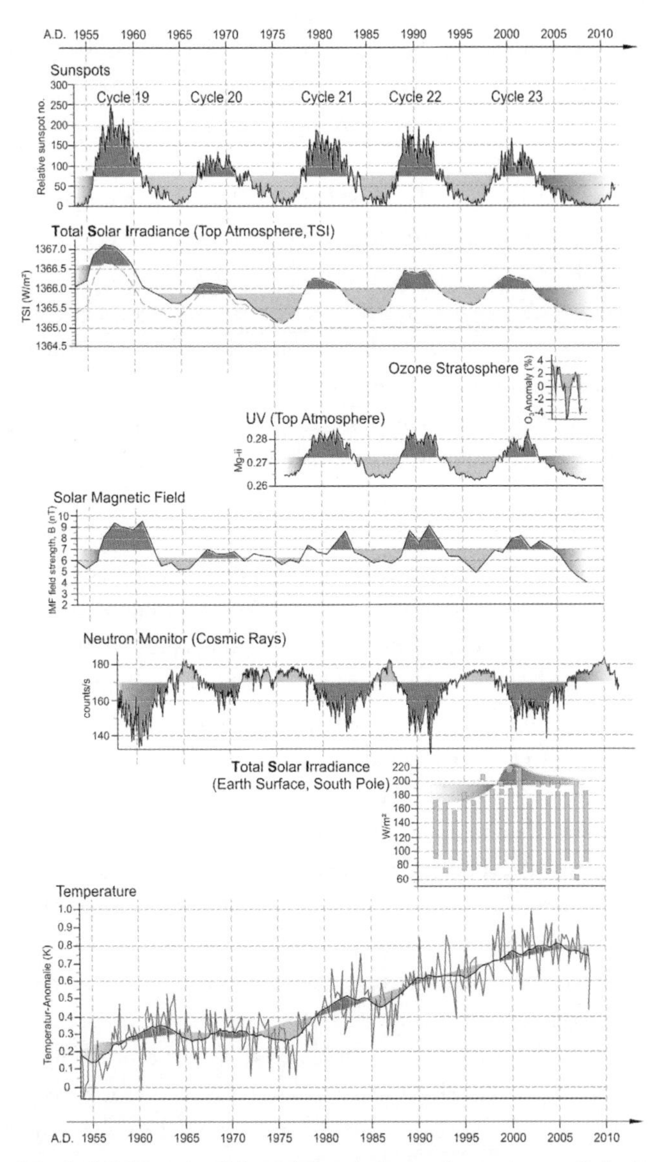

Figure 3.5 Cyclical changes in radiation and atmospheric parameters are in sync with the 11-year solar activity cycles. Sources: sunspots [38]; total solar irradiance (TSI) [39]; stratospheric ozone concentration [30]; UV at the top of the atmosphere [26]; solar magnetic field [40]; neutrons (cosmic radiation) [41]; total solar irradiance at the earth's surface (South Pole) [42]; global surface temperature [43, 33].

since observations were first made by satellite. The changes in density exceeded expectations based on conventional models by 30 per cent [44–48]. The solar cycle plays an important role here as well.

In summary: Above the earth's atmosphere solar irradiance fluctuates during the 11-year solar cycle by 0.1 per cent when you take the entire spectrum into account, but by a few per cent in the UV range.

Now let's return to the earth's surface and look at what fluctuation ranges arrive after solar radiation penetrates the earth's atmosphere. As we have seen, solar radiation is obstructed in the atmosphere before it finally completes its journey to earth. During this transit the magnitudes of irradiation fluctuations can also change.

Once again we are taken by surprise. The changes in total solar irradiance are ten times greater than those at the upper edge of the atmosphere. Over the course of an 11-year cycle the values fluctuate over a scale graduated in full percentage points [42, 49] (Figure 3.5). It seems that something is happening in the earth's atmosphere that significantly amplifies the small decimal fluctuations of the solar power plant (see Weber in Chapter 6).

But that's not all. Yet another parameter changes over the course of the 11-year cycle, and that at an astonishing 10–15 per cent [50] (Figure 3.5). That parameter is the cosmic ray intensity, which fluctuates above the earth's atmosphere 100 times more than the sun's total irradiance. The term 'cosmic radiation' is historical and does not characterize an electromagnetic radiation, but rather high-energy charged particles that strike the earth from outer space. Near the earth these particles consist of about 98 per cent atomic nuclei and about 2 per cent electrons, which for the most part were created during supernova star explosions far away from the solar system.

So how can we have large changes in cosmic rays with only very small changes in solar irradiance? Solar activity not only affects the output of electromagnetic radiation, but also the strength of the sun's magnetic field. They pulsate together, synchronously (Figure

3.5). The solar magnetic field forms a protective shield against the bombardment of cosmic particles streaming in from space. The stronger the sun's magnetic field, the fewer cosmic rays that can reach the earth. It is apparent that this magnetic process generates far more intensive changes in the atmosphere than primary solar irradiation does. Solar radiation and cosmic radiation are analogous to two boats bobbing up and down on the sea. Both boats rock synchronously, yet the rocking of one boat does not the cause the other boat to rock. It is the waves that cause them both to rock. But if the boats are very different sizes, for example, a massive supertanker and a small raft, then the supertanker rocks far less than the small raft does.

So why should we be interested in cosmic rays in the first place? There is an important reason. Cosmic rays are the most important suspect in the search for the sun's ominous amplification mechanism as a climate driving factor. The IPCC fleetingly mentioned that a strong sun led to reduced cosmic ray intensity reaching the earth. However, they deemed it unnecessary to report on the significant 10–20 per cent fluctuation [51].

Today we know that the cosmic ray intensity over the last 150 years has declined considerably and has been increasing only since the beginning of the twenty-first century. If there had been feedback effects, for example, cosmic rays influencing cloud formation, then this would have had a marked effect on our climate. But none of this is of interest to the IPCC, for it would refute its claim that the current warming is almost exclusively due to an increase in greenhouse gas concentrations. (More on this in Chapter 6 and Svensmark in Chapter 5.)

Documented climatic impacts of the solar base cycles

Now we are compelled to ask whether we can find an imprint of the Schwabe cycles in the historical temperature data sets or other climate parameters. The answer is yes. Henrik Svensmark and Eigil Friis-Christensen were able to show that the temperature development of the lower atmosphere and the oceans over the last 50 years correlated well with the 11-year solar activity (Figure 3.5) [52]. Other teams came up with similar results [43, 53–55].

However, in order to obtain the solar signal, the scientists first had to filter out other climate effects such as El Niño warm-ups, the cooling of ash and sulphur from large volcanic eruptions, and ocean-internal cycles such as the PDO. Temperatures are fundamentally generated by an entire series of climate variables and so it is always necessary to apply the filtering process in order to isolate the specific signals of individual climate factors. What is more, not all regions react to solar irradiation changes in the same way [48, 56]. Analyses show that the 11-year solar cycle has particular climatic impacts at the middle latitudes and in the Tropics [57].

We also find an impressive imprint of the Schwabe cycle at an unexpected location, namely the second largest freshwater body on earth, Lake Victoria. Over much of the twentieth century, its level fluctuated in line with the 11-year solar cycles [58–60] (Figure 3.6). Evidently the sun has an impact on precipitation over the drainage basin of this East African lake. But rainfall elsewhere follows the behaviour of the sun. Indeed, we find an 11-year rhythm in the water feed-in for the Mississippi [61], precipitation amounts in the north-west of the United States [62] and in the Tropics [63–64]. Even tree growth in Scotland [65] and other parts of Europe [66] are aligned with the Schwabe cycles.

Other solar cycles have left a permanent imprint in the climate archives [58]. For example, the 22-year Hale cycle influenced the temperature development and the tree-ring growth pattern at different locations [66–70]. The cycle is found, for example, in temperature curves in England [71], Nebraska [72] and global data sets [73], and can be shown in tree-ring records along the Russian Arctic coast [74]. Precipitation amounts also appear to be connected to the Hale cycle, which was also recently discovered in Brazilian rain archives [75–76] and in the discharge rates of the River Po in Italy [77].

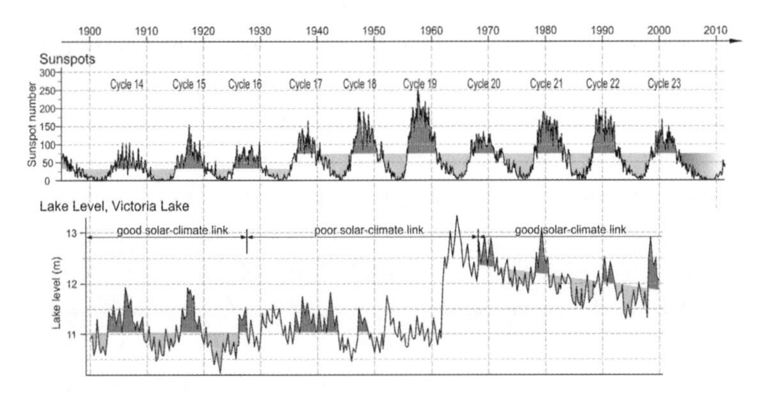

Figure 3.6 The level of Lake Victoria in East Africa fluctuated during 1896–1928 and 1968–2005 in line with the 11-year solar cycles. In the interim phase, the linkage was interrupted [58–60].

The 87-year Gleissberg cycle influences the development of sea surface temperatures [67, 78–79, 80] and also plays a role in oceanographic changes in the north Atlantic [81–82] and the Iberian Atlantic coast [83]. And temperatures in central England [84] and China [85–86] appear unable to ignore the Gleissberg cycle.

The 210-year Suess/De Vries cycle shows up in moisture fluctuations on the Tibetan plateau and elsewhere in China over the last millennia [87–88] and has an influence on tree growth on the Tibetan plateau, in Tien Shan [89] and other parts of the world [90], plays a role in the drought cycles over the American Great Plains [91] and the Mexican Yucatan peninsula [92], controls the deposits of lake sediment in East Africa [93], regulates the glacier lengths in Alaska [94] and is seen in the dust concentration curves in one Greenland ice core [50] and two Antarctic [95] ice cores.

In addition to the solar-driven basic cycles, independent oscillations come into play in the climate system and have an impact on the climate. Such internal oscillations with periods of decadal scales are found, for example, in the thermohaline circulation of the oceans [96] and the coupled atmosphere-ocean systems [97] (see Chapter 7). Solar cycles and natural earthly oscillations must be clearly distinguished. Both naturally [98–99], so if one study shows that a particular internal cycle of the climate system had nothing to do with the sun, then it certainly does not cast doubt on the existence of solar-driven cycles. Just because one happens to be standing in front of a pear tree, one does not conclude that apple trees don't exist. Identifying the physical nature (solar, lunar, astronomical or internal) of all observed cycles and natural variations of the climate system and their relative magnitude is fundamental to properly interpreting climate change.

We have seen that the 11-year cycle and the other natural oscillations are firmly etched everywhere in the climate archives and have left an indelible mark on the earth's climate development. Moreover, if we differentiate among processes, we see that solar activity variations are larger and more powerful than previously thought. This includes measuring different radiation wavelengths, distinguishing among the various measurement locations above, within and below the atmosphere, and separating the effects of electromagnetic radiation and cosmic high-energy particle streams.

When an effort is made to establish a clear and fundamental link between the sun and climate, a picture emerges that is completely different from that which certain organizations stubbornly cling to. Simple statements like 'The intensity of solar irradiation within the 11-year solar cycle fluctuates a mere 0.1 per cent and thus is too weak to have any significant impact on the climate' appear in a completely different light and raise the question of just how much we should trust such analyses, let alone the conclusions drawn from them [100].

Millennium cycles: underestimated climate drivers?

The solar 11-year cycle and the related Hale, Gleissberg and Suess/de Vries cycles are, however, only the tip of the iceberg. As we shall see, long-term climate development was mainly affected by cycles with a thousand-year timescale, and the course of these cycles correlates surprisingly well with solar activity. The degree of change that these long cycles bring with them is far more pronounced than the shorter cycles just discussed.

A better understanding of long solar activity cycles is a prerequisite for answering the question of just how much the sun's contribution is to the temperature increases we have seen since the industrial era began in around 1850. This is not about fundamentally refuting CO_2 as a climate gas, but about improving the quantitative estimate of the climate impact of individual control factors. The main objective here is to ascertain if there are signs that the assumed CO_2 climate sensitivity of the IPCC could be inflated. In simple terms: We have 0.8° C of warming that has to be distributed over a number of climate factors. If the sun is shown to be more important, then CO_2's importance has to be reduced.

There are no systematically recorded observations of sunspots available for the period before the seventeenth century. The reconstruction of solar activity for this earlier period is done using so-called cosmogenic nuclides (^{14}C, ^{10}Be, ^{36}Cl), which are generated by cosmic rays [101]. The stronger the cosmic rays, the higher are the concentrations of the cosmogenic nuclides, which means the sun was less active because the solar magnetic field shielded the earth from the galactic particle showers.

On numerous occasions, when analysing data sets of the historical development of solar activity, scientists have detected a characteristic fluctuation lasting about 2300 years [19, 21, 102–105]: the Hallstatt cycle. Its cause is unknown. Some hypothesize that the cause is the recurring position of the large planets (Jupiter, Saturn, Uranus and Neptune), which acts to shift the sun by twice its diameter at regular intervals [106]. This could have an impact on the effectiveness of the solar power plant. The last minimum of the Hallstatt cycle occurred between 1300 and 1800, the period known as the Little Ice Age (LIA) (Figure 3.7). If one projects the Hallstatt cycle into the future, then we can expect the next Hallstatt solar irradiation minimum to occur in about 1500 years' time [105]. Also, the solar 88-year Gleissberg cycle and the 208-year Suess/de Vries cycle are assumed to be related to planetary gravity effects [107].

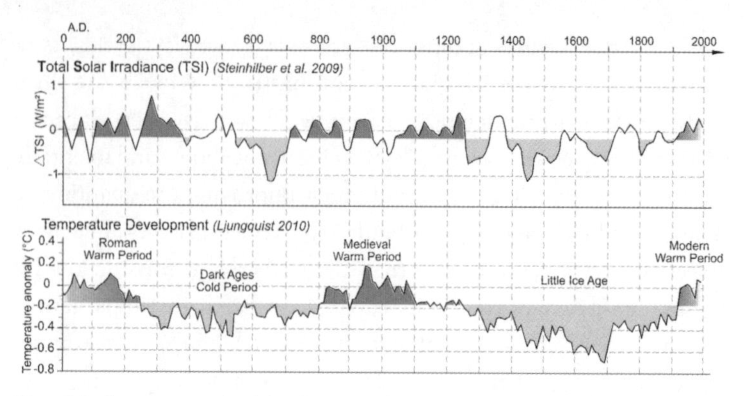

Figure 3.7 Long-term synchronicity of solar activity and temperature development (extra-tropical Northern Hemisphere) over the last 2000 years.

A closer look at the Hallstatt cycle in the natural solar archives shows a double peak about 1000 years apart [105]. Therefore, the Hallstatt cycle can be divided once again into two single oscillations, each about 1000 years long [110]. The 'semi-Hallstatt cycle' was recently dubbed the 'Eddy cycle' [101] in honour of the American astronomer John Eddy, who passed away in 2009 and who made an outstanding contribution to research on the weak sun during the Little Ice Age. The 1000-year Eddy cycle will reach its next minimum in about 500 years, which is why we currently find ourselves on a plateau area of a solar irradiation high point (Figure 3.7). We now see that the Eddy cycle had a great impact on the post-ice age climate development of the earth during the last 10,000 years and was the principal player in the global warming of the last 150 years.

What impact has the sun had on the climate over the last 1000 years? The Medieval Warm Period, the Little Ice Age and the Modern Warm Period

Reconstructions of solar activity show that the sun had a similar maximum about 1000 years ago, just as we saw over the last decades of the twentieth century [28, 111, 64–112]. This corresponds precisely with the 1000-year Eddy cycle. And between these two solar irradiation peak phases, the sun took it easy for several hundred years and was significantly less active [108, 113–116] (Figure 3.7). During this low activity phase, a 3579-day dry spell occurred as the sun went spot-free from 15 October 1661 to 2 August 1671.

To test the solar influence we compared solar activity history during this period to climate development. The result is striking. The temperature reconstruction curve runs almost completely parallel to the solar activity curve [109, 117–118] (Figure 3.7). The irradiation maximum about 1000 years ago occurred at the same time as the Medieval Warm Period, a time when higher temperatures made it possible to cultivate vineyards in southern Scotland and wheat as far north as Trondheim. Arctic sea ice melted so much that the Vikings were able to make expeditions to Iceland and Greenland in the ninth century before settling there. The glaciers in the Alps, North America and other regions retreated immensely [119].

Then, in the fifteenth century, the climate began to change. Arctic sea ice once again reclaimed the vast expanses of the Arctic, crops that Arctic settlers had planted failed and supply ships from the Scandinavian motherland were encased in ice. Because of the cooling climate, the settlers were forced to abandon their villages in Greenland. This cold phase went down in history as the Little Ice Age. It became bitterly cold on the European continent and there were extensive crop failures. Hunger, poverty and disease became rampant. The Thames froze over repeatedly. The Caspian Sea rose due to frequent rainfalls over the Caspian basin [120].

The Little Ice Age and the associated pronounced irradiation minimum in the middle of the last millennium corresponded to the low point of the Hallstatt cycle. This minimum also marks a low point in the Eddy cycle. A series of prominent low irradiation intervals developed during the Little Ice Age, including the Dalton (1790–1820), Maunder (1650–1719) [121–124], Spörer (1450–1550) and Wolf (1280–1350) minima (Figure 3.7). Recently, an American-Swedish team was able to demonstrate the very close relationship between climate development and solar activity. Using sediment cores off the Norwegian coast, they were able to substantiate each of these minima and found temperature fluctuations of 1–2° C within each cycle [125]. The agreement between the climate fluctuation curve and the solar activity curve is amazingly good. The relationship of the cold phases of the Little Ice Age with the solar slumber has in the meantime been confirmed by climate model simulations [126] and in principle is even recognized by the IPCC.

Finally, around the year 1800, the frosty period of the Little Ice Age ended; the cold point had been overcome. Temperatures began to rise and glaciers in the Alps started to recede once again [119]. Warming proceeded steadily as the sun and its radiation output steadily ramped up [127]. This increase in solar activity might have been far greater than what the IPCC previously assumed. Recent findings from Swiss researchers indicate that irradiation has increased since the Little Ice Age until today. They arrived at a value six times higher than the value the IPCC uses [39, 128]. These scientists also looked at the spectral range of UV radiation, which, as we described earlier, has a very high potential for significantly impacting the climate. For certain UV ranges, the Swiss team found increases of up to 26 per cent since the Little Ice Age, once again far higher than what the IPCC had earlier assumed.

The enormous increase in irradiation occurred between the exceptionally inactive phase of the Maunder minimum [33, 129–130] and the late twentieth century, which had remarkably high irradiation values (Figure 3.8). A similarly high level of solar activity of the kind we have witnessed over the past decades occurred very rarely over the last 11,000 years [111, 131–132]. Solar magnetic field strength increased enormously, parallel to the increase in solar irradiation over the last 100 years, having more than doubled [133–134] (Figure 3.8). The solar magnetic field protects the earth from cosmic rays because it forms a protective shield around it. As a result cosmic rays have decreased 9 per cent over 150 years up to 2000, when cosmic ray levels started to increase once again [135].

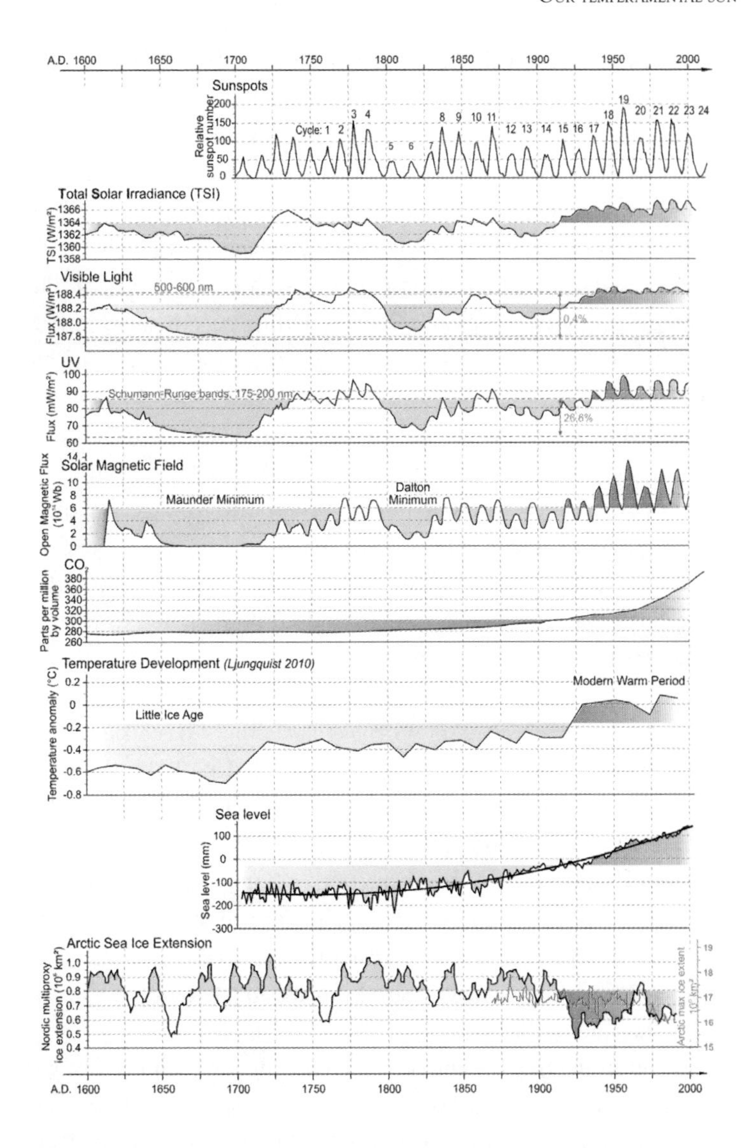

Figure 3.8 Synchronicity of solar activity, temperature, sea level and Arctic Sea ice cover during the last 400 years. Sources: sunspots [136]; total solar irradiance [39]; visible light [39]; UV [39]; solar magnetic field, [134]; CO_2 [51]; temperature (extra-tropical Northern Hemisphere) [109]; sea level [137]; Arctic sea ice cover [138].

The end of the Little Ice Age also represents the start of modern climate warming – the return from a natural cold phase, over a mean temperature and on the way to an expected warm phase, similar to the one the earth went through 1000 years ago during the Medieval Warm Period. The general warming trend of the twentieth century is without a doubt part of a natural cycle, one that has run through its complete course once over the last 1000 years and is still in play. The climate warming of the past 150 years is thus no one-time, unique anomaly. The only task that is left is to determine the extent that manmade CO_2 emissions have really contributed to the temperature increase. How would the warming appear without the anthropogenic effect? Today's global surface temperature increase is at least 1° C with respect to the cyclic low point of the Little Ice Age. Taking a neutral reference line, the zero point of the natural cycle, then the increase up to today is an easily bearable 0.4° C. How much of this very modest warming is due to mankind and how much to natural causes? That is the key question.

The positive relationship between solar activity and temperature during the last 1000 years shows that solar activity has a significant impact on the earth's climate for this period [139–141]. Even the Antarctic winds blow in sync with the variable sun [142]. Swiss lakes fill up and empty out to the same rhythm [143], as does the Indian monsoon [144]. Also the discharge rate of South America's second largest river, the Paraná, has varied in accordance with solar activity over the last 100 years [145–146] (Figure 3.9) because during times of increased solar activity, increased precipitation occurs in the vast area of the Paraná basin.

So what is the likelihood that the sun and climate would suddenly have little to do with each other in recent decades, as the IPCC vehemently postulates? Could there be a reason why the IPCC massively plays down the sun as a climate factor? Yes indeed there could: the less important the sun can be made to appear and the more benign earlier temperature fluctuations can be made to look, the more prominent the twentieth century's warming becomes and

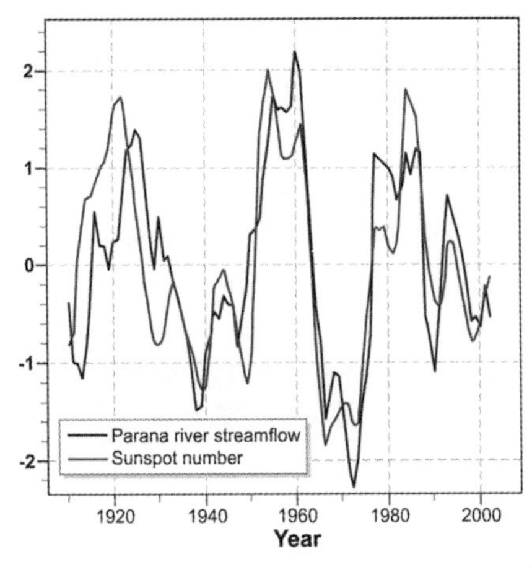

Figure 3.9 The discharge of the Paraná over the course of the last 100 years is in sync with solar activity [146].

the more it is able to steal the limelight. And if you assign the recent warming almost exclusively to the CO_2 effect, as the IPCC does, then the political message gains a lot more traction. The political aim that emerges – a transformation of the energy supply system, particularly in the OECD countries – as a consequence gains higher acceptance among the public (see Chapter 9).

The solar constant

When it comes to the irradiative power of the sun, the misleading term 'solar constant' was gladly employed in the past, and is even used at times today. The solar constant describes the theoretical average strength with which the sun irradiates the earth. As the word 'constant' implies, there is only one universal value for the solar constant. The magnitude was defined by the World Meteorological Organization in Geneva in 1982 and is set at 1367 watts per square metre. That is how much energy the sun would shine vertically on a square metre of the earth if no atmosphere existed. In fact, due to the filtering effects of the atmosphere, only 700 W/m^2 at sea level and about 1000 W/m^2 in mountains where the air is thinner are able to penetrate to the earth's surface. Also built into the value of the solar constant are a few long-term mean values, such as the changing distance between the sun and the earth, and the 11-year solar cycle of activity. As we have seen, the irradiative strength of the sun in the past has been anything but stable over the millennia. The solar constant is thus much more of a 'solar variable'. The obsolete term solar constant has to be employed with extreme caution.

What impact has the sun had on the climate over the last 10,000 years? Postglacial millennial cycles

Now we should attempt to follow the Hallstatt and Eddy millennium cycles further into the past. The last 1000 years are just enough to cover a single Eddy cycle. We now extend the period of interest back 10,000 years in order to see if the solar 1000 and 2300-year cycles of activity are connected to climatic development. Can the 1000-year climate cycle consisting of the Modern Warm Period (today), Little Ice Age and Medieval Warm Period be found further back in the past or was it just a flash in the pan? Without a doubt showing that a long-term sun–climate millennium cyclic exists would further strengthen the importance of natural factors in climate warming over the last 150 years.

At the end of the 1990s a research team led by Gerard Bond examined the postglacial seabed layers of the north Atlantic. In the seabed deposits of the last 10,000 years, the team was looking for clues about the region's climate history. When examining the sediment cores, they found unusual layers of debris deposits which were repeated at regular intervals [114]. The only means of transport for this kind of coarse material to this particular oceanic site was floating icebergs, which unloaded their freight as they slowly melted, thereby spreading their cargo over the ocean floor. Working with the diligence of detectives, the team reconstructed what had happened. Apparently, the north Atlantic system of currents took unplanned tours at regular intervals and carried cool, ice-driving surface water masses from the Arctic to southern regions as far down as the latitudes of Great Britain. Bond's team also reconstructed the intervals between these climatic special events. The result should appear familiar to us: the cycle length corresponded to the 1000–2300-year millennial cycles.

But Bond's team went a step further. To shed more light on natural climate cycles, they compared the pattern of the debris deposits to solar activity over the same period. The climate and sun dynamic fit together like a hand in a glove (Figure 3.10)! They found a very good correlation between the ice-rafted debris layers (a climatic cooling indicator) and the cosmic rays (measured from the ^{10}Be and ^{14}C concentrations). The intensity of the cosmic rays is modulated by solar activity, as discussed earlier. In a nutshell, cold phases occurred predominantly during times of weak solar activity. If one takes a closer look at the documented cyclicality, one suspects that the result consists of the 2300-year Hallstatt cycles, whereby additional 1000-year Eddy cycles are at times glaringly, but at times only weakly, visible [147]. A synthetic mean value over all the cycles produces a period of about 1500 years [114]. (This is a theoretical cycle length that as such does not exist, but rather only illustrates a composite of both real cycle types.)

Bond continued his work until shortly before his death in 2005, and was co-author of a study about the climatic millennial cycles of the last interglacial in southern Germany [148]. Two years earlier he had co-authored a study on Alaska, where scientists had researched abundance changes of diatoms in postglacial sea deposits [115]. This study provided solid confirmation of the north Atlantic results. It is as if a metronome had set the beat, with phases changing periodically from vigorous to subdued diatom growth during the sea's history and with a cycle length of 1000 years (Figure 3.11). These climate cycles run parallel to the millennium cycles and solar activity.

About 6000 km further south, off the western coast of Mexico, yet another research team took sediment cores which also contained a postglacial climate archive [149]. Using the magnesium content of the shells of calcareous protozoa, the scientists were able to reconstruct a temperature curve for the region. The 1000-year cycle was once again found to be at work.

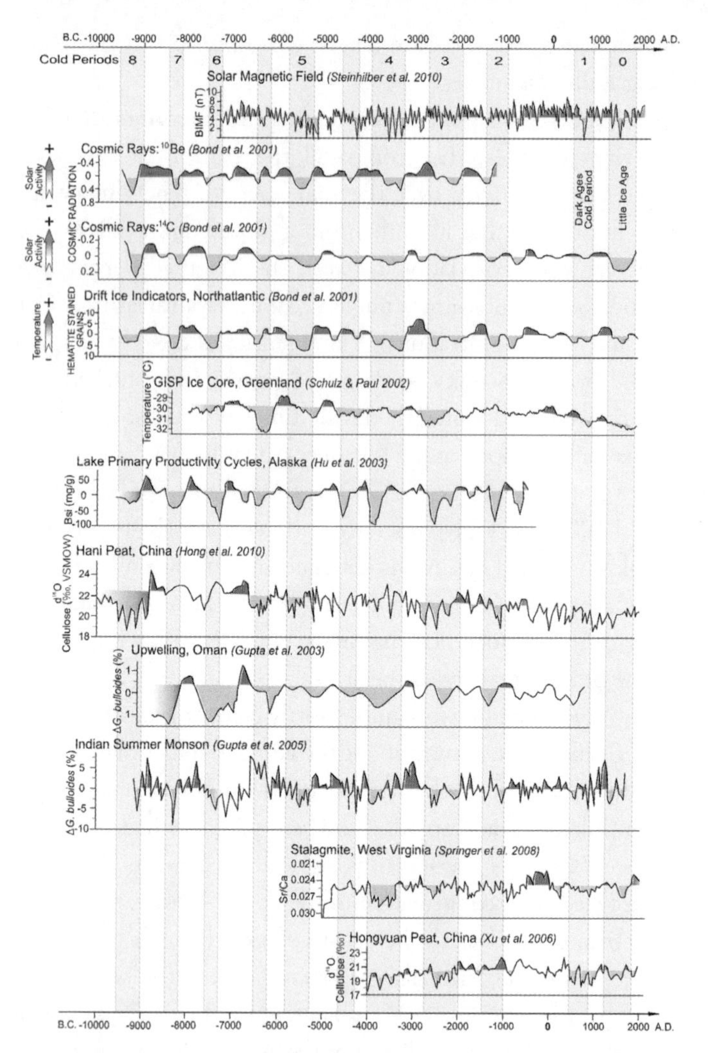

Figure 3.10 Solar activity (the upper three curves) and climate (the other curves) pulsated in concert according to the millennial beat during the entire 10,000-year postglacial period. Characteristic cold intervals (grey) occurred simultaneously over many different parts of the planet during phases of weak solar activity. Numbering of the cold phases according to Bond and colleagues [114]. Sources: Solar magnetic field [105]; ^{10}Be and ^{14}C as an indicator of solar activity [114]; iceberg sediment deposits north Atlantic [114]; Greenland GISP ice core [150]; Alaska sea sediment deposit [115]; China Hani peat [86]; Oman stalagmite zone [151]; Indian monsoon [1]; West Virginia stalagmite [153]; China Hongyuan peat [154]. In some regions the link between the sun and climate was interrupted. To provide a better overview, the peaks of these phases are unmarked.

The temperatures fluctuated reliably according to the millennium beat, in harmony with solar activity.[5] A similar picture is found in New Mexico. Here, during times of reduced solar activity, precipitation declined [155].

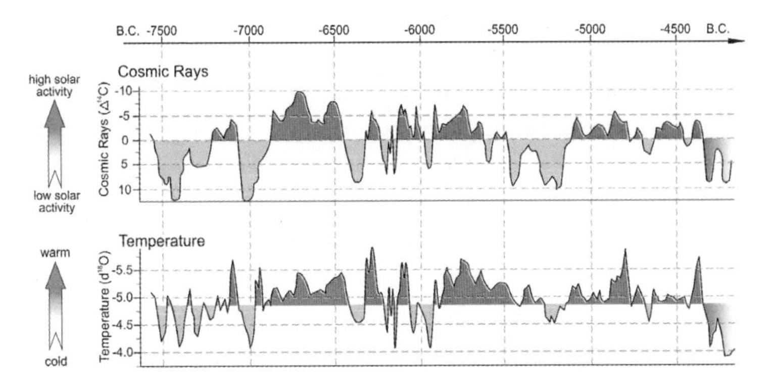

Figure 3.11 Studies of dripstones in Oman for the period 7500–4500 BC show a high degree of synchronicity between solar activity and temperature development. Source: modified from [156].

Another good example of solar-driven climate cycles on a millennium scale from the north Atlantic comes from the sea off the coast of Newfoundland. Using protozoan organisms, two American researchers reconstructed the temperature development of the last 4000 years [157]. Here they found temperature fluctuations on a 1000-year scale. The cold phases occurred simultaneously with the deposit of the ice-rafted debris layers found by Bond [114]. In addition, characteristic glacier advances took place in the surrounding Arctic regions. In other areas of North and South America, as well as in Scandinavia, Iceland, Greenland and the Mediterranean, characteristic climatic millennium cycles were discovered in the postglacial period [153, 158–159, 160–168].

Now let's jump across the Atlantic to the Mauritanian coast in subtropical West Africa. Here a group of researchers extracted

5 At this location the active sun apparently triggered a cold La Niña ocean circulation, while a weak sun caused warm El Niño conditions.

a sediment core from the upper seabed layers which also covered the postglacial period [169]. From the composition of the protozoa, they were able to reconstruct the temperature development of the past 10,000 years. It showed that the climate curve is shaped by characteristic temperature fluctuations with amplitudes of several degrees Celsius, which are repeated every 1000–2000 years. The chronology is very similar to the climate cycles that Bond found in the north Atlantic and occurred parallel to the changes in solar activity. Particularly interesting in the Mauritanian core is the era of the Little Ice Age. Here the scientists were able to find evidence of two distinct cold events with a cooling of 3–4° C. The preceding Medieval Warm Period revealed temperatures for the West African region that were slightly above today's values [169]. A very similar millennial cycle with a distinct temperature curve and good time agreement was also found in an Atlantic sediment core off the north-west coast of Spain [170]. And the climate events in the Alps over the last 7000 years are oriented about the solar millennium rhythms and oscillate in sync with the 2300-year Hallstatt and 1000-year Eddy cycles [171–172].

Further east we find a similar picture, and in Oman the postglacial temperatures pulsated according to the distinct 1000-year cycle [156] (Figure 3.11). Scientists from Heidelberg studied stalagmites from which they extracted the temperature signal for the period of 6000–9000 years before the present. It should come as no surprise that here again a very good correlation between temperature and solar activity was identified.

A research group of the Ocean Drilling Programme (ODP) arrived at a similar result after examining a core that had been extracted off the coast of Oman [152] (Figure 3.11). Based on the abundance patterns of a calcareous foraminifera species, they were able to reconstruct fluctuations in the strength of the summer monsoons for the postglacial phase. As is the case in all the other examples, the curve is characterized by strong millennium cycles of 1000–2000 years.

If we continue our journey a few thousand kilometres to the east, we find the same picture. In a south China cave scientists studied a stalagmite that documents the climate's history over the last 9000 years. The temperature curve is characterized by a series of extraordinarily warm periods during which the Asian monsoon was especially strong [173]. The warm phases occurred about every 1200 years and also coincide with times of increased solar activity. They also correlate well with the sediment deposit layers from the north Atlantic that Bond found. In China [85, 154, 174], Taiwan [175] and Korea [176] a temperature cyclicality predominated over the last 6000 years, which for the most part proceeded in step with solar activity.

Climate cycles with millennial scale periods of 1000–2300 years are well established for the entire postglacial period of the last 12,000 years. The examples studied originate from various oceans and several continents. The cycle is found in the upper, middle and lower latitudes and encompasses the different climate zones, from the Arctic to the Tropics. The temperature fluctuations at times are several degrees Celsius, and thus have a similar or even greater range than the more than 1 ° C of warming that we have experienced since the Little Ice Age. The parallel between millennial climate cycles and solar activity is a primary feature of this variability [177]. The sun's control of the cycles is thus highly probable, particularly because the opposite influence, when solar activity is driven by climate cycles, can be excluded. The cycle length depends on the long solar basic cycles, namely the 2300-year Hallstatt cycle and the 1000-year Eddy cycle. One also has to consider that determining the mean over multiple millennium cycles can result statistically in cycle lengths that are between 2300 and 1000 years, as some Eddy cycles may be too weak to be discernible, and after taking the error margin of the analytical methods and potential climatic disturbance processes into account [178].

The comprehensive collection of case studies we have presented strongly shows that the 1000-year climate cycle that comprises the Medieval Warm Period–Little Ice Age–Modern Warm Period is the continuation of a natural, solar-driven millennial cycle that characterized the postglacial climate development at many locations on earth. Thus suspicions are hardening that only a limited amount of the 0.8° C or so of warming since the end of the Little Ice Age can be attributed to CO_2. Consequently, in its reports the IPCC appears to have grossly overstated CO_2's climate impact and severely neglected natural factors. If CO_2 is truly such a potent climate driver, and if the sun is a far more powerful climate driver than previously assumed, then the current warming since the Little Ice Age should have been far greater. Shouldn't we be concerned that the IPCC climate models are unable to reconstruct the solar-synchronous millennial scale climate cycles of the past 10,000 years [179–181]? Yet they claim to be capable of projecting our future climate.

What impact has the sun had on climate over the last 150,000 years? Millennium cycles during the last Ice Age and the last interglacial

Let's now go even further back into the past to the last ice age, which began 115,000 years ago and ended 10,000 years ago. That ice age is part of an entire series of ice ages, each lasting approximately 100,000 years, interrupted by shorter warm periods, called interglacials (Figure 3.12). The current postglacial is the latest interglacial and it too will eventually come to an end and enter a new 100,000-year long ice age – maybe in a few thousand years' time. The trigger for these changes between ice ages and interglacial periods has to do with the cyclically changing orbital parameters of the earth, which lead to variable conditions for incoming sunlight, so-called insolation (more on this below).

What do we know about the climate fluctuations during this period? What role did the primary solar activity cycles play? First and foremost, we need to recognize that the initial situation was not quite as simple as the sedate postglacial time, our current Holocene. The last ice age was characterized by much more powerful

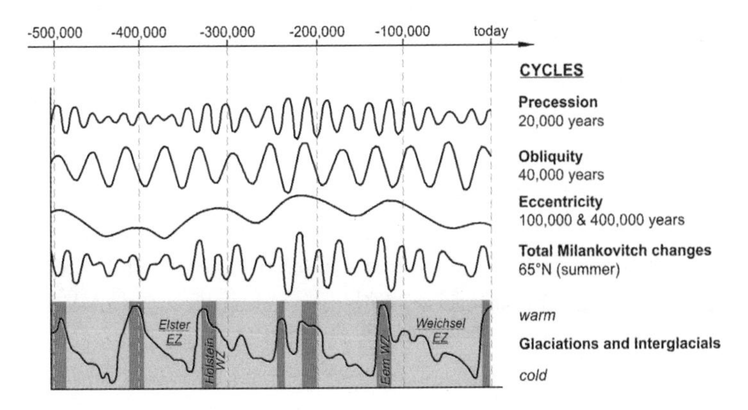

Figure 3.12 Cyclical changes in the earth's orbital parameters (Milankovitch cycles) are thought to have triggered the successive ice ages (EZ) and interglacial periods (WZ) over the last 500,000 years [182].

climate dynamics. Unstable conditions predominated with, at times, extremely rapid changes in average annual temperatures of up to 10° C within just a few decades [183]. Therefore, it would be little wonder if other processes had taken charge of climate events and had the ability to push the few degrees Celsius solar Hallstatt and Eddy cycles into the background.

Ice cores extracted from the Greenland ice sheets provide an important window into the climate archives of the last ice age. By examining the oxygen isotopes in the ice, it is possible to produce an accurate temperature reconstruction. In one of these ice cores, the period of 20,000–60,000 years is contained. Once again one finds the millennial cycles. But are they really the same cycles found in the post-ice age Holocene?

The millennium cycles of the last ice age follow a rather unusual progression. The warming phase is very abrupt. Within just a few decades the average annual temperature in the Arctic rises 6–10° C [183]. However, the subsequent cooling phase lasts many centuries until the cycle starts up again. This saw tooth pattern is the distinguishing feature of these fluctuations, the so-called Dansgaard-Oeschger cycles (named after the Danish and Swiss scientists who discovered them) [184]. Some of the cycles are stronger, while others are weaker. These single cycles in turn group together to form bundles with cycles having a uniform longer-term trend [185].

During Dansgaard-Oeschger cycles with exceptionally low temperatures and strong subsequent warming, massive numbers of icebergs broke off the North American ice sheets and made their way across the ocean [186]. The sediment material rained down from the melting icebergs and spread across large areas of the north Atlantic to form layers on the seabed. These so-called Heinrich layers can be up to 0.5 metre thick and contain deposits that had been scraped off the ground in North America by ice age glaciers before being carried out to sea [187–193]. Interestingly, during the Dansgard-Oeschger cyclic, the temperatures of the Antarctic appear to have mirrored those in Greenland [185], which underscores the complexity of the ice age climate system.

The Dansgaard-Oeschger climate cycles last approximately 1500 years [194]. In this case it is not only a statistical mean of various individual periods, but a natural climate cycle length that varies only slightly [150]. We can guess the next step. We have to compare the climate cycles with the solar activity curve in order to ascertain if the processes are related. The reconstruction of solar irradiative strength is done via cosmic rays, particularly through the 10Be concentration, and that using the same ice cores from which the temperature signal had been extracted. The reconstruction of solar activity during the ice age is not trivial as the 10Be concentration is affected by the powerful ice age climate variations; therefore, this effect has to be deduced in order to reach a value for solar irradiance strength.

An American and a Swiss took on this task in 2006 by examining the relationship between the climatic Dansgaard-Oeschger cycles and solar activity [195]. Their result is sobering. They could find no stable relationship between these two processes. But this is not surprising, as we have determined that no independent 1500-year solar cycle exists [194]. As we have seen, the solar Hallstatt and Eddy basic oscillations are known to be 2300 years and 1000 years long, respectively. Why should they deviate from this during the ice age? We do not gain much enlightenment from the millennium cycles of the last ice age. Scientists are still puzzled over what might be the exact driver behind these abrupt climate changes in connection with the Dansgaard-Oeschger cyclic.

However, we cannot rule out that the sun had a hand in the ice age Dansgaard-Oeschger cyclic too. The Potsdam Institute for Climate Impact Research (PIK) climatologist Stefan Rahmstorf favours an extra-terrestrial cause for the cycles because of the regularity of the fluctuations [196]. In 2005 he co-authored a study that showed that the 1470-year Dansgaard-Oeschger periodicity could have its origin in a combination of the 210-year Suess/de Vries cycle and the 87-year Gleissberg cycle [194, 197–202]. Mathematically, it seems quite plausible as after 1470 years the 210-year de Suess/Vries cycle occurs seven times and the Gleissberg cycle occurs seventeen times when the latter is defined as 86.5 years. However, this is just one model among many, and so the scientific dispute over the exact causes of the ice age Dansgaard-Oeschger-cyclic remains unresolved.

Why is everything so complicated? Why is there no direct linear correlation between the millennium climate cycles and the Hallstatt and Eddy fluctuations of the sun, as is the case in the postglacial period? Clearly, entirely different climate processes operated during the ice age from those in the postglacial period. Therefore, it is assumed that the thermohaline ocean circulation in the ice age climate system reacted to the sun only above certain threshold values [203]. Perhaps there were processes that enhanced the effectiveness of the solar 87-year Suess/de Vries and the 210-year Gleissberg cycles while at the same time reducing the importance of the 1000- and 2300-year Eddy and Hallstatt millennium cycles. There is still a lot research to do. The unclear situation that prevailed during the last ice age should not cause us very big headaches. What is clear is that the postglacial situation is far more relevant to understanding our current climate processes than are the extreme and complex conditions of the ice age era. Regardless of glacial or postglacial, one thing is undeniable: the climate undergoes changes, which at times are extreme and dramatic, and does so naturally.

Let's go back further into climatically calmer times, to the interglacial before the last ice age, the so-called Eem interglacial which started 126,000 years ago. Here once again the world appears to be in order. A German-American research team analysed sediment deposits extracted from a former lake in the south-west foothills of the Alps, 40 km north-west of Memmingen [148]. There they found millennium cycles of between 1000 and 1500 years in the pollen flora. Unfortunately, the precise period length cannot be determined because of the reconstruction method's resolution limits. Also, the reconstruction of solar activity going back that far using traditional methods is not possible without great difficulty, and so a comparison of the climate cycles with solar activity is problematic.

Overall we cannot expect any other high resolution, fully quantitative data sets from previous interglacial phases that would allow comparison with modern climate data. But in the future even qualitative observations from these periods could bring us further and allow us to determine how a normal natural climate dynamic of a typical interglacial is characterized. Studies of sediment deposits from a lake in the Lüneburger Heide showed that in the next to last interglacial (the Holsteinian) 330,000 years ago, the climate was significantly co-determined by the Gleissberg, Hale and Schwabe solar cycles [204].

Long-period Milankovitch earth orbital cycles

Taking into account that the total solar irradiance (TSI) is variable, the term 'solar constant' is somewhat misleading. TSI is the solar energy that strikes one square metre of the earth's outer atmosphere in one second. Here the energy of all wavelengths is summed up. TSI is expressed in watts per square metre.

Changes in TSI are produced by two factors. The first is the change in primary energy emission by the solar power plant. This is the dominant factor when considering relatively short periods (a few thousand years), and thus concerns the postglacial period and the current climate events, as we have seen.

The second major factor that influences the solar energy reaching the earth is the position and movement of the earth during its annual orbit round the sun [205]. Here changes in the distance between the sun and the earth and the tilting of the irradiation angle of incidence play a decisive role. These fluctuations exhibit characteristic cyclical behaviour, but their periods are vastly longer (lasting tens or even hundreds of thousands of years) than the known solar activity cycles. The astronomical cycles of the changing earth's orbital parameters are called Milankovitch cycles after Serbian mathematician Milutin Milankovitch (1879–1958), who came up with the theory of astronomical cycles in the 1920s [206].

Three basic magnitudes are involved here (Figure 3.12). The first Milankovitch basic magnitude is the shape of the earth's orbit, which fluctuates between an ellipse and a circle (eccentricity). These changes occur in 100,000 and 400,000-year cycles. The more eccentric (non-circular) the earth's orbit, the greater the distance to the sun varies over the course of one year. And it means the sun's irradiation intensity striking the earth fluctuates more over the course of the year. Currently, the earth has an orbit whose shape is only slightly elliptical and so its distance to the sun varies by about 3 per cent over the year; this leads to fluctuations in solar irradiance of almost 7 per cent.

The second Milankovitch fundamental magnitude is the axial tilt (obliquity) of the earth's axis, which varies over a period of 41,000 years. Here the climatic relevance is the enhancement of seasons. The more the earth's axis is tilted, the greater is the difference between summer and winter. Currently, the axis tilt is 23.44°, which is approximately halfway between the two extremes. The next tilt minimum is expected to be reached in about 8000 years' time.

Finally, the earth's axis spins with a gyroscopic motion (precession) about the vertical with a period of about 20,000 years.[6] As a result the seasons do not always occur at the same point of the earth's elliptical orbit. Currently, the earth reaches its closest point to the sun in January, in the middle of the northern winter. However, in 11,000 years' time the closest point to the sun will occur in the middle of the northern summer and thus northern winters will become harsher.

Taken together, the Milankovitch cycles lead to changes in solar irradiation of the earth with magnitudes of single to low double-digit percentages, and are therefore serious climate drivers over the long run [207]. This explains why Milankovitch cycles have played a leading role in the changes between the ice ages and the interglacials over the last 1.5 million years [208–209]. Primary solar irradiation variability here plays only a subordinate role, unlike during the climatically stable postglacial period.

We find numerous examples of the influence of astronomical cycles impacting climate deep into the earth's geological history going back hundreds of millions of years [210–211]. But because of the long period lengths, the importance of the Milankovitch cycles for variability during the current interglacial and today's climate is minimal.

The sun's role in climate history over the last 500 million years

Here we mention briefly another much longer climate cycle which has been suggested by the astrophysicist Nir Shaviv and the isotope geochemist Jan Veizer [212–214]. According to their model the solar system travels across the spiral arms of the Milky Way every 140 million years as it journeys through the galaxy, which in turn results in increased cosmic irradiation of the earth. The cosmic rays cause greater amounts of condensation nuclei for cloud formation, which

6 26,000 years with respect to the stars, and 21,600 years when one takes into account that the earth's orbit too is shifting.

then blocks out the sun and reflects it back into space. This has a cooling effect on the earth. A temperature cycle of similar periodicity can be shown from the geological record. This finding may appear to be of little importance to today's climate, but the interrelationship of cosmic rays and cloud formation identifies another decisive piece of the puzzle in explaining the solar-induced warming from 1977 to 2000, a period when cosmic rays were largely blocked out by intense solar activity.

Last but not least: the significant 60-year cycle

As we have seen, the earth's climate is profoundly impacted by a number of oscillations of different lengths. Oscillation is a typical behaviour for natural systems, with some oscillations being induced by external factors, while others occur independently. In the global temperature measurement series, as well as in some of the larger oceanic systems, a clear cycle with a period length of approximately 60 years is found [215], including the Atlantic AMO, PDO and certain parts of the NAO [216] (see Chapter 4). The strength of the Indian monsoons [217] and the Arctic temperature [218] appear to follow the 60-year cycle.

So how does this 60-year cycle come about? It doesn't appear to be because of a primary solar cycle, as it lies somewhere between the 22-year Hale and the 87-year Gleissberg cycles. Or does it? It is known that the length of the Gleissberg cycle fluctuates between 50 and 140 years and thus it can at times also cover the 60-year mark [13]. This would make it a sort of trump card that could be used to cover everything and nothing in this range. More interesting is the observation that the Northern Lights appearing at the middle latitudes in the eighteenth and nineteenth centuries had a length of 62 years [219]. Also cosmic radiation appears to follow a 65-year cycle, as 10Be measurement curves of the Greenland and Antarctic ice cores have shown. Perhaps here too we find a poorly understood and indeterminate solar activity cycle [220–222].

A Norwegian research team led by Odd Helge Otterå recently found that the AMO with its typical 6-year cyclicality must be significantly pulsed by external drive systems [223]. They were able to demonstrate that over the last 600 years the phases of the AMO were driven above all by solar activity fluctuations and large volcanic eruptions. The AMO is the main magnitude of influence for north Atlantic sea surface temperatures, which are connected to the global temperature curve.

It cannot be excluded that the two largest planets in our solar system play a role in the 60-year cycle. It takes almost 30 years for Saturn to orbit the sun; Jupiter, which is closer to the sun, takes only 12 years. Every 20 years Jupiter catches up with Saturn and thus both gas planets are aligned with the sun and together exert a gravitational pull on our mother star. After a total of 60 years, both planets are again in alignment at the original starting point of their astro ballet. The cyclic distortions of the planetary system possibly lead to slight changes in the earth and lunar orbits, as well as to disturbances in the inner workings of the solar power plant, which may be indirectly discerned in the earth's climate system [150, 215, 224–229]. Noteworthy is that shorter-term movements of the sun round the gravity centre of the solar system with periods of 7–9 years also appear to have an effect on the climate [230].

Even though we do not know much about the exact causes of these planetary tidal effects, the 60-year cycle influences our climate via natural climatic oscillations of the world's oceans in a not negligible manner (see Chapter 4 and Scafetta in that chapter). Climate models that ignore this mechanism cannot be considered complete.

References

1. Schröder, K.-P. and R. C. Smith (2008) Distant future of the sun and earth revisited. *Mon. Not. R. Astron. Soc.* 386 (1), 155–63.
2. Bounama, C., W. von Bloh and S. Franck (2004) *Das Ende des Raumschiffs Erde. Spektrum der Wissenschaft* 10, 52–9.
3. Haigh, J. D. (2009) Mechanisms for solar influence on the earth's climate. In T. Tsuda, R. Fujii, K. Shibata and M. A. Geller (eds.), *Climate and Weather of the*

Sun–Earth System (CAWSES). Selected Papers from the 2007 Kyoto Symposium (TERRAPUB Tokyo), 231–56.

4. Wikipedia (2010) Sonnenstrahlung. http://de.wikipedia.org/wiki/Sonnenstrahlung (modified).
5. Whitehouse, D. (2006) *The Sun – A Biography*. Wiley, Chichester.
6. Marusek, J. A. (2007) Solar Storm Threat Analysis. http://www.breadandbutterscience.com/SSTA.pdf.
7. Earth Observatory (2011) Sunspots. http://earthobservatory.nasa.gov/IOTD/view.php?id=37575.
8. Lockwood, M. and C. Fröhlich (2007) Recent oppositely directed trends in solar climate forcings and the global mean surface air temperature. *Proceedings of the Royal Society* A 463, 2447–60.
9. Wenzler, T., S. K. Solanki and N. A. Krivova (2009) Reconstructed and measured total solar irradiance: is there a secular trend between 1978 and 2003? *Geophysical Research Letters* 36, 1–4.
10. Arendt, T. (2008) *Heinrich Samuel Schwabe – Der Entdecker der Sonnenfleckenperiode. Reprint nach dem Original das 1927 im C. Dünnhaupt Verlag Dessau erschien.* Funk Verlag Bernhard Hein e.K., Dessau.
11. Demetrescu, C. and V. Dobrica (2008) Signature of Hale and Gleissberg solar cycles in the geomagnetic activity. *Journal of Geophysical Research* 113.
12. McCracken, K. G., G. A. M. Dreschhoff, D. F. Smart and M. A. Shea (2001) Solar cosmic ray events for the period 1561–1994. 2. The Gleissberg periodicity. *Journal of Geophysical Research* 106 (A10), 21599–609.
13. Ogurtsov, M. G., Y. A. Nagovitsyn, G. E. Kocharov and H. Jungner (2002) Long-period cycles of the sun's activity recorded in direct solar data and poxies. *Solar Physics* 211, 371–94.
14. Peristykh, A. N. and P. E. Damon (2003) Persistence of the Gleissberg 88-year solar cycle over the last 12,000 years: evidence from cosmogenic isotopes. *Journal of Geophysical Research* 108 (A1).
15. Dergachev, V. A. and O. M. Raspopov (2009) Long-term solar activity as a controlling factor for global warming in the 20th century. *Geomagnetism and Aeronomy* 49 (8), 1271–4.
16. Ma, L. H. (2009) Gleissberg cycle of solar activity over the last 7000 years. *New Astronomy* 14, 1–3.
17. Garcia, A. and Z. Mouradian (1998) The Gleissberg cycle of minima. *Solar Physics* 180, 495–8.
18. Waple, A. M., M. E. Mann and R. S. Bradley (2002) Long-term patterns of solar irradiance forcing in model experiments and proxy based surface temperature reconstructions. *Climate Dynamics* 18, 563–78.
19. Suess, H. E. (1980) The radiocarbon method in tree rings of the last 8000 years. *Radiocarbon* 20, 200–9.
20. Sonett, C. P. (1984) Very long solar periods and the radiocarbon record. *Rev. Geophys. Spa. Phys.* 22 (3), 239–54.
21. Damon, P. E. and C. P. Sonett (1991) Solar and terrestrial components of the atmospheric 14C variation spectrum In C. P. Sonett, M. S. Giampapa and M. S. Mathews (eds.), *The Sun in Time*. University of Arizona Press, Tuscon, 360–88.
22. Stuiver, M. and T. F. Braziunas (1993) Sun, ocean, climate and atmospheric $14CO_2$: an evaluation of causal and spectral relationships. *The Holocene* 3, 289–305.

23. Wagner, G., J. Beer, J. Masarik and R. Muscheler (2001) Presence of the solar de Vries cycle (~205 years) during the last ice age. *Geophysical Research Letters* 28 (2), 303–6.
24. Knudsen, M. F., P. Riisager, B. H. Jacobsen, R. Muscheler, I. Snowball and M.-S. Seidenkrantz (2009) Taking the pulse of the sun during the Holocene by joint analysis of 14C and 10Be. *Geophysical Research Letters* 36, L16701.
25. Richards, M. T., M. L. Rogers and D. S. P. Richards (2009) Long-term variability in the length of the solar cycle. *Publications of the Astronomical Society of the Pacific* 121 (881), 797–809.
26. Gray, L. J., J. Beer, M. Geller, J. D. Haigh, M. Lockwood, K. Matthes, U. Cubasch, D. Fleitmann, G. Harrison, L. Hood, J. Luterbacher, G. A. Meehl, D. Shindell, B. van Geel and W. White (2010) Solar influences on climate. *Reviews of Geophysics* 48, 1–53.
27. Babcock, H. W. (1961) The topology of the sun's magnetic field and the 22-year cycle. *Astrophys. J.* 133, 572–87.
28. Bard, E. and M. Frank (2006) Climate change and solar variability: what's new under the sun? *Earth and Planetary Science Letters* 248, 1–14.
29. Meehl, G. A., J. M. Arblaster, K. Matthes, F. Sassi and H. von Loon (2009) Amplifying the Pacific climate system response to a small 11-year solar cycle forcing. *Science* 325, 1114–18.
30. Haigh, J. D., A. R. Winning, R. Toumi and J. W. Harder (2010) An influence of solar spectral variations on radiative forcing of climate. *Nature* 467, 696–9.
31. Wintoft, P. (2011) The variability of solar EUV: a multiscale comparison between sunspot number, 10.7 cm flux, LASP MgII index, and SOHO/SEM EUV flux. *Journal of Atmospheric and Solar–Terrestrial Physics* 73, 1708–14.
32. Woods, T. N. and G. J. Rottman (2002) Solar ultraviolet variability over time periods of aeronomic interest. In M. Mendillo, A. Nagy and J. H. Waite (eds.), *Atmospheres in the Solar System: Comparative Aeronomy. Geophysical Monograph* 130, American Geophysical Union, Washington, DC, 221–33.
33. Krivova, N. A., L. E. A. Vieira and S. K. Solanki (2010) Reconstruction of solar spectral irradiance since the Maunder minimum. *Journal of Geophysical Research* 115, 1–11.
34. Ningombam, S. S. (2011) Variability of sunspot cycle QBO and total ozone over high altitude western Himalayan regions. *Journal of Atmospheric and Solar–Terrestrial Physics* 73, 2305–13.
35. Merkel, A. W., J. W. Harder, D. R. Marsh, A. K. Smith, J. M. Fontenla and T. N. Woods (2011) The impact of solar spectral irradiance variability on middle atmospheric ozone. *Geophysical Research Letters* 38, 1–6.
36. Gabriel, A., H. Schmidt and D. H. W. Peters (2011) Effects of the 11-year solar cycle on middle atmospheric stationary wave patterns in temperature, ozone, and water vapour. *J. Geophys. Res.* 116 (D23), D23301.
37. Lean, J. (2000) Evolution of the sun's spectral irradiance since the Maunder minimum. *Geophysical Research Letters* 27 (16), 2425–8.
38. NGDC (2011) Sun spots (monthly averages). ftp://ftp.ngdc.noaa.gov/stp/solar_data/sunspot_numbers/international/monthly/monthly.plt.
39. Shapiro, A. I., W. Schmutz, E. Rozanov, M. Schoell, M. Haberreiter, A. V. Shapiro and S. Nyeki (2011) A new approach to long-term reconstruction of the solar irradiance leads to large historical solar forcing. *Astronomy & Astrophysics* 529, 1–8.

40. Svalgaard, L. (2011) Long-term reconstruction of solar and solar wind parameters. http://www.leif.org/research/Svalgaard_ISSI_Proposal_Base.pdf.
41. NMBD (2011) Kiel Neutronen-Monitor. http://www.nmdb.eu/nest/search.php.
42. Frederick, J. E. and A. L. Hodge (2011) Solar irradiance at the earth's surface: long-term behavior observed at the South Pole. *Atmos. Chem. Phys.* 11, 1177–89.
43. Scafetta, N. (2009) Empirical analysis of the solar contribution to global mean air surface temperature change. *Journal of Atmospheric and Solar–Terrestrial Physics* 71, 1916–23.
44. Solomon, S. C., T. N. Woods, L. V. Didkovsky, J. T. Emmert and L. Qian (2010) Anomalously low solar extreme-ultraviolet irradiance and thermospheric density during solar minimum. *Geophysical Research Letters* 37.
45. Solomon, S. C., L. Qian, L. V. Didkovsky, R. A. Viereck and T. N. Woods (2011) Causes of low thermospheric density during the 2007–2009 solar minimum. *Journal of Geophysical Research* 116, 1–14.
46. Hunt, L. A., M. G. Mlynczak, B. T. Marshall, C. J. Mertens, J. C. Mast, R. E. Thompson, L. L. Gordley and J. M. Russell III (2011) Infrared radiation in the thermosphere at the onset of solar cycle 24. *Geophysical Research Letters* 38, 1–5.
47. Mlynczak, M. G., F. J. Martin-Torres, B. T. Marshall, R. E. Thompson, J. Williams, T. Turpin, D. P. Kratz, J. M. Russell III, T. Woods and L. L. Gordley (2007) Evidence for a solar cycle influence on the infrared energy budget and radiative cooling of the thermosphere. *Journal of Geophysical Research* 112, 1–7.
48. Beig, G. (2011) Long-term trends in the temperature of the mesosphere/lower thermosphere region: 2. Solar response. *Journal of Geophysical Research* 116, 1–7.
49. Weber, W. (2010) Strong signature of the active Sun in 100 years of terrestrial insolation data. *Annalen der Physik* 522 (6), 372–81.
50. Ram, M., M. R. Stolz and B. A. Tinsley (2009) The terrestrial cosmic ray flux: its importance for climate. *Eos* 90 (44), 397–8.
51. IPCC (2007) *Climate Change 2007: The Physical Science Basis. Contribution of Working Group I to the Fourth Assessment Report of the Intergovernmental Panel on Climate Change.* Cambridge University Press, Cambridge and New York.
52. Svensmark, H. and E. Friis-Christensen (2007) Reply to Lockwood and Fröhlich – the persistent role of the sun in climate forcing. *Danish National Space Center, Scientific Report* 3.
53. Shaviv, N. J. (2008) Using the oceans as a calorimeter to quantify the solar radiative forcing. *Journal of Geophysical Research* 113, 1–13.
54. Barnhart, B. L. and W. E. Eichinger (2011) Empirical mode decomposition applied to solar irradiance, global temperature, sunspot number, and CO_2 concentration data. *Journal of Atmospheric and Solar–Terrestrial Physics* 73 (13), 1771–9.
55. Camp, C. D. and K. K. Tung (2007) Surface warming by the solar cycle as revealed by the composite mean difference projection. *Geophys. Res. Lett.* 34 (14), L14703.
56. Pišoft, P., E. Holtanová, P. Huszár, Jiří Mikšovský and M. Žák (2012) Imprint of the 11-year solar cycle in reanalyzed and radiosonde data sets: a spatial frequency analysis approach. *Climate Change* 110, 85–99.
57. Gleisner, H. and P. Thejll (2003) Patterns of tropospheric response to solar variability. *Geophysical Research Letters* 30 (13), 44/41–44/44.
58. Alexander, W. J. R., F. Bailey, D. B. Bredenkamp, A. van der Merwe and N. Willemse (2007) Linkages between solar activity, climate predictability and water

resource development. *Journal of the South African Institution of Civil Engineering* 49 (2), 32–44.

59. Stager, J. C., A. Ruzmaikin, D. Conway, P. Verburg and P. J. Mason (2007) Sunspots, El Niño, and the levels of Lake Victoria, East Africa. *Journal of Geophysical Research* 112, 1–13.
60. Mason, P. J. (2006) Lake Victoria: a predictably fluctuating resource. *International Journal on Hydropower Dams* 13 (3), 118–20.
61. Perry, C. A. (2007) Evidence for a physical linkage between galactic cosmic rays and regional climate time series. *Advances in Space Research* 40, 353–64.
62. van Loon, H., G. A. Meehl and D. J. Shea (2007) Coupled air–sea response to solar forcing in the Pacific region during northern winter. *Journal of Geophysical Research* 112, 1–8.
63. van Loon, H., G. A. Meehl and J. M. Arblaster (2004) A decadal solar effect in the tropics in July–August. *Journal of Atmospheric and Solar–Terrestrial Physics* 66, 1767–78.
64. Meehl, G. A., J. M. Arblaster, G. Branstator and H. Van Loon (2008) A Coupled air–sea response mechanism to solar forcing in the Pacific Region. *Journal of Climate* 21, 2883–97.
65. Dengel, S., D. Aeby and J. Grace (2009) A relationship between galactic cosmic radiation and tree rings. *New Phytologist* 184, 545–51.
66. Dobrica, V., C. Demetrescu and G. Maris (2010) On the response of the European climate to solar/geomagnetic long-term activity. *Annals of Geophysics* 53 (4), doi: 10.4401/ag-4552.
67. Lohmann, G., N. Rimbu and M. Dima (2004) Climate signature of solar irradiance variations: analysis of long-term instrumental, historical, and proxy data. *International Journal of Climatology* 24, 1045–56.
68. Rigozo, N. R., H. E. d. Silva, D. J. R. Nordemann, E. Echer, M. P. d. S. Echer and A. Prestes (2008) The medieval and modern maximum solar activity imprints in tree ring data from Chile and stable isotope records from Antarctica and Peru. *Journal of Atmospheric and Solar–Terrestrial Physics* 70, 1012–24.
69. Miyahara, H., Y. Yokoyama and K. Masuda (2008) Possible link between multi-decadal climate cycles and periodic reversals of solar magnetic field polarity. *Earth and Planetary Science Letters* 272, 290–5.
70. Prestes, A., N. R. Rigozo, D. J. R. Nordemann, C. M. Wrasse, M. P. Souza Echer, E. Echer, M. B. da Rosa and P. H. Rampelotto (2011) Sun–earth relationship inferred by tree growth rings in conifers from Severiano De Almeida, Southern Brazil. *Journal of Atmospheric and Solar–Terrestrial Physics* 73, 1587–93.
71. King, J. W., A. J. Hurst, A. J. Slater, P. A. Smith and B. Tomkin (1974) Agriculture and sunspots. *Nature* 252, 2–3.
72. Willet, H. C. (1974) Recent statistical evidence in support of the predictive signifcance of solar–climatic cycles. *Monthly Weather Review* 102, 679–86.
73. Souza Echer, M. P., E. Echer, N. R. Rigozo, C. G. M. Brum, D. J. R. Nordemann and W. D. Gonzalez (2012) On the relationship between global, hemispheric and latitudinal averaged air surface temperature (GISS time series) and solar activity. *Journal of Atmospheric and Solar–Terrestrial Physics* 74, 87–93.
74. Raspopov, O. M., V. A. Dergachev and T. Kolström (2004) Hale cyclicity of solar activity and its relation to climate variability. *Solar Physics* 224, 455–63.
75. King, J. W. (1975) Sun–weather relationships. *Aeronautics and Astronautics* 13, 10–19.

76. Pugacheva, G., A. Almeida, A. Gusev, I. Martin, H. Pinto, V. Pankov and W. Spjeldvik (2001) New evidences of space weather impact on weather and climate in southern hemisphere. Proceedings of ICRC 2001, *Copernicus Gesellschaft*, 4153–6.
77. Zanchettin, D., A. Rubino, P. Traverso and M. Tomasino (2008) Impact of variations in solar activity on hydrological decadal patterns in northern Italy. *Journal of Geophysical Research* 113, D12102.
78. Reid, G. C. and K. S. Gage (1988) The climatic impact of secular variations in solar irradiance. In F. R. Stephenson and A. W. Wolfendale (eds.), *Secular Solar and Geomagnetic Variations in the Last 10,000 Years*. Kluwer Academic, Norwell, MA, 225–43.
79. Reid, G. C. (1991) Solar total irradiance variations and the global sea surface temperature record. *J. Geophys. Res.* 96, 2835–44.
80. Mufti, S. and G. N. Shah (2011) Solar-geomagnetic activity influence on earth's climate. *Journal of Atmospheric and Solar–Terrestrial Physics* 73, 1607–15.
81. Andrews, J. T., J. Hardadottir, J. S. Stoner, M. E. Mann, G. B. Kristjansdottir and N. Koc (2003) Decadal to millennial-scale periodicities in North Iceland shelf sediments over the last 12 000 cal yr: long-term North Atlantic oceanographic variability and solar forcing. *Earth and Planetary Science Letters* 210, 453–65.
82. Sejrup, H. P., H. Haflidason and J. T. Andrews (2011) A Holocene North Atlantic SST record and regional climate variability. *Quaternary Science Reviews* 30, 3181–95.
83. Santos, F., M. Gómez-Gesteira, M. deCastro and I. Álvarez (2011) Upwelling along the western coast of the Iberian peninsula: dependence of trends on fitting strategy. *Climate Research* 48 (2–3), 213–18.
84. Burroughs, W. J. (1992) *Weather Cycles Real or Imaginary?* Cambridge University Press, Cambridge.
85. Xu, H., Y. Hong, Q. Lin, B. Hong, H. Jiang and Y. Zhu (2002) Temperature variations in the past 6000 years inferred from δ 18O of peat cellulose from Hongyuan, China. *Chinese Science Bulletin* 47 (18), 1578–84.
86. Hong, B., M. Uchida, X. T. Leng and Y. T. Hong (2010) Peat cellulose isotopes as indicators of Asian monsoon variability. *PAGES News* 18, 18–20.
87. Zhao, C., Z. Yu, Y. Zhao and E. Ito (2009) Possible orographic and solar controls of late Holocene centennialscale moisture oscillations in the northeastern Tibetan plateau. *Geophysical Research Letters* 36.
88. Faurschou Knudsen, M., B. H. Jacobsen, P. Riisager, J. Olsen and M.-S. Seidenkrantz (2012) Evidence of Suess solar-cycle bursts in subtropical Holocene speleothem δ18O records. *The Holocene* 22 (5), 597–602.
89. Raspopov, O. M., V. A. Dergachev, J. Esper, O. V. Kozyreva, D. Frank, M. Ogurtsov, T. Kolström and X. Shao (2008) The influence of the de Vries (~200-year) solar cycle on climate variations: results from the Central Asian mountains and their global link. *Palaeogeography, Palaeoclimatology, Palaeoecology* 259, 6–16.
90. Breitenmoser, P., J. Beer, S. Brönnimann, D. Frank, F. Steinhilber and H. Wanner (2012) Solar and volcanic fingerprints in tree-ring chronologies over the past 2000 years. *Palaeogeography, Palaeoclimatology, Palaeoecology* 313–14, 127–39.
91. Yu, Z. and E. Ito (1999) Possible solar forcing of century-scale drought frequency in the northern Great Plains. *Geology* 27 (3), 263–6.
92. Hodell, D. A., M. Brenner, J. H. Curtis and T. Guilderson (2001) Solar forcing of drought frequency in the Maya lowlands. *Science* 292, 136–70.

93. Castañeda, I. S., J. P. Werne, T. C. Johnson and L. A. Powers (2011) Organic geochemical records from Lake Malawi (East Africa) of the last 700 years, part II: biomarker evidence for recent changes in primary productivity. *Palaeogeography, Palaeoclimatology, Palaeoecology* 303, 140–54.
94. Wiles, G. C., R. D. D'Arrigo, R. Villalba, P. E. Calkin and D. J. Barclay (2004) Century-scale solar variability and Alaskan temperature change over the past millennium. *Geophysical Research Letters* 31, 1–4.
95. Delmonte, B., J. R. Petit, G. Krinner, V. Maggi, J. Jouzel and R. Udisti (2005) Ice core evidence for secular variability and 200-year dipolar oscillations in atmospheric circulation over East Antarctica during the Holocene. *Climate Dynamics*.
96. Casford, J. S. L., R. Abu-Zied, E. J. Rohling, S. Cooke, K. P. Boessenkool, H. Brinkhuis, C. D. Vries, G. Wefer, M. Geraga, G. Papatheodorou, I. Croudace, J. Thomson and V. Lykousis (2001) Mediterranean climate variability during the Holocene. *Mediterranean Marine Science* 2 (1), 45–55.
97. Dima, M., G. Lohmann and I. Dima (2005) Solar-induced and internal climate variability at decadal timescales. *International Journal of Climatology* 25, 713–33.
98. Debret, M., V. Bout-Roumazeilles, F. Grousset, M. Desmet, J. F. McManus, N. Massei, D. Sebag, J.-R. Petit, Y. Copard and A. Trentesaux (2007) The origin of the 1500-year climate cycles in Holocene North Atlantic records. *Clim. Past* 3, 569–75.
99. Wanner, H., O. Solomina, M. Grosjean, S. P. Ritz and M. Jetel (2011) Structure and origin of Holocene cold events. *Quaternary Science Reviews* 30, 3109–23.
100. Archer, D. and S. Rahmstorf (2010) *The Climate Crisis*, 1st edn. Cambridge University Press, Cambridge.
101. Abreu, J. A., J. Beer and A. Ferriz-Mas (2010) Past and Future solar activity from cosmogenic radionuclides. In S. R. Cranmer, J. T. Hoeksema and J. L. Kohl (eds.), SOHO-23: *Understanding a Peculiar Solar Minimum. ASP Conference Series*, 428, 287–95.
102. Sonett, C. P. and S. A. Finney (1990) The spectrum of radiocarbon. *Philosophical Transactions of Royal Society of London* A330, 413–26.
103. Vasiliev, S. S. and V. A. Dergachev (2002) The ~2400-year cycle in atmospheric radiocarbon concentration: bispectrum of 14C data over the last 8000 years. *Annales Geophysicae* 20, 115–20.
104. Dreschhoff, G. A. M. (2008) Paleo-strophysical data in relation to temporal characteristics of the solar magnetic field. In R. Caballero, J. C. D'Olivo, G. Medina-Tanco, L. Nellen, F. A. Sánchez and J. F. Valdés-Galicia (eds.), *Proceedings of the 30th International Cosmic Ray Conference*, Universidad Nacional Autónoma de México, Mexico City, 541–4.
105. Steinhilber, F., J. A. Abreu, J. Beer and K. G. McCracken (2010) Interplanetary magnetic field during the past 9300 years inferred from cosmogenic radionuclides. *Journal of Geophysical Research* 115, A01104.
106. Charvátová, I. (2000) Can the origin of the 2400-year cycle of solar activity be caused by solar inertial motion? Ann. *Geophysicae* 18, 399–405.
107. Abreu, J. A., J. Beer, A. Ferriz-Mas, K. G. McCracken and F. Steinhilber (2012) Is there a planetary influence on solar activity? *A&A* 548, A88.
108. Steinhilber, F., J. Beer and C. Fröhlich (2009) Total solar irradiance during the Holocene. *Geophysical Research Letters* 36, L19704 (modified).
109. Ljungqvist, F. C. (2010) A new reconstruction of temperature variability in the extra-tropical northern hemisphere during the last two millennia. *Geografiska Annaler*: Series A 92 (3), 339–51 (modified).

110. Ma, L. H. (2007) Thousand-year cycle signals in solar activity. *Solar Physics* 245, 411–14.
111. Solanki, S. K., I. G. Usoskin, B. Kromer, M. Schüssler and J. Beer (2004) Unusual activity of the sun during recent decades compared to the previous 11,000 years. *Nature* 431, 1084–7.
112. Muscheler, R., F. Joos, J. Beer, S. A. Müller, M. Vonmoos and I. Snowball (2007) Solar activity during the last 1000 years inferred from radionuclide records. *Quaternary Science Reviews* 26, 82–97.
113. Bond, G., W. Showers, M. Cheseby, R. Lotti, P. Almasi, P. deMenocal, P. Priore, H. Cullen, I. Hajdas and G. Bonani (1997) A pervasive millennial-scale cycle in North Atlantic Holocene and glacial climates. *Science* 278, 1257–66.
114. Bond, G., B. Kromer, J. Beer, R. Muscheler, M. N. Evans, W. Showers, S. Hoffmann, R. Lotti-Bond, I. Hajdas and G. Bonani (2001) Persistent solar influence on North Atlantic climate during the Holocene. *Science* 294, 2130–6.
115. Hu, F. S., D. Kaufman, S. Yoneji, D. Nelson, A. Shemesh, Y. Huang, J. Tian, G. Bond, B. Clegg and T. Brown (2003) Cyclic variation and solar forcing of Holocene climate in the Alaskan subarctic. *Science* 301, 1890–3.
116. Singer, S. F. and D. T. Avery (2008) *Unstoppable Global Warming – Every 1,500 Years*. Rowan & Littlefield, Lanham, MD.
117. Chambers, F. M., M. I. Ogle and J. J. Blackford (1999) Palaeoenvironmental evidence for solar forcing of Holocene climate: linkages to solar science. *Progress in Physical Geography* 23 (2), 181–204.
118. Stuiver, M., T. F. Braziunas and P. M. Grootes1 (1997) Is there evidence for solar forcing of climate in the GISP2 oxygen isotope record? *Quaternary Research* 48, 259–66.
119. Grove, J. M. and R. Switsur (1994) Glacial geological evidence for the medieval warm period. *Climatic Change* 26, 143–69.
120. Leroy, S. A. G., H. A. K. Lahijani, M. Djamali, A. Naqinezhad, M. V. Moghadam, K. Arpe, M. Shah-Hosseini, M. Hosseindoust, C. S. Miller, V. Tavakoli, P. Habibi and M. N. Beni (2011) Late Little Ice Age palaeoenvironmental records from the Anzali and Amirkola lagoons (south Caspian Sea): vegetation and sea level changes. *Palaeogeography, Palaeoclimatology, Palaeoecology* 302, 415–34.
121. Miyahara, H., D. Sokoloff and I. G. Usoskin (2006) The solar cycle at the Maunder minimum epoch. In M. Duldig (ed.), *Advances in Geosciences, Vol. 2: Solar Terrestrial*, World Scientific Co., Singapore, 1–20.
122. Luterbacher, J., R. Rickli, E. Xoplaki, C. Tinguely, C. Beck, C. Pfister and H. Wanner (2001) The late Maunder minimum (1675–1715) – a key period for studying decadal scale climatic change in Europe. *Climatic Change* 49, 441–62.
123. Morellón, M., A. Pérez-Sanz, J. P. Corella, U. Büntgen, J. Catalán, P. González-Sampériz, J. J. González-Trueba, J. A. López-Sáez, A. Moreno, S. Pla, M. Á. Saz-Sánchez, P. Scussolini, E. Serrano, F. Steinhilber, V. Stefanova, T. Vegas-Vilarrúbia and B. Valero-Garcés (2012) A multi-proxy perspective on millennium-long climate variability in the southern Pyrenees. *Climate Past* 8, 683–700.
124. Soon, W. W.-H. and S. H. Yaskell (2003) *The Maunder Minimum and the Variable Sun–Earth Connection*. World Scientific Publishing, Singapore.
125. Sejrup, H. P., S. J. Lehman, H. Haflidason, D. Noone, R. Muscheler, I. M. Berstad and J. T. Andrews (2010) Response of Norwegian Sea temperature to solar forcing since 1000 A.D. *Journal of Geophysical Research* 115, C12034.

126. Zinke, J., C. Dullo, H. von Storch, B. Müller, E. Zorita, B. Rein, B. Mieding, H. Miller, A. Lücke, G. Schleser, M. Schwab, J. Negendank, U. Kienel, G. Ruoco and A. Eisenhauer (2004) Evidence for the climate during the late Maunder minimum from proxy data and model simulations available within KIHZ. In H. Fischer, T. Kumke, G. Lohmann, G. Flöser, H. Miller and H. von Storch (eds.), *The Climate in Historical Times: Towards a Synthesis of Holocene Proxy Data and Climate Models*, Springer, New York, 397–414.
127. Hathaway, D. H. and R. M. Wilson (2004) What the sunspot record tells us about space climate. *Solar Physics* 224, 5–19.
128. Lockwood, M. (2011) Shining a light on solar impacts. *Nature Climate Change* 1, 98–9.
129. Vaquero, J. M., M. C. Gallego, I. G. Usoskin and G. A. Kovaltsov (2011) Revisited sunspot data: a new scenario for the onset of the Maunder minimum. *The Astrophysical Journal Letters* 731, 1–4.
130. Vieira, L. E. A., S. K. Solanki, N. A. Krivova and I. Usoskin (2011) Evolution of the solar irradiance during the Holocene. *Astronomy & Astrophysics* 531 (A6), 20.
131. Usoskin, I. G., S. K. Solanki and M. Korte (2006) Solar activity reconstructed over the last 7000 years: the influence of geomagnetic field changes. *Geophysical Research Letters* 33, 1–4.
132. Russell, C. T., J. G. Luhmann and L. K. Jian (2010) How unprecedented a solar minimum? *Reviews of Geophysics* 48, 1–16.
133. Lockwood, M., R. Stamper and M. N. Wild (1999) A doubling of the sun's coronal magnetic field during the past 100 years. *Nature* 399, 437–9.
134. Balmaceda, L., N. A. Krivova and S. K. Solanki (2007) Reconstruction of solar irradiance using the group sunspot number. *Advances in Space Research* 40, 986–9.
135. Rao, U. R. (2011) Contribution of changing galactic cosmic ray flux to global warming. *Current Science* 100 (2), 223–5.
136. NGDC (2011) Sonnenflecken, Jahresmittelwerte. ftp://ftp.ngdc.noaa.gov/stp/solar_data/sunspot_numbers/international/yearly/yearly.plt.
137. Jevrejeva, S., J. C. Moore, A. Grinsted and P. L. Woodworth (2008) Recent global sea level acceleration started over 200 years ago? *Geophysical Research Letters* 35, 1–4.
138. Polyak, L., R. B. Alley, J. T. Andrews, J. Brigham-Grette, T. M. Cronin, D. A. Darby, A. S. Dyke, J. J. Fitzpatrick, S. Funder, M. Holland, A. E. Jennings, G. H. Miller, M. O'Regan, J. Savelle, M. Serreze, K. St. John, J. W. C. White and E. Wolff (2010) History of sea ice in the Arctic. *Quaternary Science Reviews* 29, 1757–78.
139. Hoyt, D. V. and K. H. Schatten (1997) *The Role of the Sun in Climate Change*. Oxford University Press, Oxford and New York.
140. Soon, W. W.-H. (2005) Variable solar irradiance as a plausible agent for multidecadal variations in the Arctic-wide surface air temperature record of the past 130 years. *Geophysical Research Letters* 32, L16712.
141. Versteegh, G. J. M. (2005) Solar forcing of climate. 2: evidence from the past. *Space Science Reviews* 120, 243–86.
142. Mayewski, P. A., K. A. Maasch, Y. Yan, S. Kang, E. A. Meyerson, S. B. Sneed, S. D. Kaspari, D. A. Dixon, E. C. Osterberg, V. I. Morgan, T. Van Ommen and M. A. J. Curran (2005) Solar forcing of the polar atmosphere. *Annals of Glaciology* 41, 147–54.
143. Magny, M., E. Gauthier, B. Vannière and O. Peyron (2008) Palaeohydrological changes and humanimpact history over the last millennium recorded at Lake Joux in the Jura mountains, Switzerland. *The Holocene* 18 (2), 255–65.

144. Agnihotri, R., K. Dutta and W. Soon (2011) Temporal derivative of total solar irradiance and anomalous Indian summer monsoon: an empirical evidence for a sun-climate connection. *Journal of Atmospheric and Solar–Terrestrial Physics* 73 (13), 1980–7.
145. Mauas, P. J. D., E. Flamenco and A. P. Buccino (2008) Solar forcing of the stream flow of a continental scale South American river. *Physical Review Letters* 101, 1–4.
146. Mauas, P. J. D., A. P. Buccino and E. Flamenco (2010) Long-term solar activity influences on South American rivers. *Journal of Atmospheric and Solar–Terrestrial Physics* 73 (2–3), 377–82.
147. Debret, M., D. Sebag, X. Crosta, N. Massei, J.-R. Petit, E. Chapron and V. Bout-Roumazeilles (2009) Evidence from wavelet analysis for a mid Holocene transition in global climate forcing. *Quaternary Science Reviews* 28, 2675–88.
148. Müller, U. C., S. Klotz, M. A. Geyh, J. Pross and G. C. Bond (2005) Cyclic climate fluctuations during the last interglacial in Central Europe. *Geology* 33 (6), 449–52.
149. Marchitto, T. M., R. Muscheler, J. D. Ortiz, J. D. Carriquiry and A. v. Geen (2010) Dynamical response of the tropical Pacific Ocean to solar forcing during the early Holocene. *Science* 330, 1378–81.
150. Schulz, M. (2002) On the 1470-year pacing of Dansgaard-Oeschger warm events. *Paleoceanography* 17 (2).
151. Gupta, A. K., D. M. Anderson and J. Overpeck (2003) Abrupt changes in the Asian southwest monsoon during the Holocene and their links to the north Atlantic Ocean. *Nature* 421, 354–7.
152. Gupta, A. K., M. Das and D. M. Anderson (2005) Solar influence on the Indian summer monsoon during the Holocene. *Geophysical Research Letters* 32, L17703.
153. Springer, G. S., H. D. Rowe, B. Hardt, R. L. Edwards and H. Cheng (2008) Solar forcing of Holocene droughts in a stalagmite record from West Virginia in east-central North America. *Geophysical Research Letters* 35, 1–5.
154. Xu, H., Y. Hong, Q. Lin, Y. Zhu, B. Hong and H. Jiang (2006) Temperature responses to quasi-100-year solar variability during the past 6000 years based on δ18O of peat cellulose in Hongyuan, eastern Qinghai–Tibet plateau, China. *Palaeogeography, Palaeoclimatology, Palaeoecology* 230, 155–64.
155. Asmerom, Y., V. Polyak, S. Burns and J. Rassmussen (2007) Solar forcing of Holocene climate: new insights from a speleothem record, southwestern United States. *Geology* 35 (1), 1–4.
156. Neff, U., S. J. Burns, A. Mangini, M. Mudelsee, D. Fleitmann and A. Matter (2001) Strong coherence between solar variability and the monsoon in Oman between 9 and 6 kyr ago. *Nature* 411, 290–3.
157. Marchitto, T. M. and P. B. deMenocal (2003) Late Holocene variability of upper North Atlantic deep water temperature and salinity. *Geochemistry, Geophysics, Geosystems* 4 (12).
158. Viau, A. E., K. Gajewski, P. Fines, D. E. Atkinson and M. C. Sawada (2002) Widespread evidence of 1500 year climate variability in North America during the past 14,000 years. *Geology* 30 (5), 455–8.
159. Willard, D. A., C. E. Bernhardt, D. A. Korejwo and S. R. Meyers (2005) Impact of millennial-scale Holocene climate variability on eastern North American terrestrial ecosystems: pollen-based climatic reconstruction. *Global and Planetary Change* 47, 17–35.
160. Varma, V., M. Prange, F. Lamy, U. Merkel and M. Schulz (2011) Solar-forced shifts of the Southern Hemisphere westerlies during the Holocene. *Clim. Past* 7, 339–47.

161. Denton, G. H. and W. Karlén (1973) Holocene climatic variations – their pattern and possible cause. *Quaternary Research* 3, 155–205.
162. Mayewski, P. A., E. E. Rohling, J. C. Stager, W. Karlén, K. A. Maasch, L. D. Meeker, E. A. Meyerson, F. Gasse, S. van Kreveld, K. Holmgren, J. Lee-Thorp, G. Rosqvist, F. Rack, M. Staubwasser, R. R. Schneider and E. J. Steig (2004) Holocene climate variability. *Quaternary Research* 62, 243–55.
163. Berner, K. S., N. Koç, D. Divine, F. Godtliebsen and M. Moros (2008) A decadal-scale Holocene sea surface temperature record from the subpolar North Atlantic constructed using diatoms and statistics and its relation to other climate parameters. *Paleoceanography* 23 (2), PA2210.
164. Thornalley, D. J. R., H. Elderfield and I. N. McCave (2009) Holocene oscillations in temperature and salinity of the surface subpolar North Atlantic. *Nature* 457, 711–14.
165. D'Andrea, W. J., Y. Huang, S. C. Fritz and N. J. Anderson (2011) Abrupt Holocene climate change as an important factor for human migration in west Greenland. *PNAS* early edition, 30 May, 1–5.
166. Mernild, S. H., M.-S. Seidenkrantz, P. Chylek, G. E. Liston and B. Hasholt (2012) Climate-driven fluctuations in freshwater flux to Sermilik Fjord, East Greenland, during the last 4000 years. *The Holocene* 22 (2), 155–64.
167. Helama, S., M. M. Fauria, K. Mielikäinen, M. Timonen and M. Eronen (2010) Sub-Milankovitch solar forcing of past climates: mid and late Holocene perspectives. *Geological Society of America Bulletin.*
168. Nieto-Moreno, V., F. Martinez-Ruiz, S. Giralt, F. Jimenez-Espejo, D. Gallego-Torres, M. Rodrigo-Gamiz, J. Garcia-Orellana, M. Ortega-Huertas and G. J. de Lange (2011) Tracking climate variability in the western Mediterranean during the late Holocene: a multiproxy approach. *Climate of the Past* 7, 1395–414.
169. deMenocal, P., J. Ortiz, T. Guilderson and M. Sarnthein (2000) Coherent high- and low-latitude climate variability during the Holocene warm period. *Science* 288, 2198–202.
170. Pena, L. D., G. Francés, P. Diz, M. Esparza, O. Grimalt, M. A. Nombela and I. Alejo (2010) Climate fluctuations during the Holocene in NW Iberia: high and low latitude linkages. *Continental Shelf Research* 30, 1487–96.
171. Magny, M. (1993) Solar influences on Holocene climatic changes illustrated by correlations between past lake-level fluctuations and the atmospheric ^{14}C record. *Quaternary Research* 40, 1–9.
172. Magny, M. (2004) Holocene climate variability as reflected by mid European lake-level fluctuations and its probable impact on prehistoric human settlements. *Quaternary International* 113, 65–79.
173. Wang, Y., H. Cheng, R. L. Edwards, Y. He, X. Kong, Z. An, J. Wu, M. J. Kelly, C. A. Dykoski and X. Li (2005) The Holocene Asian monsoon: links to solar changes and North Atlantic climate. *Science* 308, 854–7.
174. Tan, M., T. Liu, J. Hou, X. Qin, H. Zhang and T. Li (2003) Cyclic rapid warming on centennial-scale revealed by a 2650-year stalagmite record of warm season temperature. *Geophysical Research Letters* 30 (12), 19/11–19/14.
175. Liew, P. M., C. Y. Lee and C. M. Kuo. (2006) Holocene thermal optimal and climate variability of East Asian monsoon inferred from forest reconstruction of a subalpine pollen sequence, Taiwan. *Earth and Planetary Science Letters* 250, 596–605.

176. Lim, J., E. Matsumoto and H. Kitagawa (2005) Eolian quartz flux variations in Cheju Island, Korea, during the last 6500 years and a possible sun-monsoon linkage. *Quaternary Research* 64, 12–20.
177. Beer, J. and B. van Geel (2008) Holocene climate change and the evidence for solar and other forcings. In R. W. Battarbee and H. A. Binney (eds.), *Natural Climate Variability and Global Warming: A Holocene Perspective*. Wiley-Blackwell, Chichester, 138–62.
178. Obrochta, S. P., H. Miyahara, Y. Yokoyama and T. J. Crowley (2012) A re-examination of evidence for the North Atlantic '1500-year cycle' at site 609. *Quaternary Science Reviews* 55, 23–33.
179. Bothe, O., J. H. Jungclaus, D. Zanchettin and E. Zorita (2012) Climate of the last millennium: ensemble consistency of simulations and reconstructions. *Climate of the Past* Discussion 8, 2409–44.
180. Lohmann, G., M. Pfeiffer, T. Laepple, G. Leduc and J.-H. Kim (2012) A model–data comparison of the Holocene global sea surface temperature evolution. *Climate of the Past* Discussion 8, 1005–56.
181. Crook, J. A. and P. M. Forster (2011) A balance between radiative forcing and climate feedback in the modeled 20th century temperature response. *J. Geophys. Res.* 116 (D17), D17108.
182. Wikipedia (2011) Milanković Zyklen. http://de.wikipedia.org/wiki/Milankovi%C4%87-Zyklen.
183. Daansgard, W., S. J. Johnsen, H. B. Clausen, D. Dahl-Jensen, N. S. Gundestrup, C. U. Hammer, C. S. Hvidberg, J. P. Steffensen, A. E. Sveinbjörnsdottir, J. Jouzel and G. Bond (1993) Evidence for general instability of past climate from a 250-kyr ice core record. *Nature* 364, 218–20.
184. Mogensen, I. A. (2009) Dansgaard-Oeschger cycles. In V. Gornitz (ed.), *Encyclopedia of Paleoclimatology and Ancient Environments*. Springer, New York, 229–33.
185. Hodell, D. A., H. F. Evans, J. E. T. Channell and J. H. Curtis (2010) Phase relationships of North Atlantic ice-rafted debris and surface-deep climate proxies during the last glacial period. *Quaternary Science Reviews* 29, 3875–86.
186. Sierro, F. J., D. A. Hodell, J. H. Curtis, J. A. Flores, I. Reguera, E. Colmenero-Hidalgo, M. A. Bárcena, J. O. Grimalt, I. Cacho, J. Frigola and M. Canals (2005) Impact of iceberg melting on Mediterranean thermohaline circulation during Heinrich events. *Paleoceanography* 20, PA2019.
187. Bond, G., H. Heinrich, W. Broecker, L. Labeyrie, J. McManus, J. Andrews, S. Huon, R. Jantschik, S. Clasen, C. Simet, K. Tedesco, M. Clas, G. Bonani and S. Ivy (1992) Evidence for massive discharges of icebergs into the north Atlantic Ocean during the last glacial period. *Nature* 360, 245–49.
188. Broecker, W., G. Bond, M. Klas, E. Clark and J. McManus (1992) Origin of the northern Atlantic's Heinrich events. *Climate Dynamics* 6, 265–73.
189. Bond, G. C. and R. Lotti (1995) Iceberg discharges into the North Atlantic on millennial timescales during the last glaciation. *Science* 267, 1005–10.
190. Hulbe, C. L., D. R. MacAyeal, G. H. Denton, J. Kleman and T. V. Lowell (2004) Catastrophic ice shelf breakup as the source of Heinrich event icebergs. *Paleoceanography* 19, PA1004.
191. Bond, G. (2009) Millennial climate variability. In V. Gornitz (ed.), *Encyclopedia of Paleoclimatology nd Ancient Environments*. Springer, New York, 568–573.

192. Hemming, S. (2009) Heinrich events. In V. Gornitz (ed.), *Encyclopedia of Paleoclimatology and Ancient Environments*. Springer, New York, 409–414.
193. Marcott, S. A., P. U. Clark, L. Padman, G. P. Klinkhammer, S. R. Springer, Z. Liu, B. L. Otto-Bliesner, A. E. Carlson, A. Ungerer, J. Padman, F. He, J. Cheng and A. Schmittner (2011) Ice-shelf collapse from subsurface warming as a trigger for Heinrich events. *PNAS*, 1–5.
194. Braun, H., M. Christl, S. Rahmstorf, A. Ganopolski, A. Mangini, C. Kubatzki, K. Roth and B. Kromer (2005) Possible solar origin of the 1,470-year glacial climate cycle demonstrated in a coupled model. *Nature* 438, 208–11.
195. Muscheler, R. and J. Beer (2006) Solar forced Dansgaard-Oeschger events? *Geophysical Research Letters* 33, L20706.
196. Rahmstorf, S. (2003) Timing of abrupt climate change: a precise clock. *Geophysical Research Letters* 30 (10).
197. Braun, H., P. Ditlevsen and D. R. Chialvo (2008) Solar forced Dansgaard-Oeschger events and their phase relation with solar proxies. *Geophysical Research Letters* 35, 1–5.
198. Braun, H. (2009) Strong indications for nonlinear dynamics during Dansgaard-Oeschger events. *Clim. Past Discuss.* 5, 1751–62.
199. Braun, H., P. Ditlevsen and J. Kurths (2009) New measures of multimodality for the detection of a ghost stochastic resonance. *Chaos* 19.
200. Lombard, C. S. Q., P. Balenzuela, H. Braun and D. R. Chialvo (2010) A simple conceptual model to interpret the 100,000 years dynamics of paleo-climate records. Nonlin. *Processes Geophys.* 17, 585–92.
201. Braun, H. and J. Kurths (2010) Were Dansgaard-Oeschger events forced by the sun? Eur. Phys. J. *Special Topics* 191, 117–29.
202. Braun, H., P. Ditlevsen, J. Kurths and M. Mudelsee (2010) Limitations of red noise in analysing Dansgaard-Oeschger events. *Clim. Past* 6, 85–92.
203. Dima, M. and G. Lohmann (2008) Conceptual model for millennial climate variability: a possible combined solar–thermohaline circulation origin for the ~1,500-year cycle. *Clim Dyn* 32, 301–11.
204. Koutsodendris, A., A. Brauer, H. Pälike, J. Pross, U. C. Müller and A. F. Lotter (2011) sub-decadal to decadal-scale climate cyclicity during the Holsteinian interglacial (MIS 11) evidenced in annually laminated sediments. *Clim. Past* 7, 987–99.
205. Ruddiman, W. F. (2006) Orbital changes and climate. *Quaternary Science Reviews* 25, 3092–112.
206. Berger, A. (2009) Astronomical theory of climate change. In V. Gornitz (ed.), *Encyclopedia of Paleoclimatology and Ancient Environments*. Springer, New York, 51–6.
207. Cruz Jr, F. W., S. J. Burns, I. Karmann, W. D. Sharp, M. Vuille, A. O. Cardoso, J. A. Ferrari, P. L. Silva Dias and O. Viana Jr (2005) Insolation-driven changes in atmospheric circulation over the past 116,000 years in subtropical Brazil. *Nature* 434, 63–6.
208. Köhler, P., R. Bintanja, H. Fischer, F. Joos, R. Knutti, G. Lohmann and V. Masson-Delmotte (2010) What caused earth's temperature variations during the last 800,000 years? Data-based evidence on radiative forcing and constraints on climate sensitivity. *Quaternary Science Reviews* 29, 129–45.
209. Ganopolski, A. and R. Calov (2011) The role of orbital forcing, carbon dioxide and regolith in 100 kyr glacial cycles. *Clim. Past Discuss.* 7, 2391–411.

210. Pillans, B., J. Chappell and T. R. Naish (1998) review of the Milankovitch climatic beat: template for Plio-Pleistocene sea-level changes and sequence stratigraphy. *Sedimentary Geology* 122, 5–21.
211. Roe, G. (2006) In defense of Milankovitch. *Geophysical Research Letters* 33, L24703.
212. Shaviv, N. J. and J. Veizer (2003) Celestial driver of Phanerozoic climate? GSA *Today* 7, 4–10.
213. Shaviv, N. J. (2003) The spiral structure of the Milky Way, cosmic rays, and ice age epochs on earth. *New Astronomy* 8, 39–77.
214. Scherer, K., H. Fichtner, T. Borrmann, J. Beer, L. Desorgher, E. Flükiger, H.-J. Fahr, S. E. S. Ferreira, U. W. Langner, M. S. Potgieter, B. Heber, J. Masarik, N. J. Shaviv and J. Veizer (2006) Interstellar–terrestrial relations: variable cosmic environments, the dynamic heliosphere, and their imprints on terrestrial archives and climate. *Space Science Reviews* 127, 327–465.
215. Scafetta, N. (2010) Climate change and its causes – a discussion about some key issues. *Science & Public Policy Institute (SPPI).* Original paper.
216. Mazzarella, A. and N. Scafetta (2011) Evidences for a quasi-60-year North Atlantic Oscillation since 1700 and its meaning for global climate change. *Theor. Appl. Climatol.*
217. Agnihotri, R., K. Dutta, R. Bhushan and B. L. K. Somayajulu (2002) Evidence for solar forcing on the Indian monsoon during the last millennium. *Earth and Planetary Science Letters* 198, 521–7.
218. Klyashtorin, L., V. Borisov and A. Lyubushin (2009) Cyclic changes of climate and major commercial stocks of the Barents Sea. *Marine Biology Research* 5, 4–17.
219. Komitov, B. (2009) The 'sun–climate' relationship. II. The 'cosmogenic' beryllium and the middle latitude aurora. *Bulgarian Astronomical Journal* 12, 75–90.
220. Komitov, B. (2009) The 'sun–climate' relationship. I. The sunspots and the climate. *Bulgarian Astronomical Journal* 11, 139–51.
221. Komitov, B., S. Sello, P. Duchlev, M. Dechev, K. Penev and K. Koleva (2010) The sub- and quasi-centurial cycles in solar and geomagetic activity data series. http://arxiv.org/ftp/arxiv/papers/1007/1007.3143.pdf.
222. Scafetta, N. (2011) A shared frequency set between the historical mid latitude aurora records and the global surface temperature. *Journal of Atmospheric and Solar–Terrestrial Physics* 74, 145–63.
223. Otterå, O. H., M. Bentsen, H. Drange and L. Suo (2010) External forcing as a metronome for Atlantic multidecadal variability. *Nature Geoscience* 3, 688–94.
224. Wilson, I. R. G., B. D. Carter and I. A. Waite (2008) Does a spin–orbit coupling between the sun and the Jovian planets govern the solar cycle? *Publications of the Astronomical Society of Australia* 25 (2), 85–93.
225. Mazzarella, A. (2008) Solar forcing of changes in atmospheric circulation, earth's rotation and climate. *The Open Atmospheric Science Journal* 2, 181–4.
226. Scafetta, N. (2012) Multi-scale harmonic model for solar and climate cyclical variation throughout the Holocene based on Jupiter–Saturn tidal frequencies plus the 11-year solar dynamo cycle. *Journal of Atmospheric and Solar–Terrestrial Physics* 80, 296–311.
227. Scafetta, N. (2012) Does the sun work as a nuclear fusion amplifier of planetary tidal forcing? A proposal for a physical mechanism based on the mass–luminosity relation. *Journal of Atmospheric and Solar–Terrestrial Physics* 81–2, 27–40.
228. Scafetta, N. (2012) Testing an astronomically based decadal-scale empirical

harmonic climate model versus the general circulation climate models. *Journal of Atmospheric and Solar-Terrestrial Physics* 80, 124-37.

229. Callebaut, D. K., C. de Jager and S. Duhau (2012) The influence of planetary attractions on the solar tachocline. *Journal of Atmospheric and Solar-Terrestrial Physics* 80, 73-8.
230. Antico, A. and D. M. Kröhling (2011) Solar motion and discharge of Paraná River, South America: evidence for a link. *Geophysical Research Letters* 38, 1-5.

Solar forcing and twentieth-century climate change

Nir J. Shaviv
HEBREW UNIVERSITY OF JERUSALEM

A long list of empirical results strongly suggests that solar variations play an important role in climate change. We begin by discussing why such variations are crucial if we are to understand twentieth-century climate change and how it is related to the value of the climate sensitivity, for example the amount of warming expected for a certain increase in manmade greenhouse gases. This climate sensitivity is necessary if we are to predict future climate change.

In the standard scenario advocated by the IPCC, most of the global warming observed over the twentieth century is attributed to the increase in manmade greenhouse gases. Indeed, when one considers the observed increase in temperature and the increase in manmade greenhouse gases, it is very tempting to do so. However, we have to remember that there are many uncertainties – primarily the unknown radiative forcings and unknown climate sensitivity – which imply that most of the warming is not necessarily human.

When the earth's energy budget changes, that is, when the net radiative forcing changes, so does the climate equilibrium. Loosely speaking, the temperature change over the twentieth century is the product of the changed energy balance, according to the IPCC mostly manmade greenhouse gases, and the climate sensitivity:

$$\textit{Temperature Change } (\Delta T) = \textit{Radiative Forcing Changes } (\Delta F) \times \textit{Climate Sensitivity } (S)$$

It is, though, somewhat more complicated because it takes many decades for the climate system to adjust.

Here comes the problem, we know inadequately the net radiative forcing imposed by humans over the twentieth century and we know even less about the climate sensitivity. It turns out that the Achilles' heel for both is clouds. We don't know the net radiative forcing because human activity increased the amount of atmospheric aerosols which 'seed' the clouds and cool the earth [1]. Unfortunately, there is a very large uncertainty about the size of the effect. Clouds are also very important to the determination of climate sensitivity because the climate feedback through clouds – namely, by how much the cloud cover changes when the global temperature changes – is not known. Robert Cess of the State University of New York and colleagues showed in 1989 that it is by far the biggest source of uncertainty [2]. Some twenty years later, the situation is virtually the same.

For these reasons, there is no single prediction for how large the anthropogenic warming should have been over the twentieth century – multiplying two very uncertain numbers gives an even more uncertain temperature change.

Because theory cannot uniquely predict twentieth-century warming, it can be attributed to human activity only because of *indirect* lines of argument. First, twentieth-century warming is unprecedented. In Chapter 4 we shall see that this argument holds no water. Second, climate modellers cannot explain the warming without including the anthropogenic contributions to the net radiative forcing, in particular that of the greenhouse gases.

Now we see why the role of the sun in climate is so important. Because solar activity increased over the twentieth century, if it has an effect on the climate it should have contributed a net positive forcing and it may have been responsible for some of the twentieth-century warming. This would then diminish the role of manmade activity.

Put more quantitatively, if the sun has contributed a positive radiative forcing, then the total radiative forcing change over the twentieth century is necessarily larger as well. As we shall see, the sun does have a large effect on the climate, and it is roughly twice as large as the anthropogenic forcing alone. This implies that in order to explain the same observed twentieth-century warming, we require a climate sensitivity only half the size. In fact, the range of sensitivities required to explain twentieth-century warming is just below the often quoted IPCC range of 1.5–4.5° C increase per doubling of CO_2.

Needless to say, a lower climate sensitivity is very important if we are to the predict twenty-first-century temperature increase. For a given emissions scenario, such as a 'business as usual' one, the warming should be correspondingly smaller.

Evidence for a solar–climate link

One of the most interesting aspects of the sun is that it is not constant. The variations that it exhibits appear in the total irradiance of the sun, primarily in the visible and infrared bands by as much as 0.1 per cent. But they also appear in components other than the total emitted flux. These include very large *relative* changes in the magnetic field, the number of sunspots, the strength of the solar wind and the amount of UV, to name a few.

The basic variation is an activity cycle of about 11 years, which arises from quasi-periodic reversals of the solar magnetic dipole field. Over longer timescales (of decades to millennia) there are irregular variations which modulate the 11-year cycle. For example, during the Middle Ages and again in the latter half of the twentieth century, the peaks in the 11-year cycles were strong, but were almost absent during the Maunder minimum. On the other hand, eruptions may appear on a timescale of days. Today there is evidence linking solar activity to the terrestrial climate on all of these timescales.

Since Jack Eddy published his work in the 1970s, many empirical results have shown a clear correlation between different climatic reconstructions and different solar activity proxies on the timescale of decades or longer. Eddy realized that there is a correlation between solar activity and the European climate over the past millennium [3]. For example, the Little Ice Age in Europe took place while the sun was particularly inactive, during the Maunder minimum. The Medieval Warm Period, on the other hand, occurred while the sun was as active as it was in the late twentieth century. Since then, many findings show a correlation between different climatic reconstructions and different solar activity proxies.

One of the most beautiful results is that of a multi-millennial correlation between the temperature of the Indian Ocean as mirrored in the ratio between different oxygen isotopes in stalagmites in a cave in Oman, and solar activity, as reflected in the cosmogenic carbon 14 isotope [4]. These results by Professor Mangini Heidelberg's cosmogenic isotope group are presented in Figure 3.11.

Another impressive result over the same timescale comes from Professor Bond's group, where the solar activity was compared with the northern Atlantic climate, as recorded on the ocean bed through ice-rafted debris [5] (Figure 3.10). Many other correlations exist elsewhere.

One way to see that this solar–climate link is global and that it affects the global temperature is to look at borehole data [6]. These reveal that the solar variations give rise to changes as large as 1° C between low and high solar activity.

Over the 11-year solar cycle, it is much harder to see climate variations. There are two reasons for this. First, if we study the climate over short timescales, we find that there are large annual variations (for example, due to the El Niño oscillation) which introduce cluttering 'noise', hindering the observation of solar-related signals. Second, because of the large ocean heat capacity, it takes decades before the full effects of given changes in the radiative budget, including those associated with solar variability, can be seen.

It is for this reason that the climate of continental regions is typically much more extreme than their marine counterparts.

If, for example, a given change in solar forcing is expected to give rise to a temperature change of 0.5° C after several centuries, then the same radiative forcing varying over the 11-year solar cycle is expected to give rise to temperature variations of only 0.05–0.1° C or so [7]. This is because over short timescales, most of the energy goes into heating the oceans, but because of their very large heat capacity, large changes in the ocean heat content do not translate into large temperature variations.

Nevertheless, if the global temperature is carefully analysed (for example, by folding the global temperature of the past 120 years over the 11-year solar cycle), it is possible to see variations of about 0.1° C in the land temperature and slightly less in the ocean surface temperature [7]. Moreover, as we shall demonstrate, it is possible to see the large amount of heat going into the oceans every solar cycle.

We therefore conclude that the sun has a large effect on the climate. Although the link itself is not the topic of this chapter, it should be mentioned that the leading contender is through solar modulation of the cosmic ray flux reaching the earth [8]. This is now supported by a range of empirical and experimental results, as discussed later in this work.

Quantifying the solar climate link

Having established that the sun has a large effect on the climate, we can proceed to quantify the size of the link. In particular, we are interested in the radiative forcing associated with solar variability. This is important if we are to assess its role in twentieth-century climate change.

As mentioned above, looking for the temperature response over the 11-year solar cycle is problematic because of the large heat capacity of the oceans and the climate variability over short timescales. Nevertheless, we can use the large ocean heat capacity to our advantage, since it implies that short-term variations in the energy balance will translate into heat content variations in the oceans without affecting other components, that is, without any internal feedbacks operating. This implies that the 11-year cycle variations in the heat content in the oceans can be straightforwardly used to calculate the radiative forcing imposed by the sun.

The ocean heat content can be derived from three independent data sets. First, there is the direct measurement of the heat content, as measured by small temperature changes down to depths of 700 metres since 1955, over the whole ocean; the second is the surface sea temperature; while the third are tide gauge records of the sea level. Each of the three data sets has advantages and disadvantages. The tide gauge record can be seen in Figure 3.A.1. All three records consistently reveal that the amount of heat going into the oceans every solar cycle is about 6–7 times larger than the changes expected from just the variations in the total irradiance [9]. In absolute terms, it is a variation of about 1 W/m^2.

The figure we obtain this way is very interesting. First, the leading contender to explain the solar–climate link is through cosmic ray flux modulation of the atmospheric ionization, which in turn affects the cloud cover. This implies that the radiative forcing change associated with cloud cover variations over the 11-year solar cycle should be of the order of 1 W/m^2 – the amount of heat going into the oceans. Within the radiative forcing uncertainties of clouds, this is indeed the observed variations [7]. Second, because the forcing variation is large, it is comparable to the net anthropogenic changes in the radiative forcing over the twentieth century. This implies that one has to consider solar variability when trying to understand twentieth-century global warming.

Twentieth-century climate change – the full picture

Now that we have quantified the size of the link, we can proceed to estimate the effect that the sun had over the twentieth century. Since the increased solar activity between the first half and second half of the twentieth century is comparable to the variations between solar minimum and solar maximum over the 11-year cycle, we can expect the radiative forcing to be similar, at around 1 W/m^2. For comparison, the IPCC in its 2007 report estimates the net anthropogenic forcing to have been 0.6–2.4 W/m^2, but the solar forcing that modellers typically include is only the changes in the solar irradiance, which are of order 0.1–0.2 W/m^2.

To understand the climate change better, we can employ a simple 'box' climate model, one that includes temperatures for the land, ocean mixed layer and diffusion into the deep ocean. We can then ask what the allowed ranges for the different climate variables are, including the couplings, sensitivity, radiative forcings, and so forth which can consistently explain the twentieth-century global warming. The answer is that if we allow the sun to have contributed more than changes in the solar irradiance, we find that twentieth-century warming can be much better explained than present global circulation models which exclude a large solar effect. In fact, the residual in the fit between model and observations is twice as small! This can be seen in Figure 3.A.2.

This fit gives a net solar contribution of 0.8 ± 0.4 W/m^2, and a climate sensitivity of $0.95 \pm 0.35°$ C increase per doubling of CO_2. These values are consistent with previous determinations of the solar effect and of the climate sensitivity.

Summary

We have seen that there is ample evidence to prove that the sun has a large effect on the climate. This is important because it allows us to present a much more consistent picture to explain the observed twentieth-century global warming – one in which model predictions fit the observations much better. In this picture, the sun has contributed a net radiative forcing which is comparable to the anthropogenic contribution. As a consequence, the same twentieth-century warming can be explained with a smaller climate sensitivity to CO_2. It also implies that for a given emissions scenario the predicted twenty-first-century warming should be correspondingly smaller, typically around 1–1.5° C, for a 'business as usual scenario'.

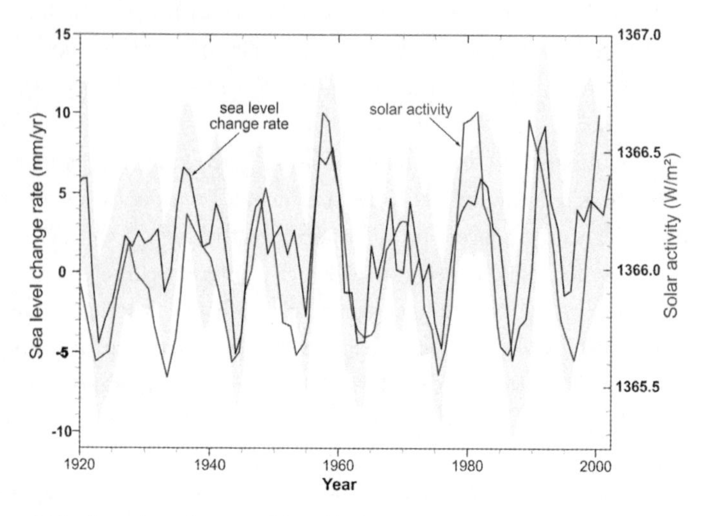

Figure 3.A.1 The sea level change rate (blue, with a hatched 1σ error) and solar constant (red line). Over short timescales, it originates predominantly from changes in the oceanic heat content. Using these data, the derived changes in the energy budget over the solar cycle correspond to 1 W/m^2, almost an order of magnitude more than can be expected from changes in the solar irradiance [9].

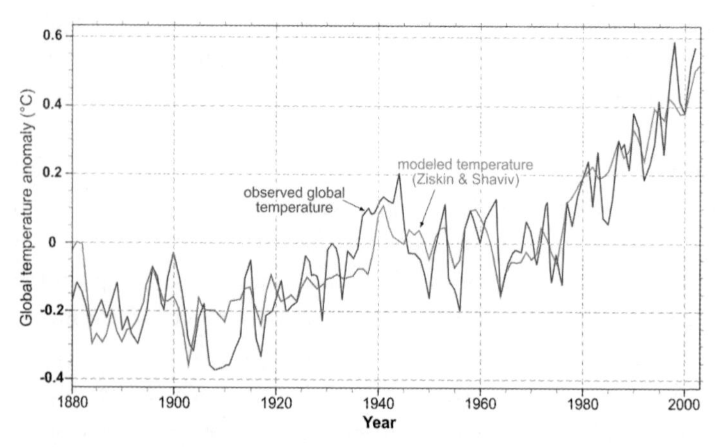

Figure 3.A.2 A comparison of the observed global temperatures (red line) and the temperatures modelled using an energy balance model with a diffusive ocean (green line). The small residual, which is twice as small than that obtained in typical global circulation models, can be obtained if we allow the sun to have large effect on the climate and the climate to have a low climate sensitivity [10].

References

1. IPCC (2007) *Climate Change 2007: The Physical Science Basis. Contribution of Working Group I to the Fourth Assessment Report of the Intergovernmental Panel on Climate Change*. Cambridge University Press, Cambridge and New York.
2. Cess, R. D., G. L. Potter, J. P. Blanchet, G. J. Boer, S. J. Ghan, J. T. Kiehl, H. Le Treut, Z.-X. Li, X.-Z. Liang, J. F. B. Mitchell, J.-J. Morcrette, D. A. Randall, M. R. Riches, E. Roeckner, U. Schlese, A. Slingo, K. E. Taylor, W. M. Washington, R. T. Wetherald and I. Yagai (1989) Interpretation of cloud–climate feedback as produced by 14 atmospheric general circulation models. *Science* 245, 513–16.
3. Eddy, J. A. (1976) The Maunder minimum. *Science* 192, 1189–202.
4. Neff, U., S. J. Burns, A. Mangini, M. Mudelsee, D. Fleitmann and A. Matter (2001) Strong coherence between solar variability and the monsoon in Oman between 9 and 6 kyr ago. *Nature* 411, 290–3.
5. Bond, G., B. Kromer, J. Beer, R. Muscheler, M. N. Evans, W. Showers, S. Hoffmann, R. Lotti-Bond, I. Hajdas and G. Bonani (2001) Persistent solar influence on North Atlantic climate during the Holocene. *Science* 294, 2130–6.
6. Huang, S., H. N. Pollack and P. Y. Shen (1997) Late Quaternary temperature changes seen in world-wide continental heat flow measurements. *Geophysical Research Letters* 24 (15), 1947–50.
7. Shaviv, N. J. (2005) On climate response to changes in the cosmic ray flux and radiative budget. *Journal of Geophysical Research* 110, 1–15.
8. Svensmark, H. (1998) Influence of cosmic rays on earth's climate. *Physical Review Letters* 81 (22), 5027–30.
9. Shaviv, N. J. (2008) Using the oceans as a calorimeter to quantify the solar radiative forcing. *Journal of Geophysical Research* 113, 1–13 (modified).
10. Ziskin, S. and N. J. Shaviv (2012) Quantifying the role of solar radiative forcing over the twentieth century. *Advances in Space Research*.

4. A brief history of temperature: our climate in the past

People love records. Anything that is special or unique seems attract us as if by magic – a new world record in the 100-metre sprint, the shortest man in the world or the hottest decade since the thermometer was invented. The notion is always fascinating. It is the thought of entering new, uncharted territory. Journalists are always grateful for the supply of new records, which serve to satisfy the public appetite. Weather records of course are part of it. In the summer of 2010, western Russia was gripped by a heatwave that lasted for weeks, with temperatures reaching 40° C. It was probably the country's most powerful heatwave in 1000 years [1]. Extreme drought also led to widespread forest and peat bog fires. Many villages were engulfed by flames. Heat and smoke-laden air led to a surge in deaths. The source of the misery and destruction was quickly found: global warming [2–5]. Surely even the last, most stubborn sceptics of the IPCC would now give up their annoying resistance for good (see Chapter 5).

In the middle of August, as temperatures steadily returned to normal levels, the fires were extinguished and the smoke gradually cleared, American scientists set out to examine what exactly had caused the heatwave [6–7]. What they found was unexpected: the heatwave had less to do with climate change and much more to do with a natural phenomenon, namely weather blocking. An atmospheric high pressure system remained stationary over western Russia for weeks and prevented the inflow of cooler air and the formation of summer storms. The researchers discovered that this sort of weather system had occurred repeatedly over the region during the last 130 years. In addition, they discovered that the climate in western Russia over the period had not warmed at all [6–7]. Well, mistakes can be made.

To a lesser extent, the same phenomenon also struck central Europe in the spring of 2011. An unusually long blocking by seven high pressure systems in succession delivered a weeks-long warm and dry period between April and early May. In the summer of the following year (2012), however, central Europe remained relatively cool. Instead, a two-month-long heatwave raged across the United States and Canada. This was an ominous sign of manmade climate change many media outlets claimed. But once again, the media had not done their science homework. A series of similar heatwaves had occurred in the 1930s in North America. Clearly, the 2012 heatwave was well within the limits of known natural climate variability.

The heat records could indeed be perplexing, had there not been a series of record cold winters in the past few years. In 2012 the coldest ever December had Russia in its grip, causing more than 100 deaths. In Siberia the temperature dropped to –60° C. The unusual cold extended to India, where 26 people died during the cold snap. In China the mercury plummeted to –37° C.

The winter of 2011–12 had its very cold moment too. By the end of October 2011, an early snowstorm plunged the US east coast into chaos. New York's Central Park experienced the heaviest October snowfall recorded there since record-keeping began in 1869. In February 2012 a deadly cold wave swept over the European continent and claimed more than 800 lives. Temperatures in several eastern and northern European countries plunged to as low as –39° C.

The winter of 2010–11 set a large number of shivery records. Great Britain and many Swedish cities recorded the coldest December since record-keeping began [8–9]. In Northern Ireland the mercury plummeted to –18.7° C – a new national record [8]. Matters were little better on the continent. In Germany the coldest December in 40 years was recorded, along with heavy snowfall [8]. The coldest May night in 50 years struck Germany on 3 May 2011, destroying parts of the wine and fruit crop [10]. Winters overseas were hardly milder. The United States suffered unusual cold and heavy snowfalls [11] and South Korea came to a standstill under the heaviest snowfall in a century [12].

The winter of 2009–10 was cold too. From the end of December 2009 until the middle of January 2010, temperatures over large parts of the globe, from North America to Europe and Asia, dropped to unexpected levels [13–15]. Snow, traffic chaos, deaths and power outages prevailed in many areas. Great Britain experienced the coldest winter in 30 years with the British media calling it 'The Big Freeze' [16]. Frost struck hard in Peking and Miami, and produced the coldest winter in 40 years [17] and since the start of weather records-keeping, respectively [15]. The cause of the intercontinental cold snap was an extremely negative NAO [15, 18], which is a climate internal fluctuation we shall discuss in more detail in Chapter 7.

Cold and heat anomalies occur regularly throughout the year. We saw them in the past and will continue to see them in the future. A look at the temperature records and other extreme weather systems reveals something 'unusual' for every region almost every year [19]. That our memories focus mainly on events of the recent years while distant events are shelved in the back of our minds appears to be a trait embedded deep in the human psyche. Our porous climate memories are partly responsible for the alarmist over-interpretation of current climate events. Single extreme weather systems neither refute nor confirm the accuracy of climate prognoses. Only a systematic and worldwide evaluation, spanning multiple decades, can uncover meaningful trends from all the statistical noise. For this reason the heat and cold waves of recent years cannot play a meaningful role in debates on either side of the climate discussion.

El Niño sets the pace

If we take a look at the global temperature curves of the last 30 years, we see peculiar warm peaks with amplitudes of 0.2–0.7° C that repeat at irregular intervals and with various intensities (Figure 4.1). After about a year or two, the heating fizzles out as quickly as it came and temperatures abruptly drop back down to normal levels. The cause of this spectacle is El Niño, which occurs in the tropical Pacific every 2–7 years, typically around Christmas time. The event is characterized by a strong warming of the upper water layer in this oceanic region. When this happens, high pressure and low pressure atmospheric systems trade places, and this leads to a partial reversal in air and ocean currents [20]. The phenomenon as a whole, comprising the El Niño and Southern oscillations, is also called the El Niño Southern Oscillation (ENSO).

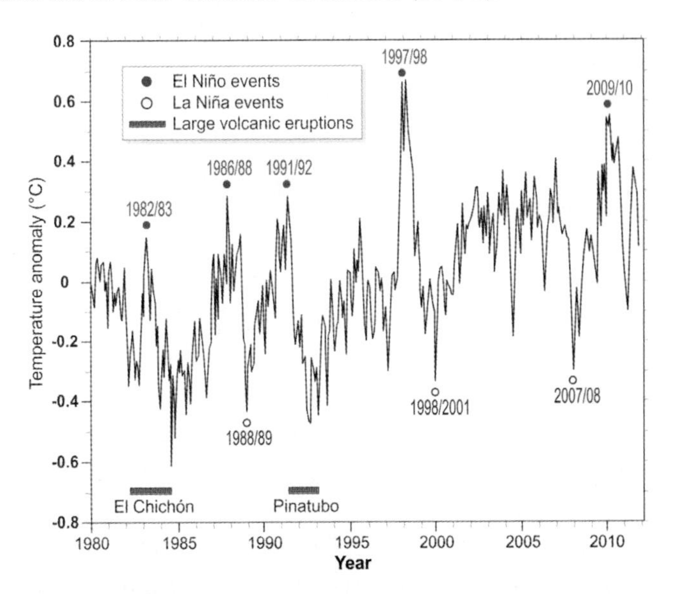

Figure 4.1 The global temperature curve spikes upwards every 2–7 years with strong warm peaks traced back to the natural El Niño phenomenon in the tropical Pacific. El Niño's cold counterpart, La Niña, pulls the curve back down with its cold spikes. Sulphur emissions from large volcanic eruptions also cause cool spikes. Source: base curve, University of Alabama-Huntsville.

The large El Niño irregularity has profound impacts on the climate, which have consequences not only for the Pacific region, but for the entire globe [21]. For example, during warm El Niño years, Southeast Asia, Australia and the Amazon area are plagued by drought, while other parts of South America are inundated by heavy rainfall. Even North America and East Africa are climatically impacted by El Niño events. The abrupt changeover in climatic processes during an El Niño also has impacts on sea life along the South American Pacific coast. Due to a lack of nutrients, algae die and shoals of fish migrate to warmer waters. With the food chain interrupted, seabirds, seals and sea reptiles draw the short straw and have to fight for survival.

El Niño events are the most powerful of the short-term internal climate fluctuations, changes that emerge spontaneously from within the climate system. During the particularly powerful El Niño of 1997–98 the global mean temperature shot up by 0.7° C. That is an enormous hike, especially if one considers that the entire climate debate comes down to the 0.8° C the earth has warmed since 1850 according to some authors [22–23]. In the tropical Pacific alone the surface water temperature increased a massive 7° C and the air temperature rose by up to 1.5° C.

Because El Niño years are special, their erratic temperature jumps have to be carefully identified and plotted on temperature curves (Figure 4.1). This is the only way of discerning the overlying climate signal. Under no circumstances can a brief El Niño temperature rise be used as confirmation of the IPCC warming prognosis because the dubious claim dissolves as soon as the El Niño fades away and the temperature collapses. A reliable forecast of the next El Niño occurrence is also impossible due to its irregularity. That means El Niño will remain one of the true Christmas surprises for some time to come. But not even El Niño is completely free of external forces. There are now good indications that the El Niño phenomenon in the tropical Pacific is influenced by fluctuating solar activity [24–29].

The stratospheric sun screen of volcanic dust: the year without a summer

In addition to the El Niño peaks, there are sharp cooling events in the temperature curve that can last up to 3 years before they return to a normal level. These short-lived cold outliers are caused by El Niño's sister La Niña (Spanish for 'little girl'). This is the counterpart of El Niño and it leads to a temporary cooling of the tropical Pacific of up to 3° C and 0.1–0.2° C for the global temperature (Figure 4.1).

However, the most distinct cold peaks are due to another climate player. The earth is punctuated with countless safety valves – volcanoes, which release huge amounts of liquid rock. Often the pressure inside a volcano is so high that it ejects great quantities of ash, gases and rocks into the air. The climatic relevance of most eruptions, however, is minor. Only when there are especially powerful and explosive volcanic eruptions, about twice a century, does enough material shoot high enough into the air that large amounts find their way into the stratosphere [30]. Once at this altitude volcanic particles are carried eastwards by the strong jet stream as if on a superhighway. Eventually, they are deposited around the entire globe. Ash and sulphur dioxide, which form into sulphuric acid aerosols, create a shadowy veil that partly absorbs solar radiation or reflects it back into space [31–33]. As a result, less solar energy reaches the earth's surface, and temperatures fall [34]. While a 'volcanic winter' [31, 35] begins abruptly, the end and transition to a normal climate is gradual because volcanic particles leave the atmosphere very slowly as they sink down or are washed out by rain and snow.

The most important volcanic eruptions impacting the climate over the last 250 years were the Icelandic Laki Crater [36–37] in 1783, the Tambora volcano [38–39] on the Indonesian island of Sumbawa in 1815, the Indonesian volcanic island of Krakatau in 1883, the Mexican El Chichon [40] in 1982 and the Philippine Pinatubo [41] in 1991 (Figure 4.1). These volcanoes each shot 20–100 million tons of sulphur dioxide and ash to altitudes of 20–45 km into the stratosphere [41–44]. The very cold winter of 1783–84 followed the Laki eruption when temperatures in the Northern Hemisphere fell an average of 1.5° C. Along the US east coast temperatures plunged by a massive 5° C [45]. North America, Central Europe and Asia experienced an extremely hard winter.

The situation was similar after Tambora erupted in April 1815. The huge clouds of ash remained so stubbornly in the atmosphere that 1816 went down in history as 'the year without a summer' [44, 46–50]. Crop failures and hunger were widespread. As a small consolation the volcanic aerosols did produce some spectacular sunsets. The Tambora cooling was exceedingly strong because the eruption occurred during the second half of a 40-year solar-related cold phase. In 1790 solar activity had dropped abruptly and marked the beginning of the so-called Dalton minimum, which persisted until 1830 (see Chapter 3). Tambora's eruption simply provided more cooling and exacerbated the cold for a number of years.

Two climate scientists working at the Helmholtz Centre, Geesthacht, recently tried to blame the entire 1790–1830 Dalton minimum cold phase on the 1815 Tambora and another ominous 1809 eruption, all in an attempt to erase the sun from the climate equation [51]. The climate model they used was not even equipped with the solar amplifier, which should have been taken into account based on the current level of knowledge (see Chapter 6). Moreover, the temperature curve had dropped more than 10 years before the eruptions occurred [52] and probably did not follow solar activity by coincidence (Figure 3.7). How can a volcano, which hasn't yet erupted, force the temperature to drop for years? The cold phase of the Dalton minimum thus has to be attributed to weak solar activity, at times made worse by one or possibly two strong volcano eruptions.

Just after Krakatau erupted in 1883, the mean temperature of the Northern Hemisphere dropped 0.5–0.8° C [53]. A cooling of similar magnitude is assumed to have occurred after the 1991 Pinatubo eruption. However, the Pinatubo temperature fall happened immediately after an El Niño year with an anomalous warming development, and so the volcanic cooling effect had an impact of only a few tenths of a degree (Figure 4.1). The 1982 El Chichon eruption falls into the category of moderate cooling and brought with it about 0.2° C of global cooling [54].

In addition to large volcanic eruptions, it appears that some middling eruptions are capable of ejecting material into the stratosphere as an American-French research team showed using satellite measurements [55]. The quantities of aerosol are certainly much less than those of major, 100-year eruptions. Nevertheless, these eruptions are more frequent and at times their cooling effect may make an appreciable contribution to climate development.

Global warming over the last 150 years

Single, anomalous hot summers and cold winters, El Niño warm peaks and volcanic dimming can produce interesting short-term temperature accents. But in the research of long-term climate development and its causes, these processes play no major role because of their transience. The starting point of the current climate discussion can be clearly identified and is not in dispute. In the last 150 years the temperature has risen 0.8° C. This is a global mean value. In some regions (e.g. the Arctic) the temperature has risen more, but less in other regions (e.g. the Tropics). This warming took place in three episodes: 1860–80, 1910–40 and 1977–2000. The rate of increase of the three episodes was similar – about 0.15° C per decade [56] (Figure 4.2). Between the warming phases the climate cooled slightly or stagnated.

What is the 0.8° C of warming based on? Is it really a suitable reference point? No, it is not. The starting point of the warming trend is the Little Ice Age (1550–1850), a natural cold period caused by the low solar activity of the 1000-year Eddy cycle. The nearly 70-year long low solar activity period of the Maunder minimum forms part of this cycle (Figure 3.7). Solar activity began to ramp up again 150 years ago, marking the end of the Little Ice Age (see Chapter 3).

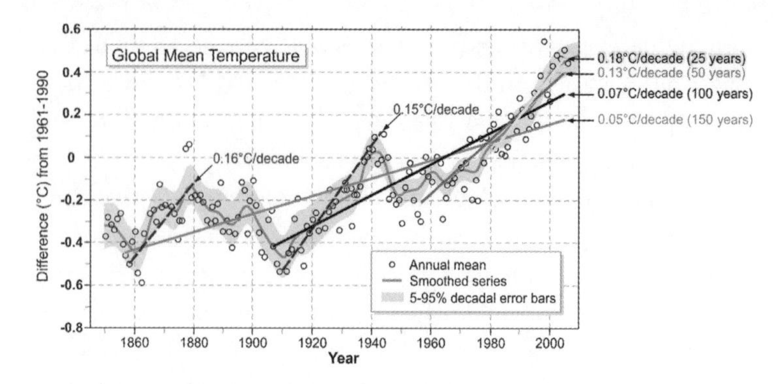

Figure 4.2 In its 2007 report the IPCC suggested that the rate of warming increased continuously over time. The further back the interval, the less was the rate of warming. However, in the IPCC's original graph the two dashed lines were missing. These clearly show that there were phases in the past which had similar warming rates as the 1977–2000 warming. The warming phases were each followed by cooling or stagnation phases. Therefore, it is unlikely that the high warming rate of the last 25 years represents a long-term trend. Source: AR4 modified [57].

From the postglacial climate archives we know that warm and cold phases have always alternated with the Eddy cycles. They oscillate about a mean level that represents a 'normal temperature' (Figures 3.9 and 3.12). It is clear that this 'normal temperature' would represent a more neutral and thus better reference level. If the climatic recovery after the Little Ice Age is 0.3° C, then the climate discussion should turn on the remaining 0.5° C of warming [58]. In addition, a part of this warming should be attributed to the expected subsequent natural Eddy warm phase. If this is also about 0.3° C, then it leaves an 'unnatural' (anthropogenic) warming of significantly less than 0.5° C which should possibly be attributed to CO_2 and other manmade factors.

The 1977–2000 temperature increase

Let's take a look at the last of the three warming phases. The temperature rise began at the end of the 1970s and lasted until about the year 2000 (Figure 4.3). The highest temperature was recorded in 1998. However, this was an El Niño year and so it is better to exclude it. In total the temperature increased about 0.5° C over a good 20 years. The exact attribution of this warming to the various potential climate factors is fiercely disputed.

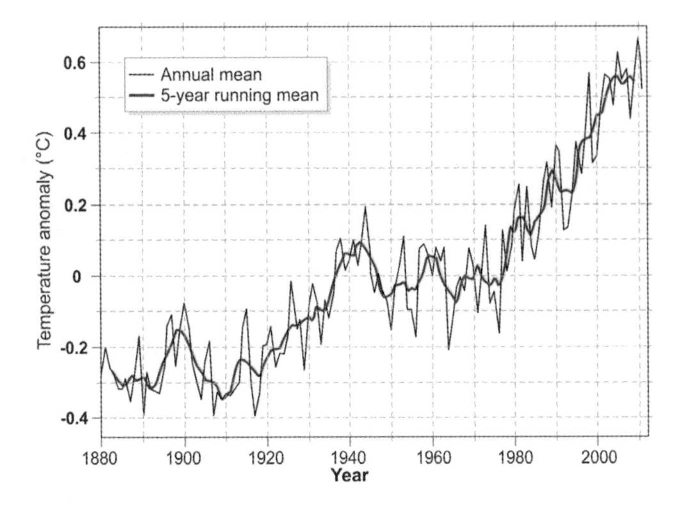

Figure 4.3 Temperature over the last 130 years. Source: GISS surface temperature data set.

Keep in mind that the IPCC likes to keep things simple and claims that CO_2 alone is to blame [57, 59]. The cause is the steady increase in the atmospheric concentrations of CO_2. The IPCC does, however, concede that the sun was very active during the 1980s and 1990s. The solar magnetic field during the 11-year solar cycles 21 and 22 reached intensities that were among the highest of the last few hundred years (Figure 3.8; see also Chapter 3). In addition, the 1977–2000 episode coincides exactly with the sharp upward flank of a natural oceanic cycle, which has a major impact on global temperatures (see more below).

Now is a good point to explain what is meant by the term 'global average temperature' and its value. This temperature is reached by a wide variety of single measured values taken from all the regions and climate zones on earth. The measurement accuracy of each station is ±0.5° C. Because of the huge quantities of data involved, differences can be reported statistically in tenths or even hundredths of a degree

Celsius. However, when one looks at the temperature development of a single measurement station, or large areas, then at times one finds surprising deviations from the global average trend.

The temperature in the Arctic during this warming episode has increased more than anywhere else and contributes disproportionately to global warming [60]. However, there are also very few measurement stations in this inhospitable region and this leads to wide area temperature generalization using very few known temperature points. That inevitably introduces possibilities for error [61]. Moreover, a part of the Arctic warming simply has to do with changed ocean current patterns. Recently, more warm water seems to have reached the Arctic region through the eastern Fram Strait, between Spitzbergen and Greenland [62]. This 'remote' warming has to be distinguished from 'local' warming.

While most of the planet has warmed significantly over the past decades, there are some regions where surface temperatures have fallen. This is especially true of Antarctica [63], the east Pacific and parts of the Indian Ocean (Figure 4.4). Greenland is another interesting example, as its inland glaciers contain enormous quantities of water which, if they were to melt completely, would cause the oceans to rise 7 metres. Here too the temperature increased during the last warming episode (1977–2000). However, in the middle of the twentieth century it was at least once as warm as it is today, and so in principle Greenland has had essentially no net warming [64–66]. The situation is similar in the United States [67–69] (Figure 4.5).

On a global scale, however, the warming trend from 1977 to 2000 is noticeable. The same conclusion was reached by a study carried out at the University of California, Berkeley (the BEST study) which in October 2011 evaluated the global land surface temperature data set [71]. The study was prompted by the fact that strong heat island effects from building construction, asphalt, concrete, heat exhaust, as well as automobile and air traffic in large urban areas, heat up urban areas by several degrees compared to the surrounding rural areas [72–74] (see Chapter 5).

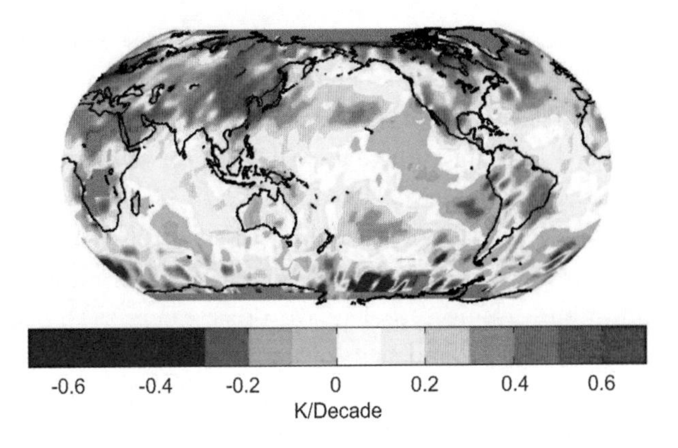

Figure 4.4 Temperature trends for 1979–2004 (HadCRUT2 temperature dataset). Many regions warmed, however others have cooled. Source: IPCC AR4 [57].

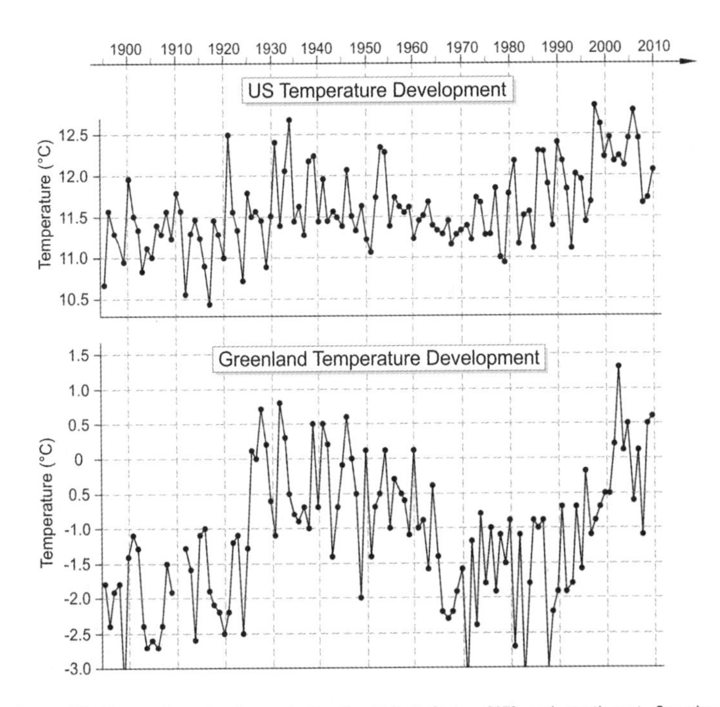

Figure 4.5 Temperature developments in the United States [67] and south-east Greenland (Angmagssalik station) [70]. In both regions the warm phase of the 1930s reached about today's levels.

The cause of the 1977–2000 warming would be relatively easy to explain if we had a network of temperature measurement stations on the other planets in our solar system. The idea is very simple: if we recorded a warming there as well, then we would know that the sun is the climate driver. Unfortunately, such a network does not exist and so we have to make do with single observations. Yet these are very interesting, even if they are not statistically sufficient and temperature changes could have other causes [75]. Various research groups have found signs of warming on Mars [76], Jupiter [77] Neptune's moon Triton [78–79] and Pluto [80]. In the future resourceful astronomers may find a way to produce a more complete temperature data set and thus shed more light on current, non-transparent planetary climate developments [81–82].

Climatic stop and go with a 60-year beat: a cycle!

Ask a friend or colleague this just for fun. Most of them won't even know that global warming actually stopped over 10 years ago. Since the year 2000, the temperature has in principle not risen [83–86] (Figure 1.2; see Chapter 1). How could this have happened? Didn't the IPCC warn us of dangerous warming that threatens to careen out of control? Or is it just a hiatus before climate warming resumes its upward death spiral?

As we have seen, for the last 150 years there have been recurrent stagnation and cooling episodes (Figure 4.2). The last one occurred between 1940 and 1977, a time when some climate scientists were fretting that a new ice age was imminent (see Chapter 7). Today's current break in warming is not unusual. The climate patterns of the past are simply repeating themselves. Still, it would be nice to know what is behind this stop and go. We have already seen with El Niño that oscillations within the climate system can have major impacts on the weather and climate. This especially applies to processes in the Pacific, whose effects can be felt all over the globe. Is there an internal climate cycle there lasting several decades that could explain the longer-term ups and downs in global temperature over the last

150 years? The answer is yes. This oceanic cycle was first discovered in 1996 by Steven Hare of the University of Washington and is called the Pacific Decadal Oscillation (PDO) [86]. Initially, it was mostly about cyclically fluctuating salmon stocks along the North American west coast [87]. Slowly, though, climate scientists became more aware of the PDO's trans-regional importance. Today it is only the IPCC that seems unwilling to accept the significance of the PDO phenomenon.

With the PDO, certain warm and cold water regions swap places in the northern Pacific Ocean (Figure 4.6), and do so every 20–30 years. A complete PDO cycle thus lasts 40–60 years [88]. As is the case with solar activity cycles, the oceanic PDO cycle is not rigid in its timing, so a perfect annual forecast is impossible. But statements on overall trends can be made and provide valuable information. The PDO Index is calculated using the north Pacific sea surface temperatures, from which the global average temperature is subtracted. Let's superimpose the PDO Index over the global temperature curve to see if El Niño's big brother, the PDO, is as strongly involved with shaping the climate as the 2–7-year El Niño oscillations are.

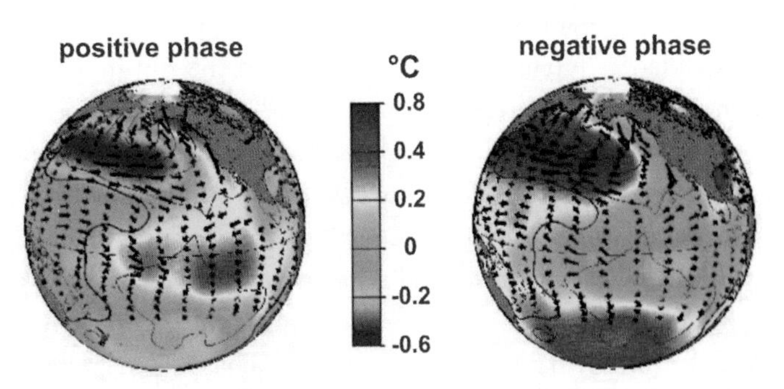

Figure 4.6 The PDO describes the position of the warm and cold water regions of the northern Pacific Ocean. They trade places every 20–30 years and have a strong impact on the global mean temperature [89].

The result is absolutely fascinating. Over the last 150 years the PDO and global temperature curves are surprisingly very well synchronized (Figure 4.9). The three warming episodes of 1860–80, 1910–40 and 1977–2000 all occur with the positive phase of the PDO, while the cooling phases occur during the negative PDO phase. So why hasn't the IPCC mentioned this? Some scientists, among them Don Easterbrook (Western Washington University), Syun-Ichi Akasofu (University of Alaska, Fairbanks) and Roy Spencer (University of Alabama, Huntsville) do not back the IPCC, Instead, they have highlighted this remarkable relationship [90–94], but have been ignored by the IPCC climate establishment.

One might suspect that the PDO is controlled by global air temperatures. However, this is not possible because the PDO only describes the distribution of cold and warm water areas, and the pattern has been repeated over the course of the climate warming of the last 150 years at various temperatures. Therefore, the PDO has to be the *trigger* of the 60-year temperature cycle, and not vice versa. In the positive PDO phases the PDO strengthens and so the global average temperature increases. In its negative phase, the PDO weakens and this leads to cooling. It is still unknown how the PDO comes about. Perhaps it is a basic internal oscillation of the Pacific climate system and the cycles occur without any external influence. Alternatively, the sun [95–96] may drive it, or the tidal forces of the giant planets Jupiter and Saturn acting on the sun in a similar fashion as lunar tidal forces on the ocean [97] (see Scafetta, below).

As we have noted, the IPCC does not take the 60-year PDO cyclic and related cycles sufficiently into account in its models [98–99]. That leads to a significant error because it causes the IPCC to assume wrongly that the rapid temperature increases during the warming episodes are the real long-term warming rates. Here the IPCC fails to note that the warming in these phases is magnified by the positive PDO. This needs to be deducted in order to arrive at the real long-term warming rate, which turns out to be much lower than what is postulated by the IPCC [100–102]. One gets the true warming rate when a line is drawn through the zero points of the PDO cycle and not along the steep positive flank of the PDO (Figure 4.7).

Owing to this misunderstanding, the IPCC manoeuvres itself into another problem, namely its explanation for the 60-year cyclically repeating cooling phases. Because the IPCC ignores the PDO cycle as a global climate driving factor along with the sun, it gets really creative and pulls aerosol clouds emitted by industry out of its magician's hat. This is what they use to explain the cooling. First, it is clear that these aerosol cooling effects do exist and that they block sunlight reaching the earth's surface. But it is rather troubling that the IPCC habitually uses aerosols as a joker in its climate models whenever they need to explain cyclic cooling phases. These cooling phases are dubbed global dimming. During the warming phases, the IPCC simply switches off this effect and claims it is due to the installation of sulphur filters on smokestacks [103]. It was not long ago that the aerosol joker was played yet again. A new round of global dimming was announced after the IPCC had become ruffled by the more than 10 years of no warming and their warming models had increasingly come under fire. On that occasion it had to be the increase in sulphur emissions by Chinese coal-fired power plants that were causing the skies to darken [83]. IPCC experts, along with Robert Kaufmann, wanted to prove that global warming had been stopped by the increased cooling effect of aerosols from China. While it is true that sulphur dioxide emissions in China increased steadily up to 2005, since then China has made enormous efforts in equipping its coal power plants with desulphurization systems. By 2006 its emissions had declined [104]. Since then three-quarters of Chinese power plants have been equipped with desulphurization units. Yet, global cooling is still with us. The attempt by Kaufmann and his proponents since then has been seriously questioned [55], for example, aerosol concentrations in the stratosphere from 2000 to 2005 had not significantly increased.

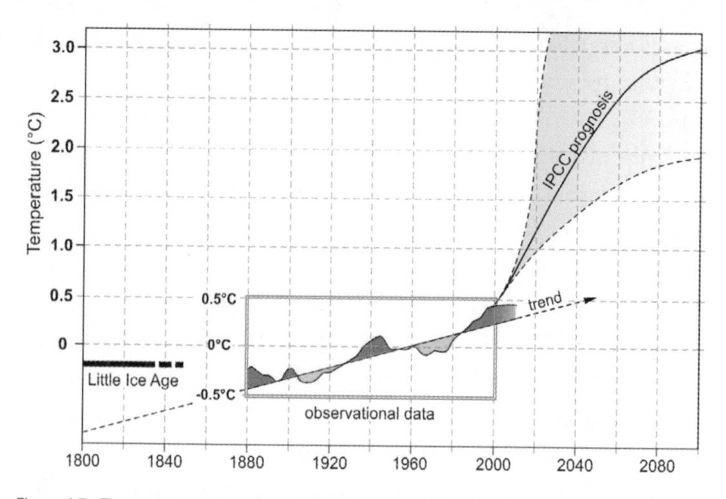

Figure 4.7 The strong warming phase of 1977–2000 coincides with the rising flank of the PDO cycle. The IPCC misinterpreted this PDO-enhanced rate of warming as a long-term global trend, then naively extended it to the year 2100. The warming trend of the twentieth century is considerably less and is dependent on a combination of the CO_2 increase and stronger solar activity. What is more, the IPCC fails to consider the likely cooling effect of the expected reduced solar activity (see Chapter 7) over the coming decades [93].

The long-term warming trend of the last 300 years

Now that we have examined the 60-year temperature oscillations, it is time to look at the overlying long-term warming trend of the last 300 years. In principle we have already described it in Chapter 3. This long-term warming is created by the sun and increased CO_2. The 1000-year Eddy solar cycle reached its irradiative minimum in the middle of the last millennium during the Little Ice Age. At the end of the Little Ice Age, solar activity increased and things warmed up again. Whereas there were no sunspots between 1661 and 1671 during the Maunder minimum, the sun was so peppered with spots during the second half of the twentieth century that one could have thought it had a bad case of chickenpox (Figures 3.3 and 3.8). This sunspot maximum marked one of the most active solar phases in the last 10,000 years. Indeed, the sun's magnetic field doubled in strength compared to the Little Ice Age (Figure 3.8). The close link between the sun and climate over the last 10,000 years is confirmed by numerous geological studies (Figure 3.12) and so it would be astonishing if that close link were to stop out of the blue (see Chapter 3). It is clear then that the enormous increase in solar activity could be responsible for a large part of the long-term warming of the last 300 years.

Parallel to solar activity, the concentration of atmospheric CO_2 has also increased due to the use of fossil fuels (Figure 3.8). Because of its greenhouse effect, CO_2 has made a contribution to the observed temperature increase, even if by how much is difficult to quantify. The apportionment of warming due to the sun and CO_2 is unclear and hotly debated. Respected scientists not affiliated to the IPCC estimate that 40–70 per cent of the observed climate warming of the last decades is caused by the sun [97, 105–112] (see Chapter 6). The IPCC's apportionment of blame for long-term warming solely to CO_2 and other anthropogenic greenhouse gases appears highly improbable. So it is reassuring when one reads in a new work by the IPCC author Stefan Rahmstorf that increased solar activity and atmospheric CO_2 together contributed to the 1910–40 warming episode [59].

The sun comes under attack

The coupling of temperature development to solar activity has been impressively shown in numerous studies examining the last 10,000 years (see Chapter 3). So what arguments do critics use to prove the ineffectiveness of our mother star on climate? Criticism focuses on irradiative development over the last 50 years. During solar cycle 19 in around 1960, solar activity accelerated to its highest level in the 400-year long measurement data series, a record that has not been exceeded. If the sun is really so influential, then one would expect that the temperature around 1960 would have reached a high point (Figure 4.8). But because we know this did not happen, some believe we can now safely cancel the sun from the climate equation beginning in the second half of the twentieth century [75]. It sounds entirely plausible at first, but it's false. The climate system isn't that simple. Let's take a look at how things are really interrelated.

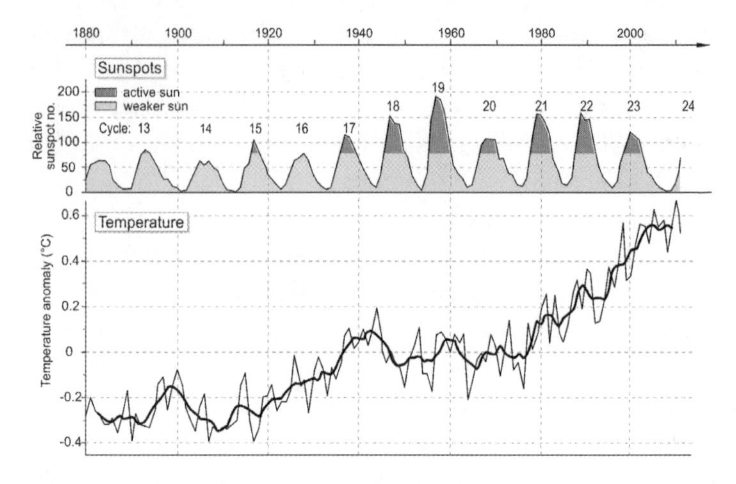

Figure 4.8 Sunspot development [113] and global temperature (GISS data set) [114] over the last 130 years.

Here we have to take a few steps back and start the story in the year 1940, when a 30-year warming had just ended. Solar activity, on the other hand, continued its rise unperturbed from 1940 (solar cycle 17) to the record year of 1960 (solar cycle 19). But we also know that the 1940–77 cooling phase was triggered by the cold phase of the PDO. Obviously, the PDO had the upper hand here and more than offset the warming effect of the strengthening sun. Even so, the record cycle 19 left its mark. If you take a closer look at the temperature curve, you'll notice there is a small 7-year warming peak of about 0.1° C around 1960 (Figure 4.3). This surely would have continued had solar cycle 20 not dropped so dramatically near 1970, when sunspots fell by half (Figure 4.8). The warming by solar cycle 19 was simply too short for the sluggish climate system to react sufficiently.

One can compare this to a pan of water on a stove. If at the start the stove is turned up to high (strong cycle 19), the water needs a few minutes to reach an equilibrium temperature with the stove. If, however, the stove is turned down before this equilibrium temperature is reached (weak cycle 20), then the water temperature will promptly decrease. If the stove is again turned up (strong cycles 21 and 22) but not as hot as at the very beginning (in cycle 19), then the temperature of the water in the pan will resume its increase until the equilibrium temperature is reached. This new equilibrium temperature may now even be higher than the water temperature during the hotter but shorter stove start phase (cycle 19) because the time was too short for the equilibrium temperature to be reached during the first heating phase. The positive PDO phase during cycles 21–22 certainly helped, as it further boosted the temperature increase (Figure 4.9). As is the case with a kettle, a flame that is consistently high for a longer time can cause the temperature to rise [115]. Over the course of strong cycles 21 and 22, the equilibrium state was finally reached. Then weaker cycle 23, in conjunction with the falling of the PDO, initiated the end of the warming period in the year 2000 (Figures 4.8 and 4.9).

It is the thermal inertia of the oceans and their deep circulation that result in the system not reacting immediately and completely as the temperature changes in response to the external climate impulses. This 'long pipeline' can lead to time lags of several years or even decades [20, 116]. A good example comes from Siberia. The temperature development for the last 700 years, reconstructed from an ice core, is closely coupled to solar activity [117]. However, at times the temperature reacted to the solar periods with a time delay of 10–30 years. Similar delays are shown in other studies [118–120].

But if climate data sets in certain cases do not oscillate synchronously, it does not necessarily mean that the processes are unrelated. Here the time for establishing equilibriums, time delays and superimposition by other climate factors can play a crucial role. When considering the temperature course over hundreds of thousands of years these short-term delays can be ignored more than when shorter time frames measured in decades are considered. Such a differentiated approach is necessary, especially for the development of temperature and solar activity over the last 70 years, but unfortunately this was not always done. Moreover, the

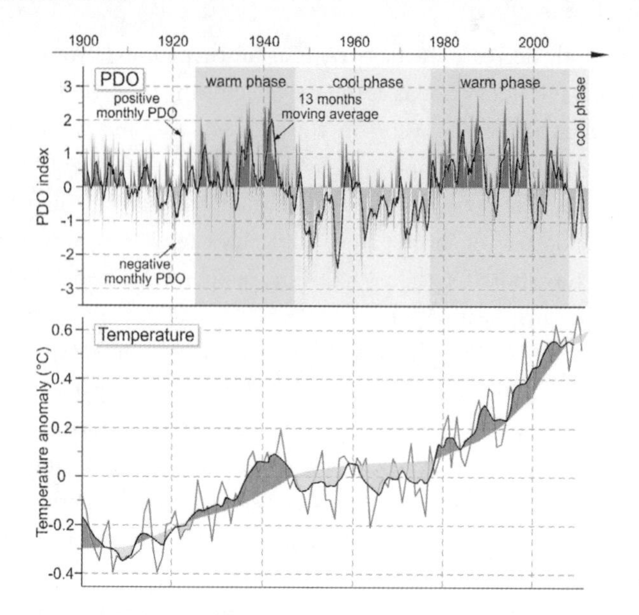

Figure 4.9 Warming and cooling episodes of the last 130 years occurred during times of positive or negative PDO phases [114, 121].

sun's impact is not of equal strength at every location, owing to the solar amplification processes (see Chapter 6). In addition, there are seasonal fluctuations. In Europe it appears that it is mainly winter temperatures that react to solar activity fluctuations [68].

In the past some authors tried to base the sun's climatic ineffectiveness on missing correlations during the past decades [75, 122]. However, this simplistic claim, as we have seen, does not go far enough and is inadequate when dealing with the climate system's complexity. We have to come to terms with reality: the sun's impact on our climate today works in the same way it has over the last 10,000 years. Why should there have been an exception over the last 50 years, or be one in the future?

Is it warming or not? A question of perspective

There has been no notable global temperature increase since the year 2000 (Figure 1.2). The temperature plateau, which is now over a decade long, marks the end of the warming of the current 60-year PDO-related cycle (Figure 4.9). Does this mean global warming has come to an end? The answer depends on how the question is framed and which perspective is chosen. Within the 60-year cycle, the warming has certainly stopped. But over the longer term we know that the cycles of the last 200 years are like staircases, where every subsequent cycle starts at a higher temperature level. We can illustrate the long-term temperature increase if we connect the zero points of the last 60-year cycles (Figure 4.7). Short cooling phases (e.g. 1880–1910, 1945–75) are natural parts of the cyclic pattern. Furthermore, the absence of warming since 2000 fits well into the picture. One could really conclude that long-term warming is still with us.

However, that the zero point of the next 60-year cycle will tend to be lower and that the long-term temperature trend is heading downwards over the next few decades is often missed. This is due to the extraordinarily high solar activity of the last decades [123], indicating that the peak of the 1000-year Eddy cycle has been reached and hence solar activity will not rise further. In addition, the fall of the 210-year Suess/de Vries and the 87-year Gleissberg cycles have already made impressive starts. And because the PDO too has passed its high point and is now on its downward side, thus adding a strong natural cooling component, rising CO_2 will not be able to offset in full the cooling over the next decades. Therefore a net cooling is anticipated (Figure 4.7; see Chapter 7). The last warming phase ended at the start of the millennium over 10 years ago. For the next few decades the earth will switch to mild cooling mode. Another oceanic internal 60-year cycle, the Atlantic Multidecadal Oscillation (AMO), will add to this development (see Chapter 7).

The climate of the last thousand years

As we have seen (Chapter 3), the 1000-year Eddy cycle has dominated the climate over the postglacial period (Holocene). It therefore makes sense to look beyond the horizon and examine our temperature history of the last one and a half centuries within a longer-term context. We have discussed the fundamental climate pattern of the last millennium in Chapter 3. A thousand years ago the Medieval Warm Period governed the earth, a time when temperatures were similar to those of today [22, 124–132]. A few centuries later the transition to the Little Ice Age was made and delivered the biting cold of the middle of the last millennium [133–142]. The transition phase between the two climate extremes lasted centuries. Globally, temperatures fell on average 0.8–1.0° C [22–23, 124–125, 143–144] and in some regions up to 2° C [145].

The fundamental scheme of this climate development had been known for some time [146–147] and was illustrated in the IPCC's First Assessment Report (FAR) of 1990 (Figure 4.10). Recent research confirms the characteristic temperature course depicting the Medieval Warm Period, Little Ice Age and the Modern Warm Period [22, 124–125]. However, after 1990 climate science proceeded down some false paths, going through one of its famous knowledge circles, only to wind up at its starting point of two decades earlier.

How could this loop back to 1990 have occurred? What had happened? A few climate scientists simply thought that the temperature dynamic of the last thousand years simply did not fit well with the IPCC's core message. The problem they saw was this: How could the temperature increase of the last 150 years be a dangerous product of rising CO_2 if a similar warm period had existed 1000 years earlier, long before atmospheric CO_2 concentrations shot up? Something didn't fit (Figure 4.10). Since CO_2 was the climate darling nobody wanted to abandon, and thus was not open to question, some fudging had to be done elsewhere. Scientists, therefore, turned their focus on revising the temperature history. Unless something there was engineered, nobody looking at the 1990 chart was going to take global warming seriously.

To the rescue came Michael Mann, who was fresh with a PhD and eager to make a rapid ascent of the academic ladder. During 1998 and 1999, working with two colleagues, he published a new

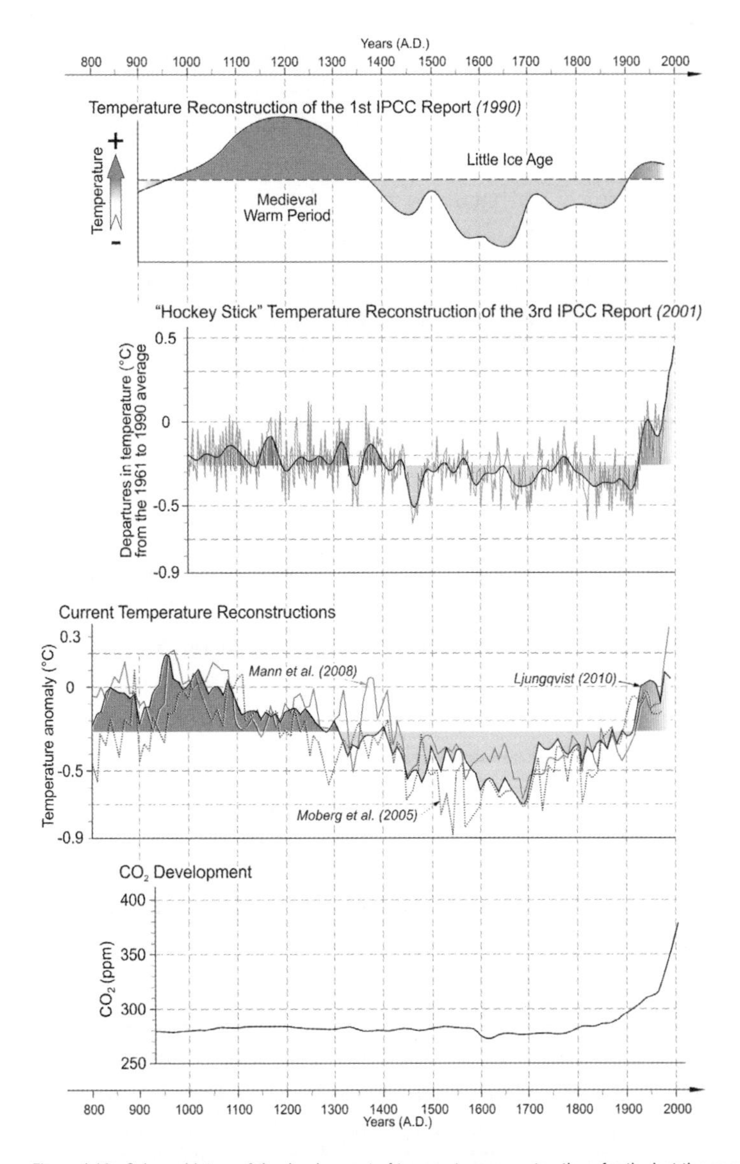

Figure 4.10 Science history of the development of temperature reconstructions for the last thousand years. In IPCC's First Assessment Report (1990), the Medieval Warm Period around the year AD 1000 was clearly shown [148]. In its Third Assessment Report (2001) the Medieval Warm Period disappeared completely because of a faulty statistical process and flawed data (the so-called hockey stick chart) [149]. Then in 2005 the Medieval Warm Period was 'rediscovered' and reappeared [22–23, 150].

temperature reconstruction of the Northern Hemisphere that included neither the Medieval Warm Period nor the Little Ice Age [151–152] (Figure 4.10). The curve resembled a hockey stick where the first 850 years were allegedly more or less climatically eventless and thus formed the handle of the stick, while the last 150 years had a rapid jump that formed the blade. The new reconstruction struck like a rapier. The chart was eagerly taken up by the IPCC and was promoted as one of the main features of its Third Assessment Report (TAR) in 2001. Mann was rewarded handsomely for his work. At a young age he was promptly appointed as an IPCC lead author on the chapter for climate variability. The hockey stick success story peaked when it featured prominently in Al Gore's blockbuster, *An Inconvenient Truth* (see Chapter 7).

The Medieval Warm Period: supposedly just an insignificant local event
Mann, of course, had been aware of the reports of low ice levels in the Arctic, which during medieval period made settlement of Greenland possible [66, 134–135, 153] and he also knew of the Thames freezing for weeks at a time during the bitter cold winters of the Little Ice Age. Even so, he went ahead and drew a more or less straight temperature line through the climatically turbulent past millennium, using a ingenious explanation to justify it: the observed climatic development of the north Atlantic region was only a local phenomenon and was not representative of the rest of the globe. There had never really been a Medieval Warm Period or a Little Ice Age elsewhere on the globe.

And the world believed him. He was the expert, after all. For a while everything went well – until opposition began to mount. Some specialists started checking the temperature history beyond the north Atlantic region for the last 1000 years. They steadily introduced data from all seven continents and in most cases their temperature curves depicted strange humps: one positive hump 1000 years ago and another negative hump 500 years later. They produced what was thought to be impossible, documenting the Medieval Warm Period and Little Ice Age, not only in the north Atlantic, but also in Africa [154–155], Antarctica [156–161], Asia [162–173], Oceania [174–175], North America [176–177] and South America [178–182]. The volume of data available today is overwhelming. Nevertheless, still acting as if nothing was wrong, scientists close to the IPCC, among them Stefan Rahmstorf (Potsdam) and Gerald Haug (Zurich), continue using the discredited argument and insist that the prominent climate peaks of the past were only a 'local phenomenon'.

Climate audit breaks the hockey stick

It took more than four years before serious resistance coalesced against Mann's hockey stick. This included an American group led by Willie Soon, who pointed out that the Medieval Warm Period and Little Ice Age were global [183–184]. Hans von Storch of the Helmholtz Centre too was not convinced by the hockey stick and calculated much stronger temperature variations for Mann's hockey stick handle [185]. But in the end it took semi-retired mining specialist Stephen McIntyre of Toronto to undertake the painstaking audit needed to see through the hockey stick's dark secrets. Not surprisingly, Mann and his colleagues were unwilling to disclose their data. Undeterred, McIntyre teamed up with economics professor Ross McKitrick and carried on. Eventually, after considerable effort, their determination paid off. The fascinating story of their journey through the hockey stick quagmire is told by A. W. Montford in his 2010 thriller, *The Hockey Stick Illusion* [186].

McIntyre and McKitrick were able to show that the statistical methodology Mann used was fundamentally flawed and tended to produce a hockey stick curve even when fed random data [187]. Using the R^2 test[7] the Mannian temperature graph failed. It was not until 2011 that statisticians took up the challenge and corrected

---

[7] This is a standard procedure for data analysis. The coefficient R^2 takes on values between 1 and 0, whereby 1 indicates a very good correlation and 0 indicates a missing correlation between the variables. All R^2 values of the hockey stick, except one, were below 0.2, according to McKintyre and McKitrick.

the errors in the statistical process [188–189]. Now an easily recognizable warm hump and a cold trough appeared from Mann's data set, namely, the Medieval Warm Period and the Little Ice Age.

In addition to mathematical and methodological errors, Mann and his colleagues committed a number of blunders when compiling their climate data sets. A large part of the input data originated from tree ring values, which are not ideal temperature proxies.[8] Tree growth reacts to a number of non-temperature influences, such as precipitation changes and insect infestation. Furthermore, in the late twentieth century some trees grew more rapidly because of the higher atmospheric CO_2 concentration, which is a temperature-independent CO_2 fertilization effect. At times Mann and his colleagues simply ignored tree ring data after the 1960s because the trend these data delivered deviated from the measurements taken by thermometer and at times even showed cooling (the so-called divergence problem). Moreover, tree rings often do not indicate winter temperatures reliably because trees for the most part are dormant during the winter. Interestingly, temperature reconstructions based on cave dripstones for the last 1000 years produce temperature fluctuations that are almost an order of magnitude greater than those of tree rings. According to the palaeo-climatologist Augusto Mangini of Heidelberg, this is because climate variability in the Northern Hemisphere occurs mostly and most distinctly in the winter, when trees are dormant [190].

There are more problems with Mann's data. Particularly with the oldest part of the hockey stick handle, the data are poor and the geographic distribution of the measured data is questionable. To some extent the data were simply extrapolated (Gaspé data) and at times they were false. Even obsolete data were used (Twisted Hill and Heartrot Hill data). Part of the error was later admitted, but they insisted that it had little effect on the overall curve. The supposedly

[8] A temperature proxy is an indirect, non-instrumental indicator of temperature. By using proxies, historical temperatures can be reconstructed from natural temperature archives such as tree rings, dripstones or sediment cores.

'independent' confirmation of the methodology and curves from Mann and his colleagues provided by Eugene Wahl and Caspar Ammann [191] from the year 2007 also has to be viewed critically. Amman was a former postgraduate of Ray Bradley, co-author of both the 1998 and 1999 hockey stick papers.

Finally, in 2008, Mann and his colleagues published a reworked version of their much criticized temperature curve [150]. In this one the Medieval Warm Period and the Little Ice Age were restored, although they were still somewhat subdued (Figure 4.10). The latter probably had something to do with the fact that although Mann used fewer tree ring data this time, he still worked with bristle cones whose strong growth over the last decades has much to do with CO_2 fertilization [187, 192]. In addition, temperature proxy data taken from a Finnish lake were incorrectly interpreted and found their way into the successor reconstruction [186].

So how did this small group of scientists manage to mislead the world for so long and so consistently? Clearing this up and looking into the exact background of the hockey stick episode is a task for future science historians. In any case, George Orwell's *1984* certainly should be among the source materials for such a study. Falsifying historical facts to ensure they always accommodated the current ideology of the state was one of the primary responsibilities of historians in the fictional state of *1984*. Orwell's Ministry of Truth provided the memorable slogan: *Who controls the past controls the future. Who controls the present controls the past.*

The last million years: a time of real climate catastrophes

The Medieval Warm Period and Little Ice Age are part of a cyclical climate dynamic with a millennial scale period. The climate forcing here stems from the sun and its fluctuations in activity, especially the 1000-year Eddy cycle and the 2300-year Hallstatt cycle. In Chapter 3 we examined case studies from various regions which impressively demonstrate the close relationship between the sun and the postglacial climate.

In addition to the important millennial cycles, there is yet another phenomenon of the postglacial period that is scarcely known to the public despite all the spectacular reports on the warming of the last 150 years. It may be hard to believe, but there was actually a phase during the past 10,000 years when temperatures over long periods and in many regions were higher than they are today. The warmest phase of the post-ice age occurred 4000–8000 years ago and is known as the Holocene Climate Optimum or the Atlantic Period [193–195] (Figure 4.11). Who would have guessed that the Alps at this time, except for a few small parts, were mostly ice free? [196]. Even the Antarctic peninsula experienced an early Holocene warm period. Unexpected but true: from about 2500–9200 years ago temperatures there were similar to current levels [197].

The largest desert in the world turned into a fertile savannah during the Holocene Climate Optimum. The Green Sahara was peppered with large lakes and populated by elephants, giraffes, hippopotamuses and crocodiles [198–200]. Due to this climate warming, the Sahel Zone and its summer rains shifted north and generously watered the Sahara. Vast quantities of water steadily seeped deep into the ground and today act as important fossil groundwater reserves, thus securing the water supply for vast regions of North Africa. Because these water reserves are not being replenished, it is foreseeable that they will eventually run dry. The causes of this postglacial climate optimum are suspected to be, for the most part, the Milankovitch earth orbital parameters (see Chapter 3). Because of geometric orbital relations, 5 per cent more solar radiation reached the earth during the summers then [200–201].

As we have seen, Milankovitch cycles are also thought to be responsible for the change between glacial phases (ice ages) and warm phases (interglacials). With impressive regularity, kilometre-thick sheets of ice ploughed their way across northern Europe and North America every 100,000 years. But once the massive ice sheets had melted, the environmental action did not quieten down at all. First, the sea level rose a massive 50–100 metres and thus submerged

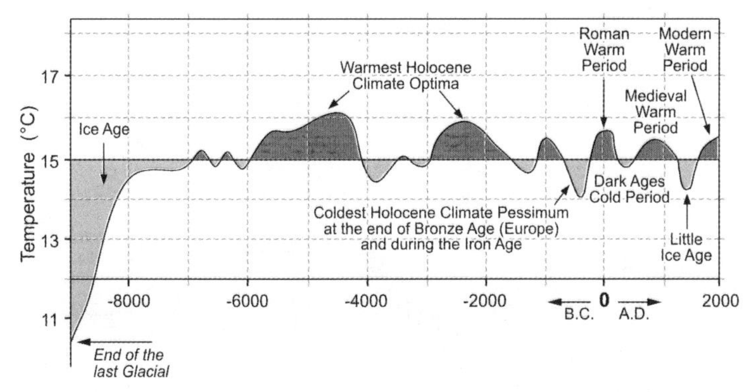

Figure 4.11 Climate history since the end of the ice age 10,000 years ago. The warmest phase of this period occurred 4000–8000 years ago and is also called Holocene Climate Optimum [202].

coastal areas, which had expanded during the preceding ice age. The average temperatures rose by about 5° C within only a few thousand years. Atmospheric CO_2 concentrations simultaneously increased, along with the global temperature. At first glance it may look as if

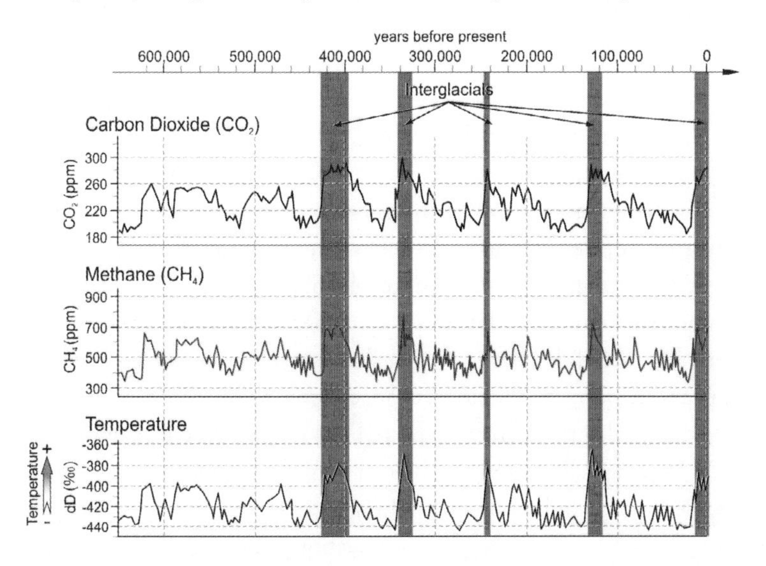

Figure 4.12 Reconstruction of temperature and atmospheric CO_2 concentrations over the last 600,000 years using the Antarctic Vostok ice core. In the transition from glacial to interglacial phases the temperature increased each time. As gas was released from the warmed ocean water, the CO_2 concentration increased. Source: AR4 [149].

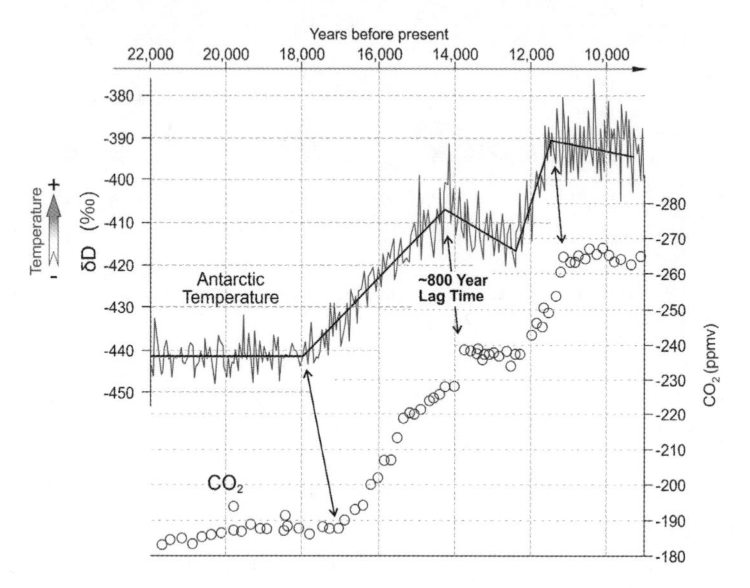

Figure 4.13 The temperature has on average an 800-year head start on CO_2 and so must be the trigger and not a consequence of CO_2 change [203].

temperatures and CO_2 moved in step whenever an ice age period changed to an interglacial (Figure 4.12). Naturally Al Gore eagerly exploited this in *An Inconvenient Truth*, using this alleged synchronicity as his star witness to prove the enormous climate potency of CO_2. At first it appears to make sense. However, Gore neglected to tell us one small detail. If you home in on the data set, you will see that the temperature curve has a head start on the CO_2 changes of up to 800 years [203–205] (Figure 4.13). On shorter timescales too, CO_2 always appears to follow temperature change [206].

Now who would have expected the temperature to conform to CO_2 in advance!? That would be real potency. Something here seems decidedly odd, and indeed something is. It is generally acknowledged that the CO_2 rise during the transition from an ice age to an interglacial is the consequence and not the trigger of the warming because less CO_2 dissolves in warm water. So as the oceans warm at the start of a warm interglacial, CO_2 simply bubbles out of the sea and into the atmosphere [20] and when it becomes colder at the start of an ice age, surplus CO_2 dissolves into the seawater and leads to a drop in atmospheric CO_2 concentration. With this sleight of hand Gore succeeded in fooling most of his audience, including the Nobel Prize committee.

A fossil precedent?

The climate sensitivity of CO_2 is very loosely defined by the IPCC, and so, according to its models, the consequences of the current CO_2 increase range from moderate to catastrophic. Everything and nothing is possible. Here it is advisable to search for comparable cases in the earth's long history, for precedents where the earth once warmed up abruptly after a gigantic injection of CO_2 into the atmosphere. Recently it was thought that such an event had been found and could serve as an analogue for today. An abrupt temperature increase occurred 55.5 million years ago at the boundary of the Paleocene–Eocene epochs – the so-called Paleocene-Eocene Thermal Maximum Event (PETM) [207]. Within just a few thousand years the global surface temperatures rose 5–9° C [208]. What triggered this? Is this geological climate catastrophe due solely to CO_2 as some claim? [209].

CO_2 must have played some role during the PETM, as it is assumed that a huge reserve of gas hydrates beneath the seabed destabilized and released massive quantities of methane, which then quickly oxidized to CO_2 [210–211]. The collapse of the gas reserve could have been caused by volcanic heating, a giant landslide or a massive meteor strike. The resulting CO_2 increase in the atmosphere has similarities in scale and course to what we are experiencing today [207, 209]. So is the PETM really comparable to today's situation? Unfortunately, it is not that simple.

During the PETM, the concentration of CO_2 increased a mere 70 per cent, from 1000 ppm to 1700 ppm [208]. Using the IPCC's own climate sensitivity calculation, only 1–3.5° C of the 5–9° C of PETM temperature increase can be explained by CO_2 [208].

The biggest part of the PETM increase had to have been caused by something other than CO_2 in the atmosphere [162, 208, 212]. Until this unknown is resolved and the reason for the abrupt warming is found, the millions of years old PETM cannot help us with our current climate questions and will remain an unsolved riddle.

Explaining the PETM mechanism and the role of CO_2 is made even more difficult because of another strange observation. In some PETM studies the CO_2 rise appears to lag behind the warming (e.g. the Bass River/New Jersey and Ocean Drilling Program Site 690) [213–214]. If this is true, how could CO_2 trigger the warming?

It is noteworthy that CO_2 and temperature were generally not coupled over the past 500 million years. New case studies confirm that the concept of a potent CO_2 climate driver has to be seriously questioned. A recent study looking at the climate in the northern Pacific 10 million years ago found that sea surface temperatures were 5–8° C warmer than today while the CO_2 concentration was lower [215].

The temperature involves many factors

As we have seen, the temperature history of the last thousand years involves a number of climate factors. The sun, volcanoes and long and short climate internal oscillations have been there from the beginning. Manmade CO_2 emissions, on the other hand, first began to get in on the climate action about 200 years ago, first only at low levels, and then taking off during the twentieth century. A solo role by CO_2, as the IPCC posits, has never been the case. Without a doubt it has played a relevant role in climate warming over the last 150 years, but it has shared the lead role with the very active sun. Today there is evidence indicating CO_2 has even been the junior partner in the warming process.

Other climate players are responsible for the smaller-scale changes within the longer-term temperature increase. Internal climate oscillations in the oceans put the temperature curves into oscillating motion of various durations. Volcanoes add short-term cooling accents. Also sulphur dioxide emitted from smokestacks adds to cooling, even if the magnitude is poorly known (see Chapter 5). The same is true for soot, which in turn has a warming effect that is not to be underestimated (see Chapter 5).

Current research shows that the climate equation is far more complicated than what was assumed a few years ago. As the IPCC managed to ignore quite a number of important climate factors in its 2007 climate modelling, Chapter 5 takes a closer look at the approach taken by the IPCC and critically examines its last report.

References

1. Focus (2010) Schlimmste Hitzewelle seit 1000 Jahren bedroht Russland. 9 August. http://www.focus.de/panorama/vermischtes/russland-schlimmste-hitzewelle-seit-1000-jahren-bedroht-russland_aid_539481.html.
2. RIA Novosti (2010) Medwedews Klimawandel-Berater prophezeit Russland neue Hitzeschläge. 16 August. http://de.rian.ru/environment_disaster/20100816/257110673.html.
3. Rahmstorf, S. (2010) Das Jahrzehnt der Wetterextreme. 1 September. http://www.wissenslogs.de/wblogs/blog/klimalounge/klimadaten/2010-09-01/das-jahrzehnt-der-wetterextreme.
4. BBC (2010) Climate change 'partly to blame' for sweltering Moscow. 10 August. http://www.bbc.co.uk/news/science-environment-10919460.
5. Rahmstorf, S. and D. Coumou (2011) Increase of extreme events in a warming world. *PNAS* 10.1073/PNAS.1101766108 1–5.
6. Dole, R., M. Hoerling, J. Perlwitz, J. Eischeid, P. Pegion, T. Zhang, X.-W. Quan, T. Xu and D. Murray (2011) Was there a basis for anticipating the 2010 Russian heatwave? *Geophysical Research Letters* 38, 1–5.
7. NOAA (2011) Natural variability main culprit of deadly Russian heatwave that killed thousands. Press release, 9 March. http://www.noaanews.noaa.gov/stories2011/20110309_russianheatwave.html.
8. Wikipedia (2011) Winter of 2010–2011 in Europe. http://en.wikipedia.org/wiki/Winter_of_2010%E2%80%932011_in_Europe#Record_low_temperatures.
9. Wikipedia (2011) Winter of 2010–2011 in Great Britain and Ireland. http://en.wikipedia.org/wiki/Winter_of_2010-2011_in_Great_Britain_and_Ireland.
10. Süddeutsche Zeitung (2011) *Nacht des Grauens*. 19 May.
11. Casey, J. L. (2011) Cold Sun. Trafford Publishing, Bloomington, IN.
12. BBC (2011) South Korea chaos after 'heaviest' snowfall. 14 February. http://www.bbc.co.uk/news/world-asia-pacific-12445509.
13. MSNBC (2009) Houston wakes up to earliest snowfall ever. 4 December. http://www.msnbc.msn.com/id/34274883/ns/weather/t/houston-wakes-earliest-snowfall-ever.
14. Wikipedia (2011) Winter of 2009–2010 in Europe. http://en.wikipedia.org/wiki/Winter_of_2009%E2%80%932010_in_Europe.

15. Wang, C., H. Liu and S.-K. Lee (2010) The record-breaking cold temperatures during the winter of 2009/2010 in the Northern Hemisphere. Atmos. Sci. Let. 11, 161–8.
16. Wikipedia (2011) Winter of 2009–2010 in the United Kingdom. http://en.wikipedia.org/wiki/Winter_of_2009%E2%80%932010_in_the_United_Kingdom.
17. Xiaoyan, Z. (2010) Extreme weather marks 2010. http://www.bjreview.com.cn/nation/txt/2011-01/23/content_328187.htm.
18. Cattiaux, J., R. Vautard, C. Cassou, P. Yiou, V. Masson-Delmotte and F. Codron (2010) Winter 2010 in Europe: a cold extreme in a warming climate. *Geophys. Res. Lett.* 37 (20), L20704.
19. von Juterczenka, W. (2007) Chronik extremer Wetterereignisse von 1900 bis 1999. In .K. G. Blüchel (ed.), *Der Klimaschwindel.* C. Bertelsmann, Munich, 250–318.
20. Latif, M. (2008) *Bringen wir das Klima aus dem Takt? Hintergründe und Prognosen.* Fischer Taschenbuch Verlag, Frankfurt a.M.
21. Alexander, M. A., I. Blade, M. Newman, J. R. Lanzante, N.-C. Lau and J. D. Scott (2002) The atmospheric bridge: the influence of ENSO teleconnections on air–sea interaction over the global oceans. *Journal of Climate* 15, 2205–31.
22. Ljungqvist, F. C. (2010) A new reconstruction of temperature variability in the extra-tropical Northern Hemisphere during the last two millennia. *Geografiska Annaler*: Series A 92 (3), 339–51.
23. Moberg, A., D. M. Sonechkin, K. Holmgren, N. M. Datsenko, W. Karlén and S.-E. Lauritzen (2005) Highly variable Northern Hemisphere temperatures reconstructed from low- and high-resolution proxy data. *Nature* 433, 613–17.
24. Nuzhdina, M. A. (2002) Connection between ENSO phenomena and solar and geomagnetic activity. *Natural Hazards and Earth System Sciences* 2, 83–9.
25. White, W. B. and Z. Liu (2008) Non-linear alignment of El Niño to the 11-year solar cycle. *Geophysical Research Letters* 35, 1–6.
26. van Loon, H. and G. A. Meehl (2008) The response in the Pacific to the sun's decadal peaks and contrasts to cold events in the Southern Oscillation. *Journal of Atmospheric and Solar–Terrestrial Physics* 70, 1046–55.
27. Meehl, G. A., J. M. Arblaster, G. Branstator and H. Van Loon (2008) A coupled air–sea response mechanism to solar forcing in the Pacific tegion. *Journal of Climate* 21, 2883–97.
28. Landscheidt, T. (1999) Solar activity controls El Niño and La Niña. http://www.john-daly.com/sun-enso/sun-enso.htm.
29. Baker, R. G. V. (2008) Exploratory analysis of similarities in solar cycle magnetic phases with Southern Oscillation Index fluctuations in eastern Australia. *Geographical Research* 46 (4), 380–98.
30. Miles, G. M., R. G. Grainger and E. J. Highwood (2004) The significance of volcanic eruption strength and frequency for climate. Q. J. R. Meteorol. Soc. 130, 2361–76.
31. Wikipedia (2011) Vulkanischer Winter. http://de.wikipedia.org/wiki/Vulkanischer_Winter.
32. Robock, A. (2000) Volcanic eruptions and climate. *Reviews of Geophysics* 38 (2), 191–219.
33. Bluth, G. J. S., W. I. Rose, I. E. Sprod and A. J. Krueger (1997) Stratospheric loading of sulphur from explosive volcanic eruptions. *The Journal of Geology* 105, 671–83.

34. Fischer, E. M., J. Luterbacher, E. Zorita, S. F. B. Tett, C. Casty and H. Wanner (2007) European climate response to tropical volcanic eruptions over the last half-millennium. *Geophysical Research Letters* 34, 1–6.
35. Rampino, M. R., S. Self and R. B. Stothers (1988) Volcanic winters. *Ann. Rev. Earth Planet. Sci.* 16, 73–99.
36. Thordarson, T. and S. Self (2003) Atmospheric and environmental effects of the 1783–1784 Laki eruption: a review and reassessment. *Journal of Geophysical Research* 108, 7/1–29.
37. Lanciki, A., J. Cole-Dai, M. H. Thiemens and J. Savarino (2012) Sulphur isotope evidence of little or no stratospheric impact by the 1783 Laki volcanic eruption. *Geophys. Res. Lett.* 39 (1), L01806.
38. Trigo, R. M., J. M. Vaquero, M.-J. Alcoforado, M. Barriendos, J. Taborda, R. Garcia-Herrera and J. Luterbacher (2009) Iberia in 1816, the year without a summer. *International Journal of Climatology* 29, 99–115.
39. Stommel, H. and E. Stommel (1983) Volcano Weather. *The Story of 1816, the Year Without a Summer*. Seven Seas Press, Newport, RI.
40. Galindo, I. (1992) Extinction of short wave solar radiation due to El Chichon stratospheric aerosol. *Atmosfera* 5, 259–68.
41. McCormick, M. P. (1995) Atmospheric effects of the Mt Pinatubo eruption. *Nature* 373 (2), 399–404.
42. Self, S., M. R. Rampino, M. S. Newton and J. A. Wolff (1984) Volcanological study of the great Tambora eruption of 1815. *Geology* 12, 659–63.
43. Sutawidjaja, I. S., H. Sigurdsson and L. Abrams (2006) Characterization of volcanic deposits and geoarchaeological studies from the 1815 eruption of Tambora volcano. *Journal Geologi Indonesia.*
44. Oppenheimer, C. (2003) Climatic, environmental and human consequences of the largest known historic eruption: Tambora volcano (Indonesia) 1815. *Progress in Physical Geography* 27 (2), 230–59.
45. Wikipedia (2011) Laki-Krater. http://de.wikipedia.org/wiki/Laki-Krater.
46. Wikipedia (2011) Jahr ohne Sommer. http://de.wikipedia.org/wiki/Jahr_ohne_Sommer.
47. Briffa, K. R. and P. D. Jones (1992) The climate of Europe during the 1810s with special references to 1816. In C. R. Harington (ed.), *The Year without a Summer? World Climate in 1816*. Canadian Museum of Nature, Ottawa, 372–91.
48. Briffa, K. R., P. D. Jones, F. H. Schweingruber and T. J. Osborn (1998) Influence of volcanic eruptions on Northern Hemisphere summer temperature over the past 600 years. *Nature* 393, 450–5.
49. Zhang, D., R. Blender and K. Fraedrich (2011) Volcanic and ENSO effects in China in simulations and reconstructions: Tambora eruption 1815. *Clim. Past Discuss.* 7, 2061–88.
50. Auchmann, R., S. Brönnimann, L. Breda, M. Bühler, R. Spadin and A. Stickler (2012) Extreme climate, not extreme weather: the summer of 1816 in Geneva, Switzerland. *Clim. Past* 8, 325–35.
51. Wagner, S. and E. Zorita (2005) The influence of volcanic, solar and CO_2 forcing on the temperatures in the Dalton minimum (1790–1830): a model study. *Climate Dynamics* 25 (2–3), 205–18.
52. Archibald, D. C. (2006) Solar cycles 24 and 25 and predicted climate response. *Energy & Environment* 17 (1), 29–35.

53. Wikipedia (2011) Krakatau. http://de.wikipedia.org/wiki/Krakatau.
54. Wikipedia (2011) El Chichón. http://de.wikipedia.org/wiki/El_Chichon.
55. Vernier, J.-P., L. W. Thomason, J.-P. Pommereau, A. Bourassa, J. Pelon, A. Garnier, A. Hauchecorne, L. Blanot, C. Trepte, D. Degenstein and F. Vargas (2011) Major influence of tropical volcanic eruptions on the stratospheric aerosol layer during the last decade. *Geophysical Research Letters* 38, 1–8.
56. Jones, P. (2010) Q & A: Professor Phil Jones 13 February. http://news.bbc.co.uk/2/hi/science/nature/8511670.stm.
57. IPCC (2007) *Climate Change 2007: The Physical Science Basis. Contribution of Working Group I to the Fourth Assessment Report of the Intergovernmental Panel on Climate Change.* Cambridge University Press, Cambridge and New York.
58. Akasofu, S.-I. (2010) On the recovery from the Little Ice Age. Natural Science 2, 1211–24.
59. Archer, D. and S. Rahmstorf (2010) *The Climate Crisis*, 1st edn. Cambridge University Press, Cambridge.
60. Graversen, R. G., T. Mauritsen, M. Tjernström, E. Källén and G. Svensson (2008) Vertical structure of recent Arctic warming. *Nature* 541, 53–7.
61. Krueger, M. (2011) Wie die NASA die Arktis warmrechnet und die Globaltemperatur zum Steigen bringt. http://www.science-skeptical.de/blog/wie-die-nasa-die-arktis-warmrechnet-und-die-globaltemperatur-zum-steigen-bringt/003815.
62. Spielhagen, R. F., K. Werner, S. A. Sørensen, K. Zamelczyk, E. Kandiano, G. Budeus, K. Husum, T. M. Marchitto and M. Hald (2011) Enhanced modern heat transfer to the Arctic by warm Atlantic water. *Science* 331, 450–3.
63. Chylek, P., C. K. Folland, G. Lesins and M. K. Dubey (2010) Twentieth century bipolar seesaw of the Arctic and Antarctic surface air temperatures. *Geophysical Research Letters* 37, 1–4.
64. Vinther, B. M., K. Andersen, P. D. Jones, K. R. Briffa and J. Cappelen (2006) Extending Greenland temperature records into the late eighteenth century. *Journal of Geophysical Research* 111, 1–13.
65. Chylek, P., M. K. Dubey and G. Lesins (2006) Greenland warming of 1920–1930 and 1995–2005. *Geophysical Research Letters* 33, 1–5.
66. Kobashi, T., K. Kawamura, J. P. Severinghaus, J.-M. Barnola, T. Nakaegawa, B. M. Vinther, S. J. Johnsen and J. E. Box (2011) High variability of Greenland surface temperature over the past 4000 years estimated from trapped air in an ice core. *Geophysical Research Letters* 38, 1–6.
67. National Climatic Data Center (2011) Climate Summary, contiguous United States. http://www.ncdc.noaa.gov/oa/climate/research/cag3/na.html.
68. Le Mouél, J.-L., E. Blanter, M. Shnirman and V. Courtillot (2009) Evidence for solar forcing in variability of temperatures and pressures in Europe. *Journal of Atmospheric and Solar–Terrestrial Physics* 71 (12), 1309–21.
69. *Der Spiegel* (1948) Es wird wärmer auf der Welt. 9/1948. http://www.spiegel.de/spiegel/print/d-44415803.html.
70. NOAA (2011) Global Historical Climatology Network. http://www.appinsys.com/GlobalWarming/climatedata.aspx?Dataset=GHCNTemp.
71. Berkeley Earth Surface Temperature (BEST) study (submitted), http://berkeleyearth.org/available-resources.
72. Wilby, R. L., P. D. Jones and D. H. Lister (2011) Decadal variations in the nocturnal heat island of London. *Weather* 66 (3), 59–64.

73. Michaels, P. J. (2008) Global warming: correcting the data. *Regulation* Fall, 46–52.
74. Fall, S., A. Watts, J. Nielsen-Gammon, E. Jones, D. Niyogi, J. Christy and R. A. Pielke Sr. (2011) Analysis of the impacts of station exposure on the US Historical Climatology Network temperatures and temperature trends. *Journal of Geophysical Research* 116, 1–15.
75. Skeptical Science (2011) Solar activity and climate: is the sun causing global warming? http://www.skepticalscience.com/solar-activity-sunspots-global-warming.htm.
76. Fenton, L. K., P. E. Geissler and R. M. Haberle (2007) Global warming and climate forcing by recent albedo changes on Mars. *Nature* 446, 646–9.
77. de Pater, I., P. Marcus and M. Wong (2006) Jupiter's New 'red oval'. In Hubble 2006 Science Year in Review, 31–41. http://hubblesite.org/hubble_discoveries/science_year_in_review/2006.
78. Elliot, J. L., H. B. Hammel, L. H. Wasserman, O. G. Franz, S. W. McDonald, M. J. Person, C. B. Olkin, E. W. Dunham, J. R. Spencer, J. A. Stansberry, M. W. Buie, J. M. Pasachoff, B. A. Babcock and T. H. McConnochie (1998) Global warming on Triton. *Nature* 393, 765–7.
79. Hammel, H. B. and G. W. Lockwood (2007) Suggestive correlations between the brightness of Neptune, solar variability, and earth's temperature. *Geophysical Research Letters* 34, 1–7.
80. Elliot, J. L., A. Ates, B. A. Babcock, A. S. Bosh, M. W. Buie, K. B. Clancy, E. W. Dunham, S. S. Eikenberry, D. T. Hall, S. D. Kern, S. K. Leggett, S. E. Levine, D.-S. Moon, C. B. Olkin, D. J. Osip, J. M. Pasachoff, B. E. Penprase, M. J. Person, S. Qu, J. T. Rayner, L. C. Roberts Jr, C. V. Salyk, S. P. Souza, R. C. Stone, B. W. Taylor, D. J. Tholen, J. E. Thomas-Osip, D. R. Ticehurst and L. H. Wasserman (2003) The recent expansion of Pluto's atmosphere. *Nature* 424, 165–8.
81. Sromovsky, L. A., P. M. Fry, S. S. Limaye and K. H. Baines (2003) The nature of Neptune's increasing brightness: evidence for a seasonal response. *Icarus* 163, 256–61.
82. Pollack, J. B., K. Rages and O. B. Toon (1980) On the relationship between secular brightness changes of Titan and solar variability. *Geophysical Research Letters* 7 (10), 829–32.
83. Kaufmann, R. K., H. Kauppi, M. L. Mann and J. H. Stock (2011) Reconciling anthropogenic climate change with observed temperature 1998–2008. *PNAS* 108 (29), 11790–3.
84. Knight, J., J. J. Kennedy, C. Folland, G. Harris, G. S. Jones, M. Palmer, D. Parker, A. Scaife and P. Stott (2009) Do global temperature trends over the last decade falsify climate predictions? In T. C. Peterson and M. O. Baringer (eds.), *State of the Climate in 2008*, Bulletin of the American Meteorological Society, Vol. 90, No. 8, August, 22–3.
85. Loehle, C. (2009) Trend analysis of satellite global temperature data. *Energy and Environment* 20 (7), 1087–98.
86. Mantua, N. J., S. R. Hare, Y. Zhang, J. M. Wallace and R. C. Francis (1997) A Pacific interdecadal climate oscillation with impacts on salmon production. *Bulletin of the American Meteorological Society* 78 (6), 1069–79.
87. Zwolinski, J. P. and D. A. Demer (2012) A cold oceanographic regime with high exploitation rates in the Northeast Pacific forecasts a collapse of the sardine stock. *Proceedings of the National Academy of Sciences* 109 (11), 4175–80.

88. MacDonald, G. M. and R. A. Case (2005) Variations in the Pacific Decadal Oscillation over the past millennium. *Geophysical Research Letters* 32, 1–4.
89. Mantua, N. (2000) The Pacific Decadal Oscillation (PDO). http://jisao.washington.edu/pdo.
90. Easterbrook, D. J. (2011) Geologic evidence of recurring climate cycles and their implications for the cause of global climate changes – the past is the key to the future. In D. J. Easterbrook (ed.), *Evidence-Based Climate Science.* Elsevier, Oxford, 3–51.
91. Easterbrook, D. Evidence of the cause of global warming and cooling: Recurring global, decadal, climate cycles recorded by glacial fluctuations, ice cores, ocean temperatures, historic measurements and solar variations. http://myweb.wwu.edu/dbunny/research/global/easterbrook_climate-cycle-evidence.pdf.
92. Akasofu, S.-I. (2009) Two Natural Components of the Recent Climate Change. http://people.iarc.uaf.edu/~sakasofu/pdf/two_natural_components_recent_climate_change.pdf.
93. Akasofu, S.-I. (2009) Natural Components of Climate Change During the Last Few Hundred Years. http://people.iarc.uaf.edu/~sakasofu/natural_components_climate_change.php (modified).
94. Spencer, R. W. (2008) Global warming as a natural response to cloud changes associated with the Pacific Decadal Oscillation (PDO). *SPPI – Science and Policy Institute,* http://scienceandpublicpolicy.org/images/stories/papers/reprint/Spencer-GW _as_Natural_response.pdf, 1–10.
95. Patterson, R. T., A. Prokoph and A. Chang (2004) Late Holocene sedimentary response to solar and cosmic ray activity influenced climate variability in the NE Pacific. *Sedimentary Geology* 172, 67–84.
96. Landscheidt, T. (2001) Trends in Pacific Decadal Oscillation subjected to solar forcing. http://www.john-daly.com/theodor/pdotrend.htm.
97. Scafetta, N. (2010) Empirical evidence for a celestial origin of the climate oscillations and its implications. *Journal of Atmospheric and Solar–Terrestrial Physics* 72, 951–70.
98. Kim, H.-M., P. J. Webster and J. A. Curry (2012) Evaluation of short-term climate change prediction in multi-model CMIP5 decadal hindcasts. *Geophys. Res. Lett.* 39 (10), L10701.
99. Robson, J. I., R. T. Sutton and D. M. Smith (2012) Initialized decadal predictions of the rapid warming of the north Atlantic Ocean in the mid 1990s. *Geophys. Res. Lett.* 39 (19), L19713.
100. Loehle, C. and N. Scafetta (2011) Climate change attribution using empirical decomposition of climatic data. *The Open Atmospheric Science Journal* 5, 74–86.
101. Wu, Z., N. E. Huang, J. M. Wallace, B. V. Smoliak and X. Chen (2011) On the time-varying trend in global-mean surface temperature. *Climate Dynamics* 37 (3-4), 759–73.
102. Zhou, J. and K.-K. Tung (2012) Deducing Multi-decadal anthropogenic global warming trends using multiple regression analysis. *Journal of the Atmospheric Sciences.*
103. Wild, M. (2009) Global dimming and brightening: a review. *Journal of Geophysical Research* 114, 1–31.
104. Lu, Z., D. G. Streets, Q. Zhang, S. Wang, G. R. Carmichael, Y. F. Cheng, C. Wei, M. Chin, T. Diehl and Q. Tan (2010) Sulphur dioxide emissions in China and sulphur trends in East Asia since 2000. *Atmos. Chem. Phys.* 10, 6311–31.

105. Scafetta, N. and B. J. West (2007) Phenomenological reconstructions of the solar signature in the Northern Hemisphere surface temperature records since 1600. *Journal of Geophysical Research* 112, D24S03.
106. Scafetta, N. and B. J. West (2008) Is climate sensitive to solar variability? *Physics Today* March, 50–1.
107. Beer, J., W. Mende and R. Stellmacher (2000) The role of the sun in climate forcing. *Quaternary Science Reviews* 19, 403–15.
108. Rao, U. R. (2011) Contribution of changing galactic cosmic ray flux to global warming. *Current Science* 100 (2), 223–5.
109. Shaviv, N. J. (2005) On climate response to changes in the cosmic ray flux and radiative budget. *Journal of Geophysical Research* 110, 1–15.
110. Shaviv, N. J. (2008) Using the oceans as a calorimeter to quantify the solar radiative forcing. *Journal of Geophysical Research* 113, 1–13.
111. El-Sayed Aly, N. (2010) Spectral analysis of solar variability and their possible role on global warming. *International Journal of the Physical Sciences* 5 (7), 1040–9.
112. De Jager, C. and S. Duhau (2011) The variable solar dynamo and the forecast of solar activity; influence on terrestrial surface temperature. In J. M. Cossia (ed.), Global Warming in the 21st Century, *Nova Science*, Hauppauge, NY, 77–106.
113. NGDC (2011) Sonnenflecken, Jahresmittelwerte. ftp://ftp.ngdc.noaa.gov/stp/solar_data/sunspot_numbers/international/yearly/yearly.plt.
114. NASA (2011) GISS Surface Temperature Analysis. http://data.giss.nasa.gov/gistemp/graphs.
115. Rawls, A. (2011) Do solar scientists still think that recent warming is too large to explain by solar activity? http://wattsupwiththat.com/2011/01/02/do-solar-scientists-still-think-that-recent-warming-is-too-large-to-explain-by-solar-activity.
116. Weber, S. L., T. J. Crowley and G. van der Schrier (2004) Solar irradiance forcing of centennial climate variability during the Holocene. *Climate Dynamics* 22, 539–53.
117. Eichler, A., S. Olivier, K. Henderson, A. Laube, J. Beer, T. Papina, H. W. Gäggeler and M. Schwikowski (2009) Temperature response in the Altai region lags solar forcing. *Geophysical Research Letters* 36, 1–5.
118. Wang, X. and Q.-B. Zhang (2011) Evidence of solar signals in tree rings of Smith fir from Sygera Mountain in southeast Tibet. *Journal of Atmospheric and Solar–Terrestrial Physics* 73, 1959–66.
119. Perry, C. A. (2007) Evidence for a physical linkage between galactic cosmic rays and regional climate time series. *Advances in Space Research* 40, 353–64.
120. Powell Jr, A. M. and J. Xu (2012) Assessment of the relationship between the combined solar cycle/ENSO forcings and the tropopause temperature. *Journal of Atmospheric and Solar–Terrestrial Physics* 80, 21–7.
121. Climate Charts & Graphs (2011) Pacific Decadal Oscillation (PDO). http://processtrends.com/images/RClimate_pdo_trend_latest.png.
122. Lockwood, M. and C. Fröhlich (2007) Recent oppositely directed trends in solar climate forcings and the global mean surface air temperature. *Proceedings of the Royal Society* A 463, 2447–60.
123. Solanki, S. K., I. G. Usoskin, B. Kromer, M. Schüssler and J. Beer (2004) Unusual activity of the sun during recent decades compared to the previous 11,000 years. *Nature* 431, 1084–7.
124. Loehle, C. (2007) A 2000-year global temperature reconstruction based on non-tree ring proxies. *Energy & Environment* 18 (7–8), 1049–58.

125. Loehle, C. and J. H. McCulloch (2008) Correction to a 2000-year global temperature reconstruction based on non-tree ring proxies. *Energy & Environment* 19 (1), 93-100.
126. Idso, C. and S. F. Singer (2009) *Climate Change Reconsidered.* Heartland Institute, Chicago.
127. Ljungqvist, F. C., P. J. Krusic, G. Brattström and H. S. Sundqvist (2012) Northern Hemisphere temperature patterns in the last 12 centuries. *Clim. Past* 8, 227–49.
128. Moschen, R., N. Kühl, S. Peters, H. Vos and A. Lücke (2011) Temperature variability at Dürres Maar, Germany during the migration period and at high medieval times, inferred from stable carbon isotopes of Sphagnum cellulose. *Clim. Past* 7, 1011–26.
129. Niemann, H., A. Stadnitskaia, S. B. Wirth, A. Gilli, F. S. Anselmetti, J. S. Sinninghe Damsté, S. Schouten, E. C. Hopmans and M. F. Lehmann (2012) Bacterial GDGTs in Holocene sediments and catchment soils of a high-alpine lake: application of the MBT/CBT-paleothermometer. *Clim. Past* 8, 889–906.
130. Surge, D. and J. H. Barrett (2012) Marine climatic seasonality during medieval times (10th to 12th centuries) based on isotopic records in Viking Age shells from Orkney, Scotland. *Palaeogeography, Palaeoclimatology, Palaeoecology* 350–352, 236-46.
131. Sicre, M.-A., J. Jacob, U. Ezat, S. Rousse, C. Kissel, P. Yiou, J. Eiríksson, K. L. Knudsen, E. Jansen and J.-L. Turon (2008) Decadal variability of sea surface temperatures off north Iceland over the last 2000 years. *Earth and Planetary Science Letters* 268 (1–2), 137–42.
132. Christiansen, B. and F. C. Ljungqvist (2012) The extra-tropical Northern Hemisphere temperature in the last two millennia: reconstructions of low-frequency variability. *Climate of the Past* 8, 765–86.
133. Fagan, B. (2000) *The Little Ice Age.* Basic Books, New York.
134. Singer, S. F. and D. T. Avery (2008) *Unstoppable Global Warming – Every 1,500 Years.* Rowan & Littlefield, Lanham, MD.
135. Plimer, I. (2009) *Heaven and Earth: Global Warming – The Missing Science.* Quartet, London.
136. Shindell, D. T. (2009) Little Ice Age. In V. Gornitz (ed.), *Encyclopedia of Paleoclimatology and Ancient Environments*, Springer, New York, 520–2.
137. Shindell, D. T., G. A. Schmidt, M. E. Mann, D. Rind and A. Waple (2001) Solar forcing of regional climate change during the Maunder minimum. *Science* 294, 2149–52.
138. Ribeiro, S., M. Moros, M. Ellegaard and A. Kuijpers (2011) Climate variability in west Greenland during the past 1500 years: evidence from a high-resolution marine palynological record from Disko Bay. *Boreas*, 1–16.
139. Perner, K., M. Moros, J. M. Lloyd, A. Kuijpers, R. J. Telford and J. Harff (2011) Centennial scale benthic foraminiferal record of late Holocene oceanographic variability in Disko Bugt, West Greenland. *Quaternary Science Reviews* 30, 2815–26.
140. Cuven, S., P. Francus and S. Lamoureux (2011) Mid to late Holocene hydroclimatic and geochemical records from the varved sediments of East Lake, Cape Bounty, Canadian High Arctic. *Quaternary Science Reviews* 30, 2651–65.
141. Larocque-Tobler, I., M. M. Stewart, R. Quinlan, M. Trachsel, C. Kamenik and M. Grosjean (2012) A last millennium temperature reconstruction using chironomids preserved in sediments of anoxic Seebergsee (Switzerland) consensus at local, regional and Central European scales. *Quaternary Science Reviews* 41, 49–56.

142. Diodato, N. and G. Bellocchi (2012) Discovering the anomalously cold Mediterranean winters during the Maunder minimum. *The Holocene* 22 (5), 589–96.
143. Hald, M., G. R. Salomonsen, K. Husum and L. J. Wilson (2011) A 2000-year record of Atlantic water temperature variability from the Malangen fjord, northeastern North Atlantic. *The Holocene* 21 (7), 1049–59
144. Morellón, M., A. Pérez-Sanz, J. P. Corella, U. Büntgen, J. Catalán, P. González-Sampériz, J. J. González-Trueba, J. A. López-Sáez, A. Moreno, S. Pla, M. Á. Saz-Sánchez, P. Scussolini, E. Serrano, F. Steinhilber, V. Stefanova, T. Vegas-Vilarrúbia and B. Valero-Garcés (2012) A multi-proxy perspective on millennium-long climate variability in the Southern Pyrenees. *Clim. Past* 8, 683–700.
145. Mangini, A., C. Spötl and P. Verdesa (2005) Reconstruction of temperature in the Central Alps during the past 2000 years from a δ18O stalagmite record. *Earth and Planetary Science Letters* 235, 741–51.
146. Lamb, H. H. (1965) The early medieval warm epoch and its sequel. Palaeogeogr. Palaeoclim. *Palaeoecol.* 1, 13–37.
147. Duphorn, K. (1976) Kommt eine neue Eiszeit? *Geologische Rundschau* 65, 845–64.
148. IPCC (1990) First Assessment Report. http://www.ipcc.ch/publications_and_data/publications_and_data_reports.shtml.
149. IPCC (2001) *Climate Change 2001: The Scientific Basis.* Cambridge University Press, Cambridge.
150. Mann, M. E., Z. Zhang, M. K. Hughes, R. S. Bradley, S. K. Miller, S. Rutherford and F. Ni (2008) Proxy-based reconstructions of hemispheric and global surface temperature variations over the past two millennia. *PNAS* 105 (36), 13252–7.
151. Mann, M. E., R. S. Bradley and M. K. Hughes (1998) Global-scale temperature patterns and climate forcing over the past six centuries. *Nature* 392, 779–87.
152. Mann, M. E., R. S. Bradley and M. K. Hughes (1999) Northern Hemisphere temperatures during the past millennium: inferences, uncertainties, and limitations. *Geophysical Research Letters* 26 (6), 759–62.
153. Behringer, W. (2007) *Kulturgeschichte des Klimas.* C. H. Beck Verlag, Munich.
154. CO_2 Science (2011) Medieval Warm Period (Regional: Africa) – Summary. http://www.co2science.org/subject/m/summaries/mwpafrica.php.
155. Powers, L. A., T. C. Johnson, J. P. Werne, I. S. Castañeda, E. C. Hopmans, J. S. S. Damsté and S. Schouten (2011) Organic geochemical records of environmental variability in Lake Malawi during the last 700 years, Part I: The TEX86 temperature record. *Palaeogeography, Palaeoclimatology, Palaeoecology* 303, 133–9.
156. CO_2 Science (2011) Medieval Warm Period (Regional: Antarctica) – Summary. http://www.co2science.org/subject/m/summaries/mwpantarctica.php.
157. Li, Y., J. Cole-Dai and L. Zhou (2009) Glaciochemical evidence in an East Antarctica ice core of a recent (AD 1450–1850) neoglacial episode. *Journal of Geophysical Research* 114, 1-11.
158. Bertler, N. A. N., P. A. Mayewski and L. Carter (2011) Cold conditions in Antarctica during the Little Ice Age – implications for abrupt climate change mechanisms. *Earth and Planetary Science Letters* 308, 41–51.
159. Orsi, A. J., B. D. Cornuelle and J. P. Severinghaus (2012) Little Ice Age cold interval in West Antarctica: evidence from borehole temperature at the West Antarctic Ice Sheet (WAIS) divide. *Geophys. Res. Lett.* 39 (9), L09710.
160. Rhodes, R. H., N. A. N. Bertler, J. A. Baker, H. C. Steen-Larsen, S. B. Sneed, U.

Morgenstern and S. J. Johnsen (2012) Little Ice Age climate and oceanic conditions of the Ross Sea, Antarctica from a coastal ice core record. *Climate of the Past* 8, 1223–38.

161. Lu, Z., R. E. M. Rickaby, H. Kennedy, P. Kennedy, R. D. Pancost, S. Shaw, A. Lennie, J. Wellner and J. B. Anderson (2012) An ikaite record of late Holocene climate at the Antarctic Peninsula. *Earth and Planetary Science Letters* 325–326, 108–15.
162. CO_2 *Science* (2011) Medieval Warm Period (China) – Summary. http://www.co2science.org/subject/m/summaries/mwpchina.php.
163. Zhu, H.-F., X.-M. Shao, Z.-Y. Yin, P. Xu, Y. Xu and H. Tian (2011) August temperature variability in the southeastern Tibetan Plateau since AD 1385 inferred from tree rings. *Palaeogeography, Palaeoclimatology, Palaeoecology* 305, 84–92.
164. Mackay, A. W., D. B. Ryves, R. W. Battarbee, R. J. Flower, D. Jewson, P. Rioual and M. Sturm (2005) 1000 years of climate variability in central Asia: assessing the evidence using Lake Baikal (Russia) diatom assemblages and the application of a diatom-inferred model of snow cover on the lake. *Global and Planetary Change* 46, 281–97.
165. Yamada, K., M. Kamite, M. Saito-Kato, M. Okuno, Y. Shinozuka and Y. Yasuda (2010) Late Holocene monsoonal climate change inferred from Lakes Ni-no-Megata and San-no-Megata, northeastern Japan. *Quaternary International* 220, 122–32.
166. Tan, L., Y. Cai, Z. An, L. Yi, H. Zhang and S. Qin (2011) Climate patterns in north central China during the last 1800 years and their possible driving force. *Clim. Past* 7, 685–92.
167. Park, J. (2011) A modern pollen–temperature calibration data set from Korea and quantitative temperature reconstructions for the Holocene. *The Holocene* 21 (7), 1125–35.
168. Kaniewski, D., E. Van Campo, E. Paulissen, H. Weiss, J. Bakker, I. Rossignol and K. Van Lerberghe (2011) The medieval climate anomaly and the Little Ice Age in coastal Syria inferred from pollen-derived palaeoclimatic patterns. *Global and Planetary Change* 78, 178–87.
169. Hao, Z.-X., J.-Y. Zheng, Q.-S. Ge and W.-C. Wang (2012) Winter temperature variations over the middle and lower reaches of the Yangtze River since AD 1736. *Climate of the Past* 8, 1023–30.
170. Selvaraj, K., K.-Y. Wei, K.-K. Liu and S.-J. Kao (2012) Late Holocene monsoon climate of northeastern Taiwan inferred from elemental (C, N) and isotopic ($\delta^{13}C$, $\delta^{15}N$) data in lake sediments. *Quaternary Science Reviews* 37, 48–60.
171. Kitagawa, H. and E. Matsumoto (1995) Climatic implications of $\delta^{13}C$ variations in a Japanese cedar (Cryptomeria japonica) during the last two millenia. *Geophys. Res. Lett.* 22 (16), 2155–8.
172. Treydte, K. S., D. C. Frank, M. Saurer, G. Helle, G. H. Schleser and J. Esper (2009) Impact of climate and CO_2 on a millennium-long tree-ring carbon isotope record. *Geochimica et Cosmochimica Acta* 73 (16), 4635–47.
173. Oppo, D. W., Y. Rosenthal and B. K. Linsley (2009) 2,000-year-long temperature and hydrology reconstructions from the Indo-Pacific warm pool. *Nature* 460, 1113–16.
174. Wilson, A. T., C. H. Hendy and C. P. Reynolds (1979) Short-term climate change and New Zealand temperatures during the last millennium. *Nature* 279, 315–17.
175. Cohen, T. J., G. C. Nanson, J. D. Jansen, L. A. Gliganic, J. H. May, J. R. Larsen, I.

D. Goodwin, S. Browning and D. M. Price (2012) A pluvial episode identified in arid Australia during the medieval climatic anomaly. *Quaternary Science Reviews* 56, 167–71.

176. *CO2 Science* (2011) Medieval Warm Period (North America) – Summary. http://www.co2science.org/subject/m/summaries/mwpnortham.php.
177. Clegg, B. F., G. H. Clarke, M. L. Chipman, M. Chou, R. Walker, W. Tinner and F. S. Hu (2010) Six millennia of summer temperature variation based on midge analysis of lake sediments from Alaska. *Quaternary Science Reviews* 29, 3308–16.
178. CO2 Science (2011) Medieval Warm Period (South America) – Summary. http://www.co2science.org/subject/m/summaries/mwpsoutham.php.
179. Bird, B. W., M. B. Abbott, M. Vuille, D. T. Rodbell, N. D. Stansell and M. F. Rosenmeier (2011) A 2,300-year-long annually resolved record of the South American summer monsoon from the Peruvian Andes. *PNAS* 108 (21), 8583–8.
180. Ledru, M.-P., V. Jomelli, P. Samaniego, M. Vuille, S. Hidalgo, M. Herrera and C. Ceron (2012) The medieval climate anomaly and the Little Ice Age in the eastern Ecuadorian Andes. *Climate of the Past Discussion* 8, 4295–332.
181. Fletcher, M.-S. and P. I. Moreno (2012) Vegetation, climate and fire regime changes in the Andean region of southern Chile (38°S) covaried with centennial-scale climate anomalies in the tropical Pacific over the last 1500 years. *Quaternary Science Reviews* 46, 46–56.
182. Neukom, R., J. Luterbacher, R. Villalba, M. Küttel, D. Frank, P. D. Jones, M. Grosjean, H. Wanner, J.-C. Aravena, D. E. Black, D. A. Christie, R. D'Arrigo, A. Lara, M. Morales, C. Soliz-Gamboa, A. Srur, R. Urrutia and L. von Gunten (2011) Multiproxy summer and winter surface air temperature field reconstructions for southern South America covering the past centuries. *Climate Dynamics* 37 (1–2), 35–51.
183. Soon, W. and S. Baliunas (2003) Proxy climatic and environmental changes of the past 1000 years. *Climate Research* 23, 89–110.
184. Soon, W., S. Baliunas, C. Idso, S. Idso and D. R. Legates (2003) Reconstructing climatic and environmental changes of the past 1000 years: a reappraisal. *Energy & Environment* 14 (2–3), 233–96.
185. von Storch, H., E. Zorita, J. M. Jones, Y. Dimitriev, F. González-Rouco and S. F. B. Tett (2004) Reconstructing past climate from noisy data. *Science* 306, 679–82.
186. Montford, A. W. (2010) *The Hockey Stick Illusion*. Stacey International, London.
187. McIntyre, S. and R. McKitrick (2003) Corrections to the Mann et al. (1988) proxy data base and northern hemispheric average temperature series. *Energy & Environment* 14 (6), 751–71.
188. McShane, B. B. and A. J. Wyner (2011) A statistical analysis of multiple temperature proxies: are reconstructions of surface temperatures over the last 1000 years reliable? *The Annals of Applied Statistics* 5 (1), 5–44.
189. McShane, B. B. and A. J. Wyner (2011) Rejoinder. *The Annals of Applied Statistics* 5 (1), 99–123.
190. Mangini, A. (2007) Ihr kennt die wahren Gründe nicht. FAZ, 5 April. http://www.faz.net/artikel/C31015/weltklimabericht-ihr-kennt-die-wahren-gruende-nicht-30096191.html.
191. Wahl, E. R. and C. M. Ammann (2007) Robustness of the Mann, Bradley, Hughes reconstruction of Northern Hemisphere surface temperatures: examination of criticisms based on the nature and processing of proxy climate evidence. *Climatic Change* 85, 33–69.

192. LaMarche Jr., V. C., D. A. Graybill, H. C. Fritts and M. R. Rose (1984) Increasing atmospheric carbon dioxide: tree ring evidence for growth enhancement in natural vegetation. *Science* 223, 1019–21.
193. Crosta, X., M. Debret, D. Denis, M. A. Courty and O. Ther (2007) Holocene long- and short-term climate changes off Adélie Land, East Antarctica. *Geochem. Geophys. Geosyst.* 8 (11), 1–15.
194. Jansen, E., C. Andersson, M. Moros, K. H. Nisancioglu, B. F. Nyland and R. J. Telford (2009) The early to mid Holocene thermal optimum in the North Atlantic. In R. W. Battarbee and H. A. Binney (eds.), *Natural Climate Variability and Global Warming: A Holocene Perspective*. Wiley-Blackwell, Chichester, 123–37.
195. Liew, P. M., C. Y. Lee and C. M. Kuo. (2006) Holocene thermal optimal and climate variability of East Asian monsoon inferred from forest reconstruction of a subalpine pollen sequence, Taiwan. *Earth and Planetary Science Letters* 250, 596–605.
196. Berner, U. and H. Streif (2004) *Klimafakten: Der Rückblick – Ein Schlüssel für die Zukunft*, 4th edn. Schweizerbart, Stuttgart.
197. Mulvaney, R., N. J. Abram, R. C. A. Hindmarsh, C. Arrowsmith, L. Fleet, J. Triest, L. C. Sime, O. Alemany and S. Foord (2012) Recent Antarctic peninsula warming relative to Holocene climate and ice-shelf history. *Nature* 489, 141–4.
198. Claussen, M., C. Kubatzki, V. Brovkin, A. Ganopolski, P. Hoelzmann and H.-J. Pachur (1999) Simulation of an abrupt change in Saharan vegetation in the mid Holocene. *Geophysical Research Letters* 26 (14), 2037–40.
199. Schulz, E. (1991) Holocene environments in the central Sahara. *Hydrobiologia* 214 (1), 359–65.
200. Lézine, A.-M., C. Hély, C. Grenier, P. Braconnot and G. Krinner (2011) Sahara and Sahel vulnerability to climate changes, lessons from Holocene hydrological data. *Quaternary Science Reviews* 30, 3001–12.
201. Ruddiman, W. F. (2008) *Earth's Climate – Past and Future*, 2nd edn. W. H. Freeman, New York.
202. Kehl, H. (2008) Das zyklische Auftreten von Optima und Pessima im Holozän. http://lv-twk.oekosys.tu-berlin.de/project/lv-twk/002-holozaene-optima-und-pessima.htm.
203. Monnin, E., A. Indermühle, A. Dällenbach, J. Flückiger, B. Stauffer, T. F. Stocker, D. Raynaud and J.-M. Barnola (2001) Atmospheric CO_2 concentrations over the last glacial termination. *Science* 291, 112–14 (modified).
204. Ahn, J., E. J. Brook, A. Schmittner and K. Kreutz (2012) Abrupt change in atmospheric CO_2 during the last ice age. *Geophys. Res. Lett.* 39 (18), L18711.
205. Pedro, J. B., S. O. Rasmussen and T. D. van Ommen (2012) Tightened constraints on the time lag between Antarctic temperature and CO_2 during the last deglaciation. *Climate of the Past* 8, 1213–21.
206. Humlum, O., K. Stordahl and J.-E. Solheim (2013) The phase relation between atmospheric carbon dioxide and global temperature. *Global and Planetary Change* 100, 51–69.
207. Schmidt, G. A. (2009) Paleocen-Eocene thermal maximum. In V. Gornitz (ed.), *Encyclopedia of Paleoclimatology and Ancient Environments*. Springer, New York, 696–700.
208. Zeebe, R. E., J. C. Zachos and G. R. Dickens (2009) Carbon dioxide forcing alone insufficient to explain Palaeocene–Eocene thermal maximum warming. *Nature Geoscience*, 1–5.

209. Lovell, B. (2010) *Challenged by Carbon*. Cambridge University Press, Cambridge.
210. Zachos, J. C., U. Röhl, S. A. Schellenberg, A. Sluijs, D. A. Hodell, D. C. Kelly, E. Thomas, M. Nicolo, I. Raffi, L. J. Lourens, H. McCarren and D. Kroon (2005) Rapid acidification of the ocean during the Paleocene–Eocene thermal maximum. *Science* 308, 1611–15.
211. Dickens, G. R. (2011) Down the rabbit hole: toward appropriate discussion of methane release from gas hydrate systems during the Paleocene–Eocene thermal maximum and other past hyperthermal events. *Clim. Past* 7, 831–46.
212. Royer, D. L., S. L. Wing, D. J. Beerling, D. W. Jolley, P. L. Koch, L. J. Hickey and R. A. Berner (2001) Paleobotanical evidence for near present-day levels of atmospheric CO_2 during part of the tertiary. *Science* 292, 2310–13.
213. Sluijs, A., H. Brinkhuis, S. Schouten, S. M. Bohaty, C. M. John, J. C. Zachos, G.-J. Reichart, J. S. Sinninghe Damsté, E. M. Crouch and G. R. Dickens (2007) Environmental precursors to rapid light carbon injection at the Palaeocene–Eocene boundary. *Nature* 450, 1218–22.
214. Kelly, D. C. (2002) Response of Antarctic (ODP Site 690) planktonic foraminifera to the Paleocene–Eocene thermal maximum: faunal evidence for ocean/climate change. *Paleoceanography* 17 (4), 23/21–13.
215. LaRiviere, J. P., A. C. Ravelo, A. Crimmins, P. S. Dekens, H. L. Ford, M. Lyle and M. W. Wara (2012) Late Miocene decoupling of oceanic warmth and atmospheric carbon dioxide forcing. *Nature* 486, 97–100.

The forgotten 60-year natural cycle and its significance for the development of our climate

Nicola Scafetta
ACTIVE CAVITY RADIOMETER IRRADIANCE MONITOR (ACRIM)
LABORATORY AND DUKE UNIVERSITY, DURHAM, NC

Several studies have highlighted that a distinct 50–70-year oscillation characterizes numerous climatic records referring to the last 10,000 years as well as during recent centuries [1–9]. These examples [3] include ice core and pine tree samples from several regions, sardine and anchovy sediment core samples, global surface temperature records, fishing catch records and in an atmospheric circulation index, length of the day index [4]. A dominant 60-year periodicity is found in oceanic Atlantic Multidecadal Oscillation (AMO) indexes throughout the last 8000 years [1], as well as in the GISP2 ice core records [5] and in instrumental temperature reconstructions since 1500 covering the European Mediterranean Basin (Spain, France and Italy) [6]. A ~60-year periodicity is found in secular monsoon rainfall records from India, the Arabian Sea sediments and east China [7]. A 60-year oscillation is found in ocean and land surface temperature records of both hemispheres [2], and not only in AMO indexes such as reflected by the abundance variations of marine protozoa in the Caribbean Cariaco Basin [8], but also in Pacific Decadal Oscillation (PDO) indexes to the west [8]. A significant 60-year oscillation is found in a reconstruction of global sea level since 1700 calculated from tide gauge records [9]. Since 1850 a quasi-60-year oscillation in numerous instrumental global surface temperature records is extremely clear, where the cyclical matching between the periods 1880–1940 and 1940–2000 is evident [2].

This quasi-60-year oscillation may have a solar origin because researchers found 60–62 year cycles in cosmogenic nuclide records in the north-east Pacific region which are clearly related to solar-controlled cosmic ray variations [10]. A similar ~60-year cyclicity was observed by scientists in numerous other solar-related historical and geological data sets [11–13]. Together with my co-author Adriano Mazzarella from the University of Naples I was recently able to confirm that a quasi-60-year oscillation also characterizes the North Atlantic Oscillation (NAO) record which has been obtained using the historical instrumental records available for Europe since 1700 [14]. This NAO oscillation correlates very well with the global sea surface temperature records since 1850 and with the 60-year cycle observed in the historical middle latitude aurora records and the daily rotation rate of the earth since 1700. These results clearly suggest that a quasi-60-year oscillation exists in the climate system and, very likely, is astronomically induced.

The 60-year cycle is probably one of several solar activity cycles [11–14] which may be triggered by astronomical mechanisms. One of these mechanisms may be the tidal forces of the two largest planets of our solar system, the gas giants Jupiter and Saturn [2]. Jupiter takes about 12 years to orbit the sun while it takes Saturn about 30 years. This means that every 20 years or so the two planets meet and align with the sun. During such an alignment, the planets pull most strongly at the sun, while this force weakens as Jupiter and Saturn transit into different parts of the solar system. After three of these so-called conjunctions, that is about 60 years, the Jupiter-Saturn conjunctions occur approximately in the same constellation. This is important, because their orbits are not exactly circular. Therefore, every 60 years there will be a conjunction where the two planets come closer to the sun than during the previous two events. It can be shown that the 60-year gravitational cycle peaked during 1880–1, 1940–1 and 2000–1, as did the global surface temperature

[2]. Thus, a quasi-60-year cycle could have an astronomical physical origin, being associated with the wobbling of the sun and to the combined tidal effects of Jupiter and Saturn. Other tidal effects originating from the gravitation of the moon and the sun are powerful processes on the earth which move the sea level up and down across the world's oceans twice a day.

The exact physical link mechanisms explaining how the solar system oscillations induced by the planets may cause a 60-year cycle in the climate system of the earth are still uncertain. However, it is possible that such gravitational oscillations produce equivalent oscillations in the solar interior where an amplifier for the planetary gravitational signal through powerful nuclear feedback mechanisms may occur [15]. Then, these oscillations could be forced in the climate through the cloud system by means of a solar modulation of the incoming cosmic ray flux [16–18] (see Chapter 6).

A 60-year natural cycle would have important implications for correctly interpreting climate change. In fact, the 60-year climate cycle was in its warm phase from about 1970 to 2000 and, in conjunction with a 20-year climate cycle, has very likely induced about 0.3° C (or 60 per cent) of the 0.5° C warming observed since 1970 [2, 19]. Moreover, because the 60-year cycle entered its cooling phase around 2000, it can easily explain the lack of warming observed since 2001, which may last until about 2030–40 [2, 19].

In fact, the IPCC [20] has claimed that the ~0.5° C warming observed from 1970 to 2000 has been induced by anthropogenic forcing alone because natural forcing on its own (solar plus volcano debris) should have caused a cooling of about 0.1–0.2° C during the same 30-year interval (Figure 5.5) [20]. The IPCC's results imply that the anthropogenic forcing from 1970 to 2000 has contributed a net warming of the observed 0.5° C plus at most another 0.2° C which have offset the alleged natural cooling, which could be as large as 0.2° C. The ~0.7° C warming in his period would correspond to a warming rate of about 2.3° C a century since 1970. This anthropogenic warming rate was projected by the IPCC to continue on average from 2000 to 2050. This conclusion, however, is based on theoretical computer modelling which is characterized by huge uncertainties. For example, and as the IPCC acknowledges, the equilibrium climate sensitivity to CO_2 doubling varies between 1 and 10° C, with a central estimate of 1.5–4.5° C and an average of about 3° C [20].

On the other hand, if ~0.3° C (60 per cent) of the warming from 1970 to 2000 has been naturally induced by the 60-year cycle during its warming phase, only about 0.2° C (up to a possible maximum of 0.4° C) [21] warming from 1970 to 2000 could have been manmade, not 0.7° C as implied in the IPCC calculations. Thus, the IPCC has used climate models that, by not reproducing the 60-year climate cycle, have almost certainly overestimated the anthropogenic contribution (due mostly to greenhouse gases plus aerosol forcings) to climate change by a factor 2–4. The warming trend for the period 2000–50 implicit in the projections should consequently also be reduced from ~2.3 ± 0.3° C per century to a value in the range 0.5–1.3° C a century [21], which would also imply a lower climate sensitivity of a doubling of CO_2 of about 1° C or less. This same result has also been obtained by using alternative methods of data analysis [8, 22–23] where the climate sensitivity for a doubling of CO_2 is estimated to be in the range of 0.5–1.3° C. These results clearly imply that the current climate models used by the IPCC to support the anthropogenic climate warming theory are seriously exaggerating climate sensitivity to CO_2 changes.

Figure 4.A.1 shows a reconstruction of the global surface temperature [24] by using a model that I developed based on the twenty- and 60-year cycles plus. For the period 1850–2000 the upward trend can be related to both secular/millennial natural solar cycles and manmade forcings [25–26], while since 2000 the upward trend in the figure is related only to a corrected anthropogenic projected warming as explained above [21]. Not included is a possible additional secular cooling trend effect caused by the forecast secular and millennial decline in solar activity in the coming few decades (see Chapter 7). The model depicted in the figure reconstructs the alternating warming–cooling 30-year patterns from 1850 to 2000 very well, and also predicts the slight cooling or the temperature plateau observed since 2000 much more convincingly than using climate models opportunely calibrated [27] to explain such cooling with anthropogenic greenhouse gas and aerosol-projected forcings. In fact, the climate sensitivity to these forcings is characterized by extremely large uncertainties and it would be relatively easy to take advantage of such uncertainties to calibrate a computer model to reproduce a climate trending for a restricted number of years. According to my proposed model, the global temperature will not increase significantly – indeed, it may even cool until 2030–40 – while the IPCC projections have claimed, on average, a warming of about 1.15° C above the 2000 temperature by 2050.

This theory appears further confirmed in my two latest publications [19, 21]. In the first paper [19] I show that the observed decadal and multi-decadal climatic cycles are linked to astronomical forcing which should cause albedo oscillations by modulating the cloud cover through electrical/magnetic forcings of the heliosphere and of the terrestrial magnetosphere, as revealed by the oscillations observed in the mid latitude aurora records. In the other [21] I show that all climate models used by the IPCC in 2007 fail to reproduce the decadal and multi-decadal cycles observed in the climate system while a harmonic model based on the astronomical cycles, depicted here in Figure 4.A.1, is able to reconstruct and forecast those oscillations.

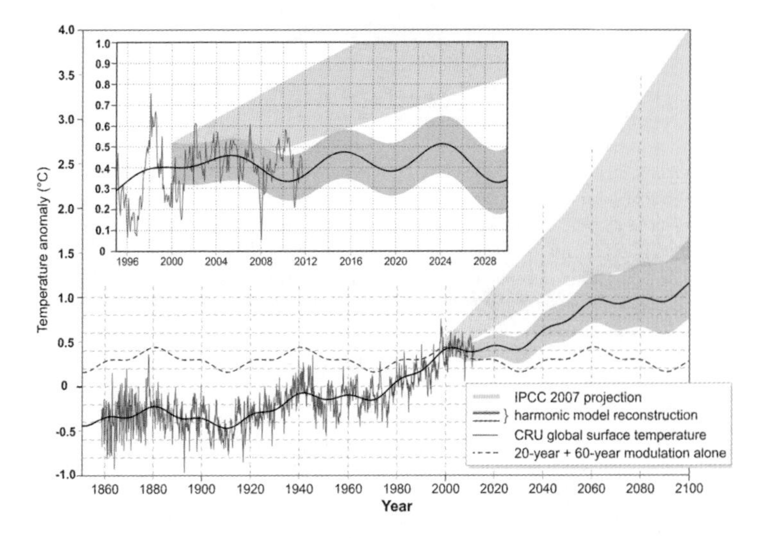

Figure 4.A.1 The Scafetta model [2] (black solid line) based on only the 20-year plus 60-year cycles plus the upward trend, against the global surface temperature (red) [24] and the IPCC [20] average manmade warming projection for the period 2000–2100 (green) of ~2.3 ± 0.3° C a century, which has been already disproved by the slight global temperature cooling observed since 2001. The blue curve depicts the projection for 2000–2100 taking into account the 60-year cyclicity plus the corrected manmade warming trend during the same period of 0.5–1.3° C a century, as explained in the text. Not included is a possible additional secular cooling trending effect caused by the forecast secular and millennial decline in solar activity which may begin in the coming few decades. The dashed curve shows the 20-year plus 60-year modulation of the temperature without the upward trend. The insert zooms in on 1995–2030 and uses an optimized harmonic model based on four astronomical cycles (9.1-year, 10.4-year, 20-year and 60-year) plus the corrected anthropogenic projected trend since 2000 as above [21]. Additional natural cycles longer than the 60-year period cycle are ignored in the 2000–2100 forecast.

References

1. Knudsen, M. F., M.-S. Seidenkrantz, B. H. Jacobsen and A. Kuijpers (2011) Tracking the Atlantic multidecadal oscillation through the last 8,000 years. *Nature Communications* 2 (178), 1–8.
2. Scafetta, N. (2010) Empirical evidence for a celestial origin of the climate oscillations and its implications. *Journal of Atmospheric and Solar–Terrestrial Physics* 72, 951–70.
3. Klyashtorin, L., V. Borisov and A. Lyubushin (2009) Cyclic changes of climate and major commercial stocks of the Barents Sea. *Marine Biology Research* 5, 4–17.
4. Mazzarella, A. (2007) The 60-year modulation of global air temperature: the earth's rotation and atmospheric circulation connection. *Theor. Appl. Climatol.* 88, 193–9.
5. Davis, J. C. and G. Bohling (2001) The search for patterns in ice-core temperature curves. In L. C. Gerhard, W. E. Harrison and B. M. Hanson (eds.), *Geological Perspectives of Global Climate Change*, *AAPG Studies in Geology*, Tulsa, OK, 213–30.

6. Camuffo, D., C. Bertolin, M. Barriendos, F. Dominguez-Castro, C. Cocheo, S. Enzi, M. Sghedoni, A. della Valle, E. Garnier, M.-J. Alcoforado, E. Xoplaki, J. Luterbacher, N. Diodato, M. Maugeri, M. F. Nunes and R. Rodriguez (2010) 500-year temperature reconstruction in the Mediterranean Basin by means of documentary data and instrumental observations. *Clim Change* 101, 169–99.
7. Agnihotri, R. and K. Dutta (2003) Centennial scale variations in monsoonal rainfall (Indian, east equatorial and Chinese monsoons): manifestations of solar variability. *Current Science* 85, 459–63.
8. Loehle, C. and N. Scafetta (2011) Climate change attribution using empirical decomposition of climatic data. *The Open Atmospheric Science Journal* 5, 74–86.
9. Jevrejeva, S., J. C. Moore, A. Grinsted and P. L. Woodworth (2008) Recent global sea level acceleration started over 200 years ago? *Geophysical Research Letters* 35, 1–4.
10. Patterson, R. T., A. Prokoph and A. Chang (2004) Late Holocene sedimentary response to solar and cosmic ray activity influenced climate variability in the NE Pacific. *Sedimentary Geology* 172, 67–84.
11. Ogurtsov, M. G., Y. A. Nagovitsyn, G. E. Kocharov and H. Jungner (2002) Long-period cycles of the sun's activity recorded in direct solar data and poxies. *Solar Physics* 211, 371–94.
12. Charvatova, I., J. Strestık and L. Krivsky (1988) The periodicity of aurorae in the years 1001–1900. *Stud. Geophys. Geod.* 32, 70–7.
13. Komitov, B. (2009) The 'sun–climate' relationship. II. The 'cosmogenic' beryllium and the middle latitude aurora. *Bulgarian Astronomical Journal* 12, 75–90.
14. Mazzarella, A. and N. Scafetta (2011) Evidences for a quasi-60-year North Atlantic oscillation since 1700 and its meaning for global climate change. *Theor. Appl. Climatol.*
15. Wolff, C. L. and P. N. Patrone (2010) A new way that planets can affect the sun. *Solar Physics* 266, 227–46.
16. Kirkby, J. (2007) Cosmic Rays and Climate. *Surveys in Geophysics* 28, 333–75.
17. Enghoff, M. B., J. O. P. Pedersen, U. I. Uggerhøj, S. M. Paling and H. Svensmark (2011) Aerosol nucleation induced by a high energy particle beam. *Geophysical Research Letters* 38, 1–4.
18. Kirkby, J., J. Curtius, J. Almeida, E. Dunne, J. Duplissy, S. Ehrhart, A. Franchin, S. Gagné, L. Ickes, A. Kürten, A. Kupc, A. Metzger, F. Riccobono, L. Rondo, S. Schobesberger, G. Tsagkogeorgas, DanielaWimmer, A. Amorim, F. Bianchi, M. Breitenlechner, A. David, J. Dommen, A. Downard, M. Ehn, R. C. Flagan, S. Haider, A. Hansel, D. Hauser, W. Jud, H. Junninen, Fabian Kreissl, A. Kvashin, A. Laaksonen, K. Lehtipalo, J. Lima, E. R. Lovejoy, V. Makhmutov, S. Mathot, J. Mikkilä, P. Minginette, S. Mogo, T. Nieminen, A. Onnela, P. Pereira, T. Petäjä, R. Schnitzhofer, J. H. Seinfeld, M. Sipilä, Y. Stozhkov, F. Stratmann, A. Tomé, J. Vanhanen, Y. Viisanen, A. Vrtala, P. E. Wagner, H. Walther, E. Weingartner, H. Wex, P. M.Winkler, K. S. Carslaw, D. R.Worsnop, U. Baltensperger and M. Kulmala (2011) Role of sulphuric acid, ammonia and galactic cosmic rays in atmospheric aerosol nucleation. *Nature* 476, 429–35.
19. Scafetta, N. (2011) A shared frequency set between the historical mid latitude aurora records and the global surface temperature. *Journal of Atmospheric and Solar-Terrestrial Physics.*

20. IPCC (2007) *Climate Change 2007: The Physical Science Basis. Contribution of Working Group I to the Fourth Assessment Report of the Intergovernmental Panel on Climate Change.* Cambridge University Press, Cambridge and New York.
21. Scafetta, N. (in press) Testing an astronomically-based decadal-scale empirical harmonic climate model vs. the IPCC (2007) general circulation climate models. *Journal of Atmospheric and Solar–Terrestrial Physics.*
22. Lindzen, R. S. and Y.-S. Choi (2011) On the observational determination of climate sensitivity and its implications. *Asia-Pacific J. Atmos. Sci.* 47 (4), 377–90.
23. Spencer, R. W. and W. D. Braswell (2011) On the misdiagnosis of surface temperature feedbacks from variations in earth's radiant energy balance. *Remote Sensing* 3, 1603–13.
24. Brohan, P., J. J. Kennedy, I. Harris, S. F. B. Tett and P. D. Jones (2006) Uncertainty estimates in regional and global observed temperature changes: a new data set from 1850. *Journal of Geophysical Research.*
25. Scafetta, N. and B. J. West (2007) Phenomenological reconstructions of the solar signature in the Northern Hemisphere surface temperature records since 1600. *Journal of Geophysical Research* 112, D24S03.
26. Scafetta, N. (2009) Empirical analysis of the solar contribution to global mean air surface temperature change. *Journal of Atmospheric and Solar–Terrestrial Physics* 71, 1916–23.
27. Kaufmann, R. K., H. Kauppi, M. L. Mann and J. H. Stock (2011) Reconciling anthropogenic climate change with observed temperature 1998–2008. *PNAS*, 1–4.

5. Has the IPCC really done its homework?

Global warming is a late discovery of the twentieth century, with the first signs of increasing global temperatures appearing in the 1980s. Yet it was just a few years earlier that public opinion had been exercised by global cooling. In its August 1974 issue, the German flagship news magazine *Der Spiegel* carried the headline: 'Is A New Ice Age Coming?' [1]. *Time* magazine in 1974 ('Another Ice Age?') and *Newsweek* in 1975 ('The Cooling World') featured similar global cooling stories at that time. *Der Spiegel* wrote:

> At the latest since 1960 meteorologists and climate scientists have been increasingly convinced that something is amiss within the complex system of global weather. The terrestrial climate, they think, is on the verge of changing ... At first the data showed an advancing cooling of the north Atlantic. Over the last 20 years the sea temperature there has dropped from an annual average of 12° C to 11.5° C. Since then, the icebergs have been drifting further south, as in the winter of 1972/73 when they were located at latitudes as far south as Lisbon – more than 400 kilometres further than in previous winters.

Editors at *Der Spiegel* were already looking back nostalgically to pleasanter, warmer times: 'There was a prolonged period of weather comparable to the first half of the twentieth century in around the years AD 1080 to 1200, when vineyards were cultivated throughout England. On Greenland, where parts of it were green and the Vikings established colonies, vegetation thrived' [1]. The article examined the topic in greater depth and reported that a drop in solar irradiance of 1 per cent or an increase in the average population would trigger a new ice age, while scientists such the American biologist Paul Ehrlich were quoted as saying that they saw no hope for a third of the population currently living in high rainfall regions because of drought and crop failure [1].

Just 12 years later, on 11 August 1986, *Der Spiegel* shocked its readers with its front cover depicting the landmark Cologne Cathedral half-submerged in seawater and accompanied by the headline: '*DIE KLIMA-KATASTROPHE*' (The climate catastrophe) [2]. James E. Hansen of NASA's Goddard Space Flight Center was reported as saying, 'In the early 21st century, global temperature will be higher than at any time in the last 100,000 years.' *Der Spiegel* based its article on information from the German Energy Working Group of the Deutsche Physikalische Gesellschaft (German Physical Society), which claimed that unless emissions of the thermal insulating trace gases were reduced drastically and immediately, grave climate change with disastrous consequences for humanity would occur within two decades.

The IPCC takes the stage

Indeed, after a cooling of a few tenths of a degree Celsius between 1940 and 1975, the earth's temperature had risen slightly (Figure 4.3). This was reason enough for the United Nations to establish the Intergovernmental Panel on Climate Change (IPCC) in November 1988. The remit of the IPCC was to assess the risks of global warming and to draw up mitigation and adaptation strategies. It expressly would not carry out any of its own studies, but would compile existing data.

The IPCC's First Assessment Report (FAR) was published in 1990 and served as the basis for the (non-binding) agreement for climate protection by 192 countries at the 1992 Rio conference, and was later signed as the Kyoto Protocol in 1997, before being implemented in 2005. The Kyoto Protocol obliged industrialized countries among the signatories to reduce greenhouse gas emissions by 5 per cent below their 1990 levels by 2012 [3]. Although the FAR concluded that it was impossible to determine that human activity was responsible for warming by the use of climate gases, it went on to predict that a warming of 0.3° C a decade would occur, and that in 2025 it would be 2° C warmer than in the pre-industrial age and 4° C warmer by 2100 [4]. As a result on average the sea level would rise 630 mm.

Even though the IPCC came to slightly different results in its 1995 Second Assessment Report (SAR), the 2001 Third Assessment Report (TAR), the 2007 Fourth Assessment Report (AR4) and the draft of the 2013 Fifth Assessment Report (AR5), its basic statements are largely unchanged. In all four reports the IPCC first described the most important climatic changes (temperature, sea level, ice and snow cover) over the last 50–100 years. These data relied on multiple measurements (Figure 5.1).

AR4 used data available up to 2005. And, contrary to their forecasts, none of the important climate trends have accelerated since then. The temperature has remained on a plateau, the sea level rise has remained constant and even appears to have slowed down, and combined Arctic and Antarctic sea ice melt has decelerated. Needless to say, this contradicts the IPCC's apocalyptic worldview, which sees an acutely threatened planet that is on the brink of tipping over the edge. The basis for the IPCC's well-known warning is the claim that the overwhelming share of the warming of the last 50 years can be attributed to manmade climate gases, with a more than 90 per cent certainty. And because the emissions of climate gases are rising, the danger continues to grow [5].

Compared to the 2001 TAR, the recorded linear temperature over the previous 100 years rose from 0.6° C to 0.74° C. This, however, is principally due to the chart's starting point, which was five years later in the 2007 report. This may seem trivial at first glance, but it means part of a cooling phase that had taken place in

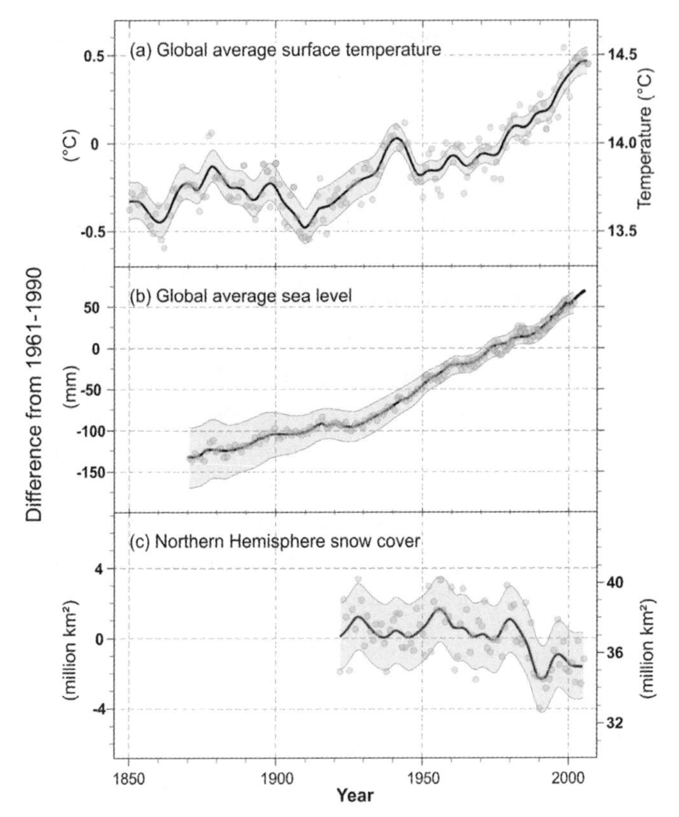

Figure 5.1 Temperature, sea level rise and the retreat of snow and ice cover over the last 150 years. Source: AR4.

the early twentieth century was no longer included in the data sets used by AR4, so thanks to this statistical sleight of hand the warming rate automatically rose. The unauthorized release of the AR5 draft shows continued temperature stagnation, a development that every climate model failed to predict.

Carbon dioxide heats up

There's no doubt that the increase in climate gases has had a detectable impact on our climate. The atmospheric concentration of CO_2 increased by over 100 parts per million (ppm) over the last 250 years, from 280 to 390 ppm today. The first 50 ppm was reached between 1760 and 1970, the second 50 ppm had been added by 2005, a mere 30 years, and emissions continue unabated today. Every year more than 7 gigatonnes of carbon (about 26 billion tons of CO_2) are released into the atmosphere by burning fossil fuels, industrial processes like cement production and land use such as deforestation. That means a growth of 4 ppm per year, of which about 2 ppm remain in the atmosphere with the other half absorbed mainly by the oceans. Measurements taken on the Mauna Loa volcano, Hawaii since 1958 confirm the trend. Keep in mind that the temperature does not increase linearly with CO_2 concentrations, but rather rises logarithmically. According to IPCC models the first 50 ppm (from 280 to 330 ppm) added 0.23° C of warming, the second 50 ppm added 0.20° C, and the next will add about 0.17° C – that is without taking any possible amplification or suppression effects into account (more on this in Chapter 6).

The climate mix according to the IPCC's formula

Before turning to the statements made by the IPCC, now is a good time to look at how the IPCC arrived at its conclusions. Scenarios for the future development of the world's population, standard of living, consumption of fossil fuels and land use changes (e.g. deforestation) serve as its basis. This in turn yields emissions projections for each climate gas (e.g. CO_2 and methane), which can contribute to warming. Cooling effects are also taken into account. The greatest contributors here are aerosols, which are fine dust particles and droplets that reflect or diffuse sunlight and thus have a cooling effect. There are other parameters that have an impact on climate and also need to be taken into account, e.g. the reflection of sunlight by snow and ice (albedo) and clouds, as well as natural changes in solar irradiation as that too is not constant.

Simulation models need not only the time and spatial development of climate gas concentrations as input parameters, but also their so-called radiative forcing. Radiative forcing is the climate gas's ability to change the amount of energy absorbed by the surface of the earth and so cause warming or cooling. The ability of radiative forcing to weaken or strengthen incoming solar energy is expressed in watts per square metre (W/m^2) and is based on the difference between today's effect and the effect it had at the beginning of the industrial era, which is set at 1750. The IPCC assumes that CO_2 has a positive radiative forcing of 1.66 W/m^2 (Figure 5.1).

Other climate gases need to be taken into account, among them CH4. Methane is not only created when producing natural gas, but is also emitted by marshlands, rice cultivation and cattle. Methane concentrations in the atmosphere are much lower than CO_2. But because the greenhouse warming effect of a methane molecule is thirty times greater than a CO_2 molecule, the IPCC believes methane accounts for a third of global warming over the last 250 years (Figure 5.1). What few among the public are aware of, however, is that methane concentrations are no longer increasing but have remained flat since the year 2000. Noteworthy too is that CH4 stays in the atmosphere for a much shorter time than CO_2. It reacts with oxygen and transforms into CO_2 over an average period of 8 years [6].

The halogen chlorofluorocarbons (CFCs), which are exclusively components of industrial refrigerants and insulation materials and were once used as a spray gas, are extremely potent greenhouse gases, their effect is 20,000 times higher than CO_2's. The large CFC emissions of the twentieth century had a positive radiative forcing of 0.34 W/m2 (Figure 5.1). But CFC emissions were curbed by the Montreal Protocol which came into effect in 1987. A significant drop in the growth rate of atmospheric CFC concentration has occurred since the 1990s, if not even a drop in CFC concentration as such. Because halogen CFCs have a half-life of 50–100 years, their reduction is slow. The Montreal Protocol to protect the ozone layer was signed by 195 countries, and has had a greater effect on limiting anthropogenic greenhouse gases than the Kyoto Protocol [6]. Today nitrous oxide, which is a major by-product of agricultural fertilizers, is steadily becoming a more important greenhouse gas.

In all four IPCC reports, the greatest uncertainty involves estimating the effects of aerosols, which are mainly dust particles that have the effect of cooling the earth because they reflect and refract sunlight. According to AR4, since 1750 aerosols have had a total radiative forcing of $-1.2\ W/m^2$. The range of uncertainty is in fact greater than the mean value itself (Figure 5.1). The total amount of the radiative forcing thus entails the same uncertainty. (We will take a closer look at aerosols later in this chapter.)

It should be noted that the models the IPCC uses assume a very low radiative forcing for solar irradiation. Beginning in the eighteenth century until today, the effect of the sun on the climate, according to the IPCC, is limited to just $0.12\ W/m^2$, and so has played a minimal role in the earth's warming over the last 250 years. The draft AR5 appears to reduce even that negligible influence to $0.04\ W/m^2$. But is the climatic impact of the sun really as modest as the IPCC claims? The relationships presented in Chapter 3 gives rise to considerable doubt [7]. In this chapter we take a closer look from different perspectives and cast more light on this obvious disagreement.

It's simulation time: virtual climate worlds from computers

Climate models are constructed at various institutes all over the world using input parameters such as those discussed above. A total of twenty-three models were used in AR4. Climate models embody physical descriptions of the energy processes in the atmosphere, hydrosphere with global water circulation and the oceans, land surface and the biosphere. The aim is to optimize the computation of the parameters using complex calculations, until the output agrees as closely as possible with the measured data of the last 100 years.

We are talking about relatively small changes to parameters when compared to the total energy budget. The earth receives at least $342\ W/m^2$ from the sun on average at the top of its atmosphere.[9] Part

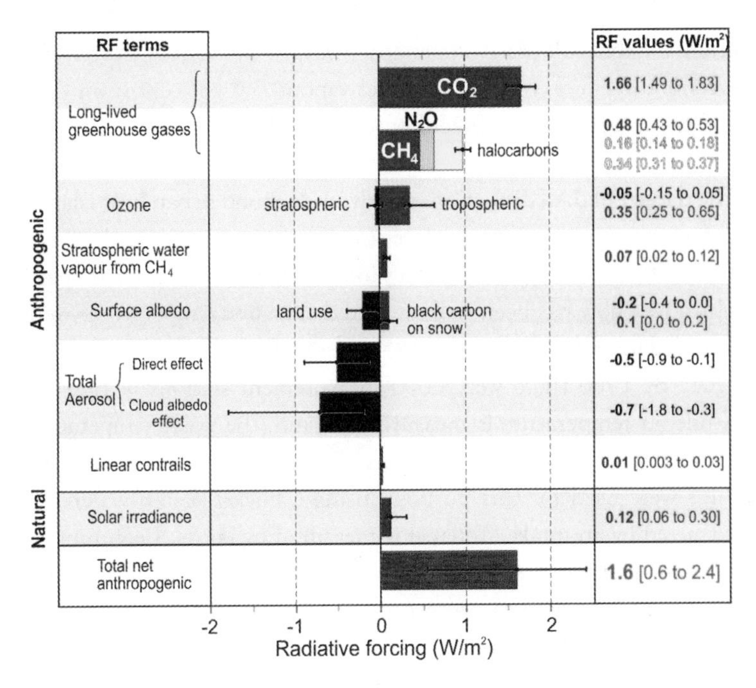

Figure 5.2 The warming and cooling effects of various climate control factors according to AR4. Shown are the so-called radiative forcings that have occurred due to changes in each factor since 1750 converted into W/m². Each conversion, however, entails many assumptions regarding climatic processes that are poorly understood, and so the real climate mix is likely to be very different.

of the solar radiation is reflected by clouds and the earth's surface, or is absorbed by the atmosphere. On the other hand, the earth emits 390 W/m², which should lead to a negative energy budget. However, because of the natural greenhouse effect, caused principally by water vapour and greenhouse gases (CO_2 and methane), 324 W/m² are returned to the earth through infrared radiation. This results in an energy budget that leads to an average global temperature of about 14° C. Without water vapour and the greenhouse gases, the earth would have a temperature of -18° C and thus would be an

[9] 342 W/m² correspond to a quarter of the total solar irradiance (TSI) of 1367 W/m², because the ratio of surface to a slice plane of the earth is 4.

uninhabitable ball of ice. Put another way, 95 per cent of the natural greenhouse effect comes from water vapour, 3.6 per cent from CO_2 and the rest from other greenhouse gases [8].

The danger of data distortion: poorly understood urban heat islands
Before attempting to make forecasts using computer models, the models first have to be verified against the climate of the recent past, which we know has been well measured. The first temperatures were recorded in central England in 1659, and we have data for Berlin since 1701. By 1960 there were 6000 measurement stations worldwide. While air temperature is measured on land, the water temperature is measured in the oceanic regions. During the twentieth century, ships were used for this purpose, using a bucket. Later, water was extracted by an intake. Today it is measured by buoys. Beginning in 1979 the first satellites began to record the temperature of the lower atmosphere. Their data are processed by two American institutes: the University of Alabama, Huntsville (UAH) and Remote Sensing Systems (RSS) in Santa Rosa, California.

Interestingly, surface measurements over the last 30 years show temperatures that are one to several tenths of a degree Celsius higher than those recorded by satellite and ocean data (Figure 1.2). The lower ocean temperature was explained by claiming the temperature of the ocean changed more slowly compared to land temperatures due to the high degree of mixing and greater heat capacity. However, another possibly still underestimated phenomenon may have played a role here. Numerous scientific studies [9–10] show a huge distortion of temperatures measured by weather stations in urban locations because of heat from the surroundings. This phenomenon is known as the urban heat island effect (UHI). Because of the heat storage capacity of asphalt roads, car parks and nearby buildings, in combination with a lack of vegetation, urban areas are notably warmer than the surrounding rural regions (Figure 5.2). For example, in Los Angeles the high temperatures over the last century increased approximately 2.5° C and the low temperatures approximately 4° C.

New York also has a night-time heat island effect of 4° C. If the asphalt surfaces and buildings in large cities were painted white, then the albedo (surface reflectivity) would be greatly increased and heat storage would be lessened. In London, peak summer heat could be reduced by as much as 10° C [11–12].

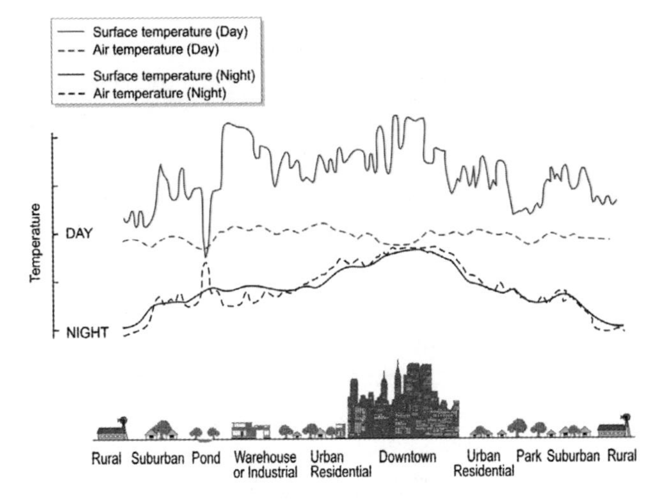

Figure 5.3 The urban heat island effect (UHI). Because of the lack of vegetation and the heat storage capability of large buildings and other infrastructure, urban areas are notably warmer than their surrounding rural areas [13].

Weather stations that were originally located in rural areas and then later swallowed up by growing cities over the course of the last century have experienced a progressive warming. In addition to global warming, these stations experience an additional temperature increase because of UHI. In a study of eastern China, scientists discovered that about one quarter of the regional warming of recent decades can be attributed to UHI [14]. In Korea, more than half of the warming of the last 55 years is attributed to UHI [15]. High UHI was found by other authors for the worldwide global warming of the last 150 years [16–19]. In western Europe the introduction of clean air policies has led to major improvements in air quality since the 1980s. Sulphur dioxide levels have declined markedly so that more sunlight now reaches the earth's surface. Scientists assume that part of the warming from 1977 to 2000 in western Europe can be attributed to this [20].

It is still unknown what role the heat generated by manmade energy production (waste heat production) [21] plays and whether UHI has any influence on the temperatures of rural surroundings [22]. Although the data from urban weather stations are an important basis for climate simulations, the IPCC discounts the importance of UHI and acknowledges a contribution in low single-digit percentage points only [6, 23].

There have also been misgivings over poorly sited measurement stations close to asphalt or concrete surfaces that could distort temperature statistics. As a result IPCC critic Anthony Watts, who runs the well-known climate-science website Watts up with That (wattsupwiththat.com), embarked on a comprehensive review of the American temperature stations network in 2007. Using 650 volunteers and the internet, 1007 of the 1221 stations were inspected according to strict guidelines over a period of two and half years. The result of the survey was sobering. Almost 90 per cent of the stations did not meet even the standard for minimum distance from heat sources such as asphalt surfaces [24]. Sixty-eight stations were sited at sewage treatment facilities and some stations sited at airports were intensely heated by the blast of jet engines and nearby asphalt and concrete surfaces (Figure 5.3). It is not known how much this has influenced the US temperature record [24–25].

The impact of closing the majority of weather stations worldwide at the start of the 1990s is also unknown. In 1990 there were 6000 weather stations worldwide; the current number is about 1500. It was precisely at this time that the statistical global temperature curve began to surge upwards [26]. Veteran meteorologists Joseph D'Aleo and Anthony Watts point out that a disproportionately large number of these decommissioned stations were located in rural areas, at higher elevations and at high northern latitudes – all stations that tended to record cold readings [16]. This is telling, to say the least.

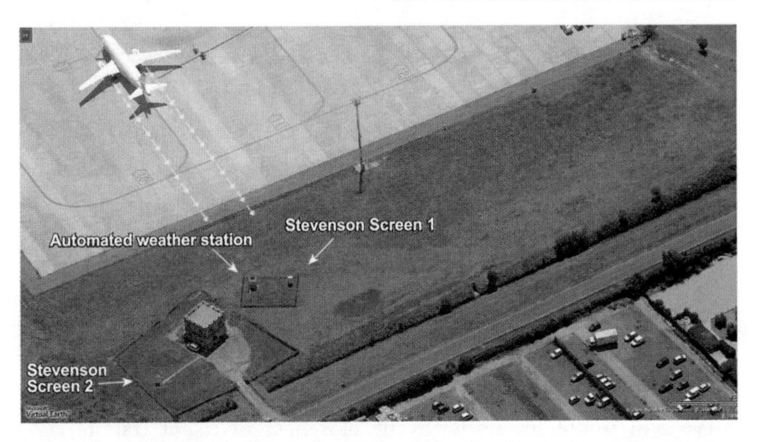

Figure 5.4 An unfavourably sited weather station at Ciampino Airport, the smaller of the two international airports in Rome. Jet engines and vast expanses of paved surfaces combine to warm up the temperature station. Source: Paolo Mezzasalma [27].

Independent of UHI and all its statistical shortcomings, it is not disputed that a significant warming occurred between 1977 and 2000. This is confirmed by satellite data. In the end it merely boils down to the question of whether or not the surface temperatures are inflated by artefacts, and if so, by how much. In view of the doubt that surrounds surface measurements, it would be appropriate to base the climate models on satellite data instead. But the IPCC does not do that. So we have to continue feeding the models with these unsatisfactory data, which takes us to the next surprise.

Our sun is climatically powerless and meaningless, says the IPCC

Some may ask why the sun appears to play no role in the climate as a whole according to the IPCC. The sun is, after all, the main source of energy and common sense would tell us that even a slight fluctuation in solar activity would lead to significant climatic effects on earth. And haven't geologists been able to show and document precisely that for the earth's history (see Chapter 3)? That's all very well, the IPCC replies. However, we have excellent climate models that can help us with this problem. So let's take two calculation scenarios and see which one best models the real temperature over the last

100 years. On one side we have the classic IPCC model where CO_2 has a very powerful climate forcing effect and the sun is ignored. On the other we have a climate model that also takes natural factors into account, namely the sun and, periodically, volcanic ash. Let the contest begin. The computers chug along and calculate their way through reams of data. Eventually, they spew out their results. The winner is ... the IPCC model! Why? The 'natural' model is unable to reproduce the warming since 1977, and simply meanders aimlessly on a plateau (Figure 5.5 top). The IPCC model, on the other hand, elegantly reproduces most of the twentieth-century temperature curve and impresses us with the closeness of its fit (Figure 5.5, bottom). 'Oh,' the reader might think. 'It really isn't such a good idea to take the sun into account. The sophisticated IPCC model has truly demonstrated its superiority.'

At this point we bring in a referee to judge how the results were reached and to rate the approach. The first low blow was that IPCC assumed the same negligible climate forcing for the sun for both scenarios. As the IPCC accepts no amplification mechanism (see Chapter 6), the sun never had a chance in this contest. It is as if the IPCC challenger had its right arm tied behind its back before the fight began. The result is unsurprising. What goes into the model also comes out. The second infraction involves the good fit between the IPCC curve and the historical temperatures. Here also, a massive manipulation is used to arrive at the result. The cooling effect of aerosols was inserted into the model at carefully chosen points to ensure that the modelled temperature curve matches the historical curve (see Chapter 7). That's like choosing the winning lottery numbers the day after the draw. How else could the sunless IPCC model have produced the cool periods as CO_2 concentration increased continuously and steadily? The IPCC's exclusion of the sun in the rigged contest is thus meaningless and does not bring us a centimetre further on the question of the impact of natural climate factors. Rather than trying to understand the role of natural factors like the sun, the IPCC is content to bury it.

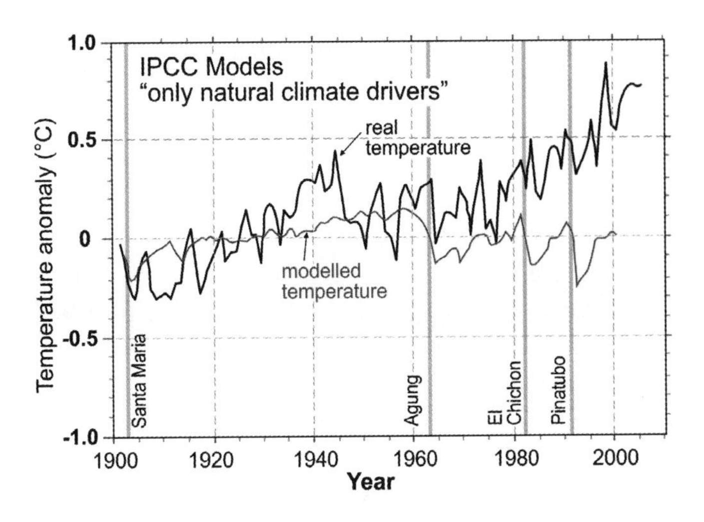

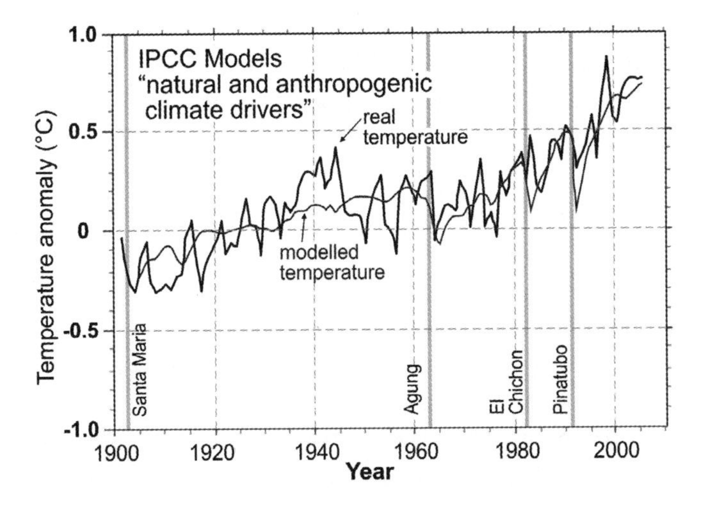

Figure 5.5 Deceptive anti-sun evidence presented by the IPCC. The IPCC compares theoretical modelled temperatures to the actual observed temperature of the last 100 years. In a direct comparison only the model with the anthropogenic components (bottom chart) appears to reproduce the measured historical temperature, while the model with the pure natural factors (top chart) shows considerable discrepancy. What is not mentioned is that the wonderful fit in the anthropogenic chart was created by throwing in an aerosol fudge factor where needed. Meanwhile, in the model with the natural factors, discrepancies occurred because amplification mechanisms and cycles were ignored. Source: modified from AR4 [28].

How the 2007 IPCC report looks at the sun

In Chapter 3 we saw how the sun affected the earth's climate in the past in a variety of ways. So it is utterly improbable that 40 years ago the sun would take early retirement and cease being one of the most important climate drivers, as the IPCC asserts. How did the IPCC reach this assumption and what is the chain of argument behind it? How could the inconvenient truth of the close historical link between the sun and climate have been marginalized so successfully in a number of climate reports and for so long?

In AR4 references to the sun are scattered across a number of chapters [29] and many of the aspects discussed in Chapter 3 are reported there. The IPCC mentions the 11-year solar cycle and relishes the fact that the changes in total solar irradiation amount to no more than about 0.1 per cent. It also mentions that the UV range fluctuates strongly, with values of 1.3 per cent (see Chapter 3). With some surprise, we read that the IPCC even considers the possibility that changes in UV may trigger an amplification mechanism and that fluctuations in solar activity may produce detectable climate signals. They also mention that UV radiation can trigger reactions in the ozone layer and thus have a clearly discernible influence on the temperature in the upper atmosphere. And then we read: a coupling of the UV effect in the upper atmosphere with the climate in the lower atmosphere is probable, but how is unknown and is still being researched. One thing, though, is certainly clear: the effect cannot be ignored. One might have expected the IPCC at least to have taken this into account in one form or another in its quantitative models. The same goes for the other important climate driver which is also the subject of much research, namely the Svensmark Mechanism, which links the solar magnetic field with cloud cover through cosmic rays, thus linking it to climate. Here too there are many positive signs (see Chapter 6), and so these processes should appear among the possibilities in the various climate models. We shall see if they are.

AR4 also recognizes that the Little Ice Age did not occur during a period of low solar activity by mere chance. Much of the temperature development in the Northern Hemisphere in the past was controlled by fluctuations in solar activity on a decadal scale, together with large volcanic eruptions. Last but not least, our millennium cycles appear in the long version of the report, where its relationship with solar activity is also suggested.

After reading all this, one might almost conclude that everything is in apple-pie order. Many important points have been raised. Solar activity cycles from every period would certainly be included in the models and various solar forcings as possible scenarios would be created. Just to be certain, one might think the IPCC would check and calibrate the models at least once using the pre-industrial, known climate curves. Then, the IPCC would introduce the CO_2 parameter and other anthropogenic elements and extend the models up to the present, working to the principle 'from easiest to more complex'. And only if the models were able to withstand these rigorous quality checks would they be deemed to be sound enough for projections into future. Surely the IPCC would have followed this process? Well, given what we already know about the IPCC, we can guess the answer: No, it did not. The IPCC deems it unnecessary. Instead, it is enough to discuss possible solar effects at length, and then simply ignore them.

The factually decoupled IPCC synthesis

Let's go a step further and consider the chapters of the IPCC report that deal with the climate models: the synthesis of the many scientific single aspects and the recommendations for policy makers. It is the nature of science that not all scientists agree on every point when you have a group that is as large as the one that worked on the IPCC report. In order to be able to put together a report that can be mutually agreed on, there has to be a core team that ultimately decides which aspects will be emphasized and pursued, and which excluded.

It is precisely at this interface that so much misfortune transpired. Nothing positive about the sun was reported in the

IPCC's model calculations, conclusions and recommendations for policy makers. It is as if the many progressive thoughts found in the earlier chapters had been swallowed up by the earth and had simply disappeared. The IPCC instead barricades itself behind the 0.1 per cent radiation variation wall of the Schwabe solar cycle and allows the climate forcing of the sun to die out with the other climate control factors. Newly uncovered possible solar amplification mechanisms, such as the UV amplifier and the cosmic ray amplifier (see Chapter 6), are excluded. The specious reason? The mechanisms are poorly understood. The aerosol effects, along with their uncertainty, however, are taken into account even though they too are poorly understood. In its quantitative estimations, the IPCC claims to cover all possibilities. Here upper and lower limits are given, along with the best anticipated and determined values. With aerosols their effect fits the picture; they serve the purpose. Possible solar amplifications, on the other hand, do not, and so they are ignored.

Ironically, the IPCC instead uses a huge CO_2-induced water vapour amplification process which inflates the rather meagre CO_2 warming effect so that it reaches dangerous levels. Here it has to be pointed out that the water vapour amplifier too is poorly understood and that some scientists, among them former IPCC author Richard Lindzen of Massachusetts Institute of Technology, consider that water vapour has a dampening effect rather than an amplifying one [30] (see below and Chapter 6). Clearly there is a growing suspicion that the IPCC is employing a double standard: thus deliberate exclusion or marginalization of inconvenient results and indications undermine the quality of IPCC's statements.

The IPCC does not waste any time on why solar activity correlated so well with temperature and other climate parameters in the past. How could this have worked then when the sun is supposedly such a weakling? No matter how powerfully the sun demonstrated its impact on the climate over the course of its different cycles during the postglacial period, the IPCC report plays it down. Even though one scientist after another has demonstrated that the sun's slowing activity is connected to the stagnation of the global temperature since 2000, we read nothing about this in the summary IPCC expert report.

In fact, behind the scenes IPCC scientists mock the solar cycle repertoire and imply that solar scientists are suffering from 'cyclomania'. The sun's clear signature on the earth's climate is dismissed in the report, based on wild reasoning. When it concerns the sun, everything has to be simple and crystal clear, otherwise it cannot be valid. But with CO_2, things are handled much more adventurously and discretely. How many among us are aware that temperatures in Antarctica, the east Pacific and parts of the Indian Ocean have dropped since 1979 while CO_2 is uniformly spread over the globe? No one can offer an explanation for that.

Is there any hope that things will improve?

It has been a number of years since AR4 was published. How has science developed since then? What new findings do we have today that we did not have in 2007? Just how probable is it that the IPCC will admit, 'Sorry, but in the 2007 report we were wrong about the sun. There's more to it than we first thought?'

Numerous publications which do not support the IPCC's solar-critical position have come out since 2007 (see Chapter 3). That there's lots of movement in the area of important fundamental data is shown by the current research results of a Swiss research team at the Davos-based World Radiation Centre and ETH, Zurich. They found that the IPCC may have understated the rise of total solar irradiation (TSI) for the period since the Little Ice Age by as much as a factor of six [31]. While the IPCC assumed an increase of merely 1 W/m^2, the Swiss team arrived at a hefty 6 W/m^2. Strangely enough, in AR4 the IPCC just halved the value given in TAR.

If you multiply these values by a factor of 0.175, then you get the magnitude of radiative forcing of the sun [33],[10] which is usually

[10] The factor 0.175 corrects the geometric effect between the earth's cross-section and its overall surface area, and also takes the albedo of 30 per cent into account.

given for the same interval. The values unfortunately also have as their units W/m^2, but must not be confused with the increase in TSI. If one uses the Swiss TSI value, then you get an updated solar forcing value for the sun of 1 W/m2 – an enormous increase compared to the 0.12 W/m2 in AR4 (Figure 5.2). That alone puts the sun near the top of the list of possible climate drivers. A recent climate modelling study using the Swiss TSI values now points to a much more significant role for the sun [34]. And this, the reader should remember, is without even considering possible solar amplification mechanisms, which probably would put the sun in pole position and CO_2 would have to be downgraded in order to keep the temperature from shooting up beyond the observed temperature values. Along the same lines, a British-American 2010 study sets the variability of UV radiation over the course of an 11-year solar cycle at five times greater than that used by the IPCC [35].

Despite this, the draft AR5 diminishes the influence of the sun even further. The IPCC argues that natural forcing would be a small fraction of anthropogenic forcing. Since 1980, satellite observations of the TSI and volcanic aerosols demonstrate a near zero (–0.04 W/m2) change in the natural forcing compared to the anthropogenic forcing increase of 1 W/m^2. This has been commented on by Alec Rawls, a climate-sceptic blogger: 'Solar activity was at historically high levels at least through the end of solar cycle 22 (1996), yet the IPCC is assuming that because this high level of solar forcing was roughly constant from 1950 until it declined during solar cycle 23 it could not have caused post-1980 warming. In effect they are claiming that you can't heat a pan of water by turning the burner to maximum and leaving it there, and that you have to keep turning the flame up to get continued warming' [36].

The IPCC's strategy for ignoring the sun

Let's take a look at the IPCC's reasoning. The starting point is that the IPCC attributes the 1977–2000 warming almost exclusively to anthropogenic greenhouse gases. Models were specially developed for this period and then simply projected into the future. Unfortunately, the models could not even reproduce the 1940–77 cooling phase and also failed for the period 2000–the present, as once again warming has stopped. If CO_2 was the only magnitude of influence acting on the climate, and other natural factors like the sun and oceanic climate oscillations played no role, then perhaps that approach could be adopted. However, we know that the sun's activity is quite erratic and shows changes ranging from minutes to thousands of years.

To gloss over all of this, the IPCC cherry-picks from the broad spectrum of available literature a solar activity curve that shows very little change over the last thousand years [32]. A supposedly minor fluctuation range disarms the sun in the most facile way. Here, possible pre-amplifiers would only be a distraction and thus have no place in the models.

If, for whatever reason, the solar fluctuations don't allow themselves to be minimized, then a safety mechanism is brought into play: temperature reconstructions of the last thousand years are statistically processed to the point where only a more or less flat curve results for the pre-industrial period [37–38]. This is the infamous hockey stick curve, the centrepiece of TAR, which has since been completely discredited. The relatively flattened temperature curve thus made it possible for scientists to downgrade the impact of natural climate forcings drastically in their models. So, no matter how much the natural processes fluctuated, the temperature curve remained more or less flat. What is really important for the IPCC is that the temperature curve started climbing powerfully just when industrial CO_2 began to increase. Only in that way could the sun's influence be kept small and CO_2's be made responsible for 95 per cent. And only in this way could political momentum be unleashed and stringent measures for protecting the earth's atmosphere demanded. However, the actual temperature development of the last thousand years gives us a very different picture. First, it distinctly shows the Medieval Warm Period (MWP), a time when Greenland was settled for generations and agriculture was practised there. It also shows the bitter cold of the Little Ice Age (LIA), which proponents of the inflated CO_2 hypothesis still stubbornly refuse to acknowledge. Stefan Rahmstorf, for example, insists that the LIA (Maunder minimum) is a regional north European event. But studies from Japan and China show that East Asia too was in the grip of a long cold period at that time (see Chapter 4).

If the solar activity curves were allowed to show their full variability and the temperature curve was again allowed to have its characteristic peaks and troughs which the first IPCC report of 1990 showed (Figure 4.10), then there would be only one retreat position left for the IPCC. The significance and effectiveness of the solar impact on climate would have to be acknowledged. However, this would have to be presented as having no serious consequence for CO_2 climate sensitivity. CO_2 would have to keep its potential for danger. But would that make any sense? Recall that the temperature has increased 0.8° C since 1850, which has to be attributed to various factors. The more important the sun becomes, the more CO_2's role has to diminish. Currently, according to the IPCC, CO_2 and other anthropogenic greenhouse gases are 95 per cent of the cause and the sun only 5 per cent. But if the pie were to get sliced up differently, then CO_2's share would have to drop accordingly to maintain the 0.8° C value. One simply cannot strengthen any desired climate factor without running into problems, even if one considers time lags in greenhouse warming. Interestingly, some scientists close to the IPCC concede that the sun has impacted the climate to some extent – but only until 1970 [6]. So just how probable is it that the sun abruptly withdrew from the climate stage and lost its influence 40 years ago?

The IPCC's distortion of scientific data

Just how scientific and serious is the IPCC's approach? How will future generations rate it? Does its stubborn refusal with respect to the sun have something to do with the fact that their scientific models, which were once believed, are now being questioned? Is it perhaps an unholy mix of science and politics? Does an alliance of institutes, NGOs, photovoltaic lobbyists and politicians fear for their privileges and benefits? Or is it that far-reaching economic and political measures that have already been implemented are making it difficult to reassert common sense and the tried and tested scientific method? The IPCC approach has to be regarded as unscientific and divorced from reality. It is unrealistic to believe that the IPCC models could describe climate dynamics without first involving solar dynamics and other natural variability.

Mojib Latif of Kiel came to the same conclusion. With an Australian colleague, Noel Keenlyside, he determined in 2011 that the models employed by the IPCC could not provide forecasts over decadal timescales. They found numerous shortcomings. The models cannot reproduce the decadal climate oscillations that are observed in the real world. Furthermore, the influence of the stratosphere is overlooked and aerosol particles are not adequately taken into account [39]. Despite these fundamental flaws, climate model calculations are still presented as sound results, which are then transformed into media reports.

Much has been reported on the IPCC's dramatization of reports, whether it's the baseless publicizing of the Himalayan glaciers melting by the year 2035, which Dr Lal justified by claiming 'We thought that if we can highlight it, it will impact policy makers and politicians and encourage them to take some concrete action' [40], or whether 40 per cent of the Amazon rainforest is threatened, or the drinking water of 4.5 billion people will be lost by 2085, or food production in North Africa dropping by 50 per cent by 2020 [41].

The way in which data can exert political pressure is also demonstrated by the IPCC's selective method for depicting temperature trends (Figure 5.1). The mean temperature rise was calculated for different periods in the past. The coloured regression lines got steeper the more recent the starting point, which implies an increase in warming. For 25 years (1981–2005), just before the editorial deadline of AR4, the IPCC calculated a warming rate of

0.177° C per decade, and thus curtly concluded that warming had accelerated [42]. Has it really? Unfortunately, the IPCC omitted two not so unimportant lines from its graph (dashed red lines in Figure 5.1, p.177). These two lines show warming in 1860–80 and 1910–40. And, guess what, they have similar slopes, or warming rates, as the supposedly unique and 'unusual' steep linear line used for the last 25-year period. At the start of 2010, IPCC's temperature specialist Phil Jones had to admit this in the wake of the Climategate scandal, saying that the rate of warming for all three episodes was statistically similar and could not be differentiated from each other [43]. In AR4 we look in vain for an explanation. Instead, the depiction of 'accelerated' warming in recent times [42] in the AR4 misleads unknowing readers. The politicians who were provided with the written summaries, in any case, deserve reliable scientific reports that can be used as a basis for sustainable planning. Since AR4 was published another 5 years of data has become available. What was merely indicated in 2005 has since been confirmed. The temperature has remained at its 2000 level; warming has stopped, at least for the time being (Figure 1.2). It is going to be very difficult for the IPCC to repeat its masterful feigning of accelerated warming in the next report.

Another method the IPCC resorts to when confronted with different scientific findings is to favour papers that minimize or exclude solar influence to the benefit of CO_2. For example, Richard Willson and his team at Columbia University showed from satellite data that the solar irradiation of solar cycles 21–23 (1978–2002) increased 0.05 per cent per decade [44], which could have had a significant impact on earth's temperature as a consequence. But that simply did not fit the IPCC's narrative. So by applying corrections to the different satellite measurements, Claus Fröhlich of the Swiss World Radiation Centre was able to generate a diagram in 2006 which showed no increase [45] (Figure 5.6). Needless to say, in AR4 the IPCC opted for Fröhlich's interpretation and detected no change in solar irradiation over the last decades either. If the original literature from Willson had been pursued, then part of the warming (30–50 per cent) since 1980 would have to be attributed to the sun [46].

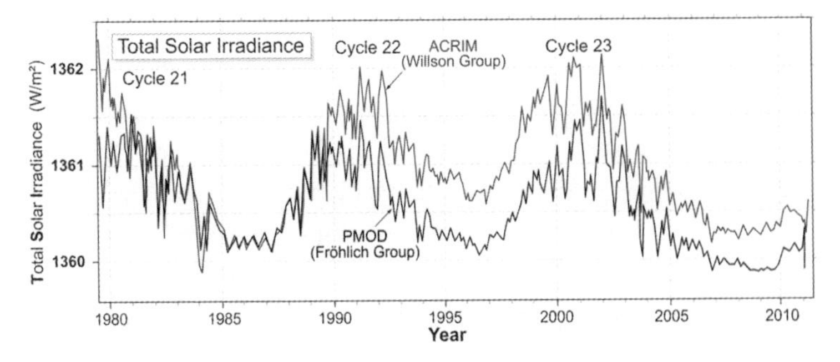

Figure 5.6 Satellite measurements of total solar irradiation (TSI). While the minima of cycle 21–23 of the ACRIM data series of the Willson group rise, they appear to be almost constant in the PMOD version of the Fröhlich group. For comparison, the PMOD is shifted down by about 5 W/m² from its native scale.

Misunderstood water vapour

Water vapour has the greatest influence on climate in the atmosphere (see Chapter 6). It is by far the most powerful 'climate gas'. First, water vapour content increases with rising temperatures as warmer air carries more water per unit volume. This water vapour amplification plays a huge role in the IPCC climate models. Without such feedback, a doubling of the atmospheric CO_2 concentrations would increase the earth's temperature by only about 1.1° C. It is only with the help of water vapour feedback that the IPCC is able to come up with its alarming up to 4.5° C temperature increase per CO_2 doubling [47]. Here the IPCC report does not base this on actual measurements, but rather on five theoretical calculations undertaken between 2003 and 2006.

Is it really correct to include the water vapour effect in the calculations? It wasn't until 2009 that two studies based on measurements were published. In one, Garth Paltridge of the University of Canberra showed from humidity measurements made by balloon probes that the humidity in the middle and upper troposphere had decreased between 1973 and 2007 [48]. In the other study, Richard Lindzen and his colleague Yong-Sang Choi of MIT evaluated data from the 1984 ERBE (Earth Radiation Budget Experiment) satellite, which recorded radiation emitted from the atmosphere into space and compared it to the change in sea temperature. They produced the surprising result that a negative feedback for water vapour is found at the tropical latitudes [30]. As the sea temperature rose, the radiation into the atmosphere increased and led to a cooling, thus constituting a negative feedback. The calculation signalled that the CO_2 effect of 1.1° C for a doubling of CO_2 concentration is reduced by the water vapour feedback to only 0.5° C.

But how can warming of the ocean, which without doubt leads to more water vapour in the atmosphere, have a neutral or even a negative feedback? The answer is that more water vapour in the atmosphere results in more cloud formation. Lindzen discovered that as ocean temperatures rise, the ratio of cooling low cumulus clouds to warming high ice crystal (cirrus) clouds goes up, thus offsetting the warming triggered by the sea surface. As water temperatures increase, more droplets form. The larger the droplets, the less they are transported into the high atmosphere where they form heat-trapping cirrus clouds. Cirrus clouds reflect less solar radiation, but effectively absorb infrared radiation reflected by the earth's surface [49]. This 'iris effect' has acted throughout the earth's history to counteract a runaway greenhouse warming by water vapour whenever the planet warmed [49]. Lindzen points out that he had only considered the Tropics and suspects that the negative feedback on a global scale is possibly less when other latitudes are considered.

To say the least, Lindzen caused outrage among some scientists because previous IPCC models claimed a strong positive feedback for CO_2, whereby CO_2 warming-induced water vapour contributed much of the global warming. Kevin Trenberth, a lead author of the IPCC's 1995, 2001, and 2007 reports, was able to show that Lindzen had made some errors [50], particularly that he had looked only at the Tropics. Roy Spencer of UAH showed that the change over the examined period could also show a zero effect (neither positive nor negative) [51]. Even so, Lindzen opened a long-overdue discussion among scientists. The IPCC's assertion that water vapour exclusively had a strong warming effect had been shown to be questionable, at least over parts of the earth.

The years since then have been marked by a bitter dispute among adversaries. Andrew Dessler [52] of Texas University joined Trenberth's cause and, based on satellite data, calculated that clouds would amplify climatic changes slightly (positive feedback). Yet at the same time he could not exclude the possibility that clouds may slightly attenuate climatic changes (negative feedback) [53]. Spencer examined satellite data and concluded that there had to be another natural relationship independent of CO_2 and its associated warming that causes cloud formation, such as the PDO [54]. Also, the effect of cosmic radiation and aerosols can lead to an independent impact on cloud formation. This could mean that the IPCC models reverse cause and effect. A CO_2-independent fluctuation in cloud cover may lead to a change in solar radiation reaching the earth's surface and thus could create the false impression that the temperature increase is caused by CO_2 [55–56]. Cloud cover in many regions of the world has declined over recent decades [57–59].

Spencer too came up with a strong negative feedback. Finally, in July 2011, Lindzen published a paper that addressed Trenberth's criticisms [60]. Here Lindzen calculated that the negative water vapour feedback leads to a CO_2 sensitivity of only 0.7° C. For a doubling of CO_2 concentration, the cloud effect would therefore reduce the temperature increase from 1.1° C to 0.7° C.

The debate over whether a positive or negative feedback exists will remain with us for quite some time. As a result, CO_2 climate sensitivities vary throughout the literature from 0.4° C [61] to 4.5° C [29]. The IPCC's very high estimates for water vapour amplification, however, can no longer be upheld because CO_2 can be responsible for only half of the warming over the last 50 years (see chapters 4 and 6). Thus a lot points to a weak positive CO_2 feedback through water vapour cloud cover at best. Even the IPCC admits that cloud feedback remains the greatest source of uncertainty in estimating the climate sensitivity of CO_2 [62]. On the other hand, 'It is still not possible to judge which of the model estimations on cloud feedback is the most reliable' [62]. The importance of clouds is obvious. On average clouds cover 65 per cent of the globe and have a net cooling effect of 30 W/m2 [63]. Compare that to the IPCC-assumed CO_2 warming effect of 1.66 W/m2 since 1750! That means a change in cloud cover of just a few per cent is enough to have the same effect as that proposed for CO_2. Clouds have an impact on our climate, says Henrik Svensmark. This is shown by the intuitive correlation between global cloud cover and global temperature (Figure 5.7). As cloud cover increases temperatures drop, the statistics show.

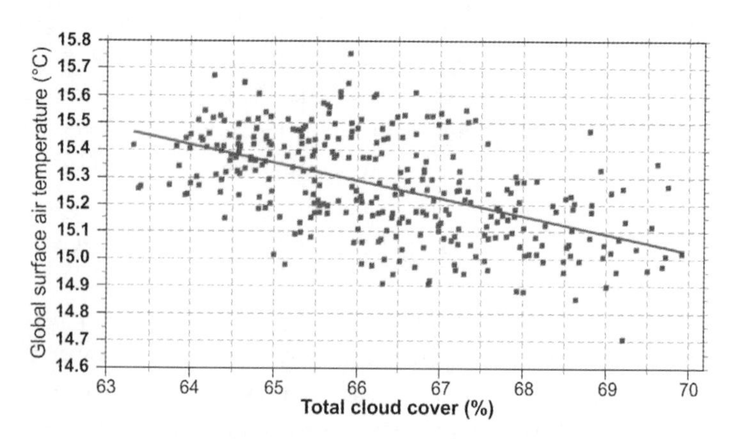

Figure 5.7 The relationship between global surface temperature and global cloud cover for 1983–2011. The trend shows that temperatures drop as cloud cover increases [64].

The CO_2/water vapour amplification model also has been shaken on the temperature side. According to IPCC models, higher specific humidity levels should have an impact, particularly in the Tropics. According to the IPCC a higher temperature has to be present at altitudes of between 3 and 7 km, where a so-called hotspot should form. This theoretical hotspot was a crucial link in AR4. On p. 675 of that report it is displayed as a model and is mostly brought in in connection with greenhouse gases like CO_2. But there's a problem: balloon measurements taken in 1979–99 show no such hotspot in the troposphere [65–67]. To the contrary, temperatures show a decided fall. So what are we supposed to think of the models when their most important hypotheses are contradicted by real observations? Not much, at least when it comes to their forecasts and to interpreting the past. If the water vapour amplification effect is excluded, then CO_2 can only be responsible for a fraction of the recent 0.8° C of warming.

In the meantime, there are now scientific opinions that claim that CO_2 does not have the lead role in the CO_2–water vapour link, but is only the junior partner. The award-winning geochemist Jan Veizer is convinced that the mighty water circulation mechanism provides the direction, whereby water vapour is controlled by solar activity [68–70]. Similar views have been expressed by Murray Salby of Macquarie University, Sydney.

Also there recently have been surprises in the stratosphere for the IPCC. In the models it is assumed that we should find cooling, unlike in the troposphere. Susan Solomon, a lead author of AR4, surprised the climate community with results from a study that supposedly showed that water vapour in the stratosphere had declined 10 per cent since 2000. With this she attempts to explain the halt in the warming since 1998. Solomon calls it the 10, 10, 10 problem: '10 per cent less water vapour at 10 km altitude has an effect on global warming over the last 10 years' [71]. Even more interesting is the finding that water vapour in the stratosphere had increased during the previous 20 years (1980–2000). Solomon estimates that about a third of the late twentieth-century warming can be attributed to this stratospheric water vapour effect [72]. In the IPCC climate models, stratospheric water vapour does not play an important role. This needs to be corrected. Indeed, it is no exaggeration to say that the formulae are currently incomplete and unable to model reality

[72]. Because it is still largely unclear which processes stratospheric water vapour fluctuations are coupled with climate, it is very difficult to determine how to account for this in the climate models. Dale Hurst of the US National Oceanic and Atmospheric Administration (NOAA) suspects that one quarter of the stratospheric water vapour increase over 1980–2000 can be explained by the increased methane concentrations in the stratosphere [73]. This is nothing new. British scientists pointed out the potential large climate impacts of water vapour in the stratosphere a decade ago [74] and so it should come as no surprise.

So what could be causing the rise of water vapour in the stratosphere 10 km above the surface of the earth? The first thing that comes to mind is that it is one of the warming effects of the rising atmospheric CO_2 level now taking place. The warming leads to more water vapour formation, which in turn is having more of an impact on the stratosphere and, unlike what IPCC's prediction, less in the troposphere. Then we would again have our water vapour amplifier. But this raises the dilemma of how water vapour emission could decrease beginning in 2000 when CO_2 was rising. Solomon's results create real headaches for the IPCC's theory. While CO_2 concentration rose continuously year after year, the stratospheric water vapour curve took a sharp turn downwards in 2000, precisely when solar activity started its slumber and the PDO began its fall (Figure 5.8).

We will refrain from pursuing in detail which impacts the different factors had on the warming process before the year 2000 and on the cooling process after 2000. By now the obvious conclusion is this: CO_2 is not the trigger of the stratospheric water vapour fluctuations. Solomon's study [72] thus strips CO_2 of a large part of the IPCC's alleged impact on the climate. The IPCC's assertion that the warming is almost exclusively attributable to anthropogenic climate gases can no longer be maintained. Instead, a host of climate factors appears to be involved. In Chapter 7 we attempt to estimate the extent each of these factors plays. The downgrading of climate

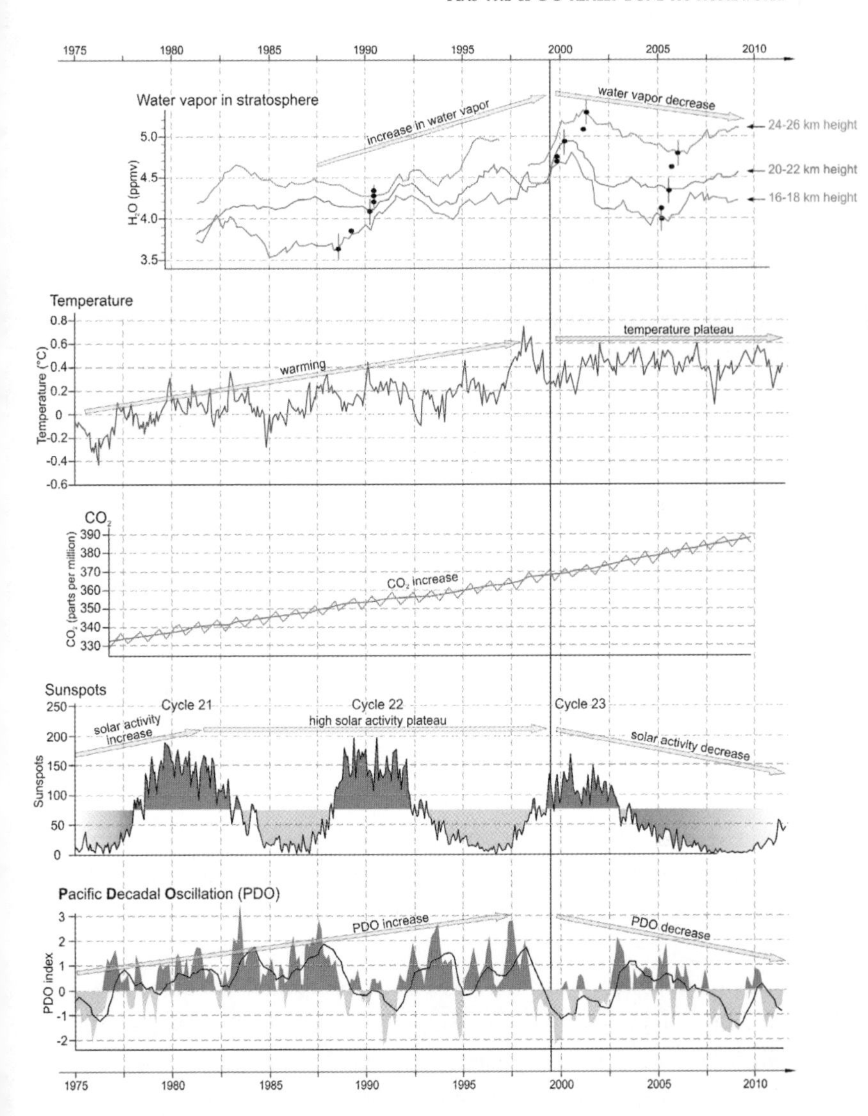

Figure 5.8 Development of water vapour content for various altitudes of the stratosphere [73]. After a protracted increase, the water vapour curves dropped abruptly around the year 2000 and then stabilized at a plateau. This reminds us of the global temperature, solar activity and the PDO, which all show a discontinuation around this time. CO2, on the other hand, continued to climb.

gases means that the warming that is expected to occur by the year 2100 will be much less than the IPCC fears.

Aerosols are the huge unknown

One thing in AR4 is clear: the IPCC admits that we really do not understand much when it comes to the impact of aerosols on the climate. According to the IPCC, the aerosols' cooling effect counteracts the proposed warming power of CO_2 by as much as two-thirds. As to the IPCC, aerosols reduce the warming caused by all the greenhouse gases combined by 45 per cent. But the uncertainty is large; it could be as little as 15 per cent or as much 85 per cent. Scientists only have 'a medium to small level of scientific understanding' of the interrelationships [75]. The impact of aerosols on cloud formation is even more poorly understood and thus can be assigned to the category 'very little scientific understanding'. So what makes it so difficult to get a grip on aerosols?

Aerosols are tiny particles of various origins. They include sulphate dust particles, soot, nitrate and mineral dust, as well as natural particles (sea salt, pollen, volcanic ash). Through diffusion and absorption, they have a direct effect on the sunlight reaching the earth. Even more important, but also more difficult to detect, are the indirect effects. Aerosol particles can act as condensation nuclei for water droplets and thus play a role in cloud formation. To put it loosely: a 'dirty' atmosphere enhances cloud formation and so indirectly cools the earth.

Naturally enough, one takes only the manmade aerosols into account when calculating climate change over the last 250 years, even when 80 per cent of all aerosols are of natural origin. In any case, total manmade emissions of aerosols have risen during this period and, at 160 million tons a year, are ten times higher today [76].

The cooling effect is of great importance for IPCC climate modellers. Only by selectively modelling the effects of aerosols were they able to explain to some extent the cooling period of 1940–70 without having to deal with natural oceanic cycles and the sun.

Rahmstorf explains that 'solar activity remained almost constant during the stagnant phase (1940–1970) as the warming caused by greenhouse gas concentrations was essentially offset by the cooling effect of aerosol pollution' [77]. After that aerosols were reduced in many parts of the world by the use of 'desulphurization systems at power plants so rising greenhouse gas concentrations again began to control the temperature beginning in 1970' [77]. These two statements contain two major errors. First, the sun, in large part, went off duty around 1970 and went into a markedly weak phase of activity, and so the notion of constant solar activity is out of the question. The other error is that Rahmstorf, as many IPCC climate scientists do, moves the desulphurization up a decade or two so that it better fits the models. For example, the first desulphurization system in Germany went into demo-operation in 1976. Desulphurization for new plants in Germany became law in 1983, with older plants being allowed a grace period for conversion until 1993. The region comprising former East Germany added the technology in the 1990s and the rest of Europe followed in line with the 2001 EU Directives. In 1970 global emissions were 60 million metric tons (calculated as sulphur), in 1990 the figure was 70 million tons and did not fall back to 60 million tons until before 2000 [78].

But none of this seems to trouble IPCC climate scientists. Some place the peak of sulphur emissions in 1960, while others peg it to 1990 – whatever suits them best. Christian Ruckstuhl of the University of California concluded that aerosols are the biggest uncertainty factor in AR4: 'Incorrect input aerosol histories are the most likely cause for the disagreement between models and observations' [79]. This is how one finds differences of a factor of 3 in the twenty studies on the climate effects of aerosols summarized by the IPCC. Some scientists simply used aerosols to bring their models into line with reality [75]. The climate effect of aerosols was often simply determined by subtracting the assumed greenhouse effect and other modelled climate effects from the measured temperature. The remaining temperature difference is then assigned to aerosols.

That is how one brings whatever value is desired for the effect of CO_2 into harmony with reality. Not all scientists did things this way, but it is odd what an analysis of the IPCC models revealed [80]. Models with high greenhouse gas sensitivities were exclusively coupled by the IPCC with high aerosol cooling, while models with minimal sensitivity had minimal aerosol cooling. And this is how IPCC scientists succeeded, without really knowing what they were doing and taking separate paths, in reaching the same target, namely a virtual temperature curve that looks as much as possible like the observed temperature development [81]. But if you couple the scenarios of low CO_2 sensitivity with high aerosol cooling, or vice versa, then the models fail. Nobody questions the relatively good agreement between models and measurements. By now it should be obvious that reliable forecasts cannot be generated with such models.

Citing aerosol emissions as an explanation for the 1945–77 cooling is not much more than an excuse. Carbon dioxide and aerosols increased in parallel from 1945 until the 1980s. Yet the temperature development switched from cooling to warming almost abruptly in 1977. According to Don Easterbrook this has nothing to do with CO_2 or aerosols, but rather with the PDO flipping from its cool phase to its warm phase in 1977 [8] (see Chapter 4).

Surely since 1990 a decoupling of CO_2 emissions from the dust and aerosol emissions has occurred. Dust and aerosol emissions have been rising less rapidly, if not at times reversing. Indeed, part of the warming since 1985 can be traced back to this decoupling. However, determining the extent of this effect borders on speculation. A radiative forcing of –0.1 to –0.9 W/m^2 is assumed for the direct aerosol effect. As for the indirect cloud forcing, the twenty-seven studies used by the IPCC show that the forcing is scattered over a range from –0.2 to –1.8 W/m^2. This uncertainty enables the IPCC to project a strong warming for the future. One can assume that the aerosol emissions of fine dust and sulphate particles will decline in Southeast Asia over the coming decades because of stringent environmental policies. According to the IPCC this is why the CO_2 effect will hit harder in the future than it currently does.

The aerosol joker is needed once again. As the global temperature has not risen since 2000 and it cannot be explained away by the IPCC scientific community, the aerosol joker is slipped into the playing deck once more. This time increased sulphur dioxide emissions in China since 2000 are claimed to be the cause of the strengthened aerosol formation and is why the warming of the last 10 years has been pushed to the side [82], according to IPCC scientist Robert Kaufmann (see Chapter 4). But these aerosol emissions were rising before 2000 and since 2006 there's been a marked reduction in China owing to a massive expansion of coal power plant desulphurization in the 2005–10 5-year plan. Moreover, Chinese emissions solely influence the Northern Hemisphere where the temperature has risen slightly. In the Southern Hemisphere the temperature has dropped slightly over the last 10 years. Immediately, two other IPCC scientists step in to explain this. Teams working with Jean-Paul Vernier and Susan Solomon discovered volcanic events that provided the urgently needed cooling potential [83–84]. Solomon found a slight aerosol increase in the stratosphere for 2000–10, which she believed accounted for 0.07° C of cooling since the end of the 1990s (i.e. about one third of the missing warming). So how were the other two-thirds neutralized? One cannot but notice how these publications are driven by the desire to explain the stagnant global temperature since 2000 without casting CO_2's importance in doubt or having to integrate the oceanic climate cycles into the models, or even to involve declining solar activity [85].

In fact, the omnipotence of CO_2 is threatened by another aerosol, which up to now has never been appropriately taken into account in any of the IPCC prognoses: soot.

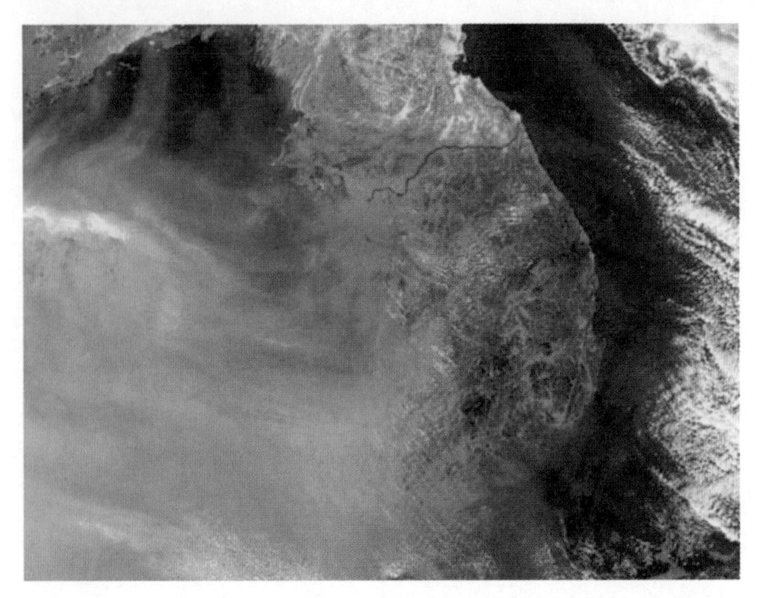

Figure 5.9 Satellite photo of the Korean peninsula darkened by eastbound industrial gases from its Chinese neighbour. Photo taken on 27 January 2006 [67].

Soot – the new star on the horizon of climate science

AR4 assigns soot a relatively weak positive radiative forcing of 0.3 W/m2. But numerous studies have been published since then which show that soot potentially has a much higher impact. Veerabhadran Ramanathan of the University of San Diego demonstrates that the radiative forcing of soot could be as high as 0.9 W/m^2, three times higher than the AR4 assumption [86]. Although soot is comparably low in emissions at 8 million tons a year and stays only a few days or at most weeks, the studies indicate soot has to be considered to have 55 per cent of the warming effect of CO_2 in the IPCC models when the past 250 years are considered. Even Greenpeace agrees that soot is second only to CO_2 as a cause of warming [87]. Because this effect has not been taken into account at this magnitude, it is essential to scale back the influence of CO_2 in the models by a corresponding amount. That would surely be an embarrassing thing to communicate to the public. Ramanathan thus resorts to claiming a large degree of uncertainty for the dampening influence of other aerosols (see above), and so simply increases the negative effect of the other aerosols by precisely the warming amount that soot is attributed to have in the latest findings. Bingo! The result is once again perfect. CO_2's effect on warming remains intact and Ramanathan is soon quoted by the IPCC community. It's a simple now-you-see-it, now-you-don't sleight of hand that is not confirmed by any new findings concerning the non-soot aerosols. It is striking that the climate-political debate so far has not acknowledged that soot could have warmed the Arctic by 0.5–1.0° C over the past decades [86, 88]. According to Mark Jacobsen of Stanford University, soot accounts for 0.25° C of the global warming so far [89]. In other regions of the world too soot accounts for a significant contribution to the warming found in past decades [90]. That means there really isn't much left for CO_2 to do. Who can honestly say they have heard of this or of any debate about a global soot reduction programme?

Soot can influence the temperature in the atmosphere in two ways: soot particles absorb sunlight and radiate warmth back into the atmosphere; and it hinders the reflection of sunlight on ice and snow (a reduced albedo). Soot has a negative effect on snow and ice regions in that it absorbs the sun's rays that otherwise would be reflected. It thus warms the surface, leading to accelerated melting. In the IPCC models soot can hardly play out climatically because it is simulated at ground level in the formulae, which has little to do with reality. The maximum concentration is at 2 km in the air where soot absorbs the sunlight reflected by low clouds and thus converts it into warmth [86].

Soot is created by the incomplete combustion of biomass, fossil fuel and combustible materials. Twenty per cent of soot is created by burning biomass, 40 per cent by fossil fuels and 40 per cent by slash-and-burn land-clearance and through forest fires. China and India are responsible for 25–35 per cent of all global soot emissions [91]. In China alone soot emissions doubled between 2000 and 2006 (Figure 5.9). A reduction through filter systems, diesel motors and switching from open fires that burn wood or charcoal to other cooking methods would quickly lead to a positive effect because soot remains in the air for a very short time. Moreover, replacing wood as a cooking fuel would significantly reduce smoke inhalation by women and children, which causes a staggering 400,000 deaths each year in poor countries, half of whom are children. By strongly scaling back soot emissions, up to 40 per cent of the observed climate warming could be eliminated [92]. From a climate-political standpoint, burning wood at home instead of burning natural gas at our middle latitudes hardly brings any advantages, unless the soot emissions are drastically reduced.

Whether it's soot or other aerosols, the IPCC's stoicism in light of all the scientific uncertainty is remarkable. Most of the climate models considered by the AR4 do not take any indirect aerosol effects into account [93]. Moreover, their fundamental assumptions vary to a great extent. In one model the maximum sulphate emissions in Europe occurred in 1960, in another they occurred in 2000. In yet another model the emissions in China have declined since 1995, in another they skyrocket after the year 2000. In one model Japanese emissions decline after 1970, but in another they steadily climb. The IPCC's adherence to unambiguous announcements is unshaken, so much so that the AR4 summary report claims a radiative forcing for CO_2 of 1.66 W/m2 since 1750, right down to a hundredth of a watt. Just the uncertainty for aerosols alone accounts for 1.5 W/m^2. One might call this barefaced cheek or even scientific irresponsibility.

What climate extremes are we threatened by?

Recently, in public discussions, anthropogenic global climate warming has been blamed for all extreme weather events. Forest fires in Russia, tornadoes in the United States, floods in Pakistan and China, hot summers and cold winters, record October snowfall in New York City, sharks off the coast of Egypt – climate scientists, journalists and politicians are quick to make a connection between unusual weather events and global warming. During the weeks-long devastating forest fires that raged round Moscow in the summer of 2010, President Medvedev declared, 'What is happening now in our central regions is evidence of global climate change' [94]. Hans-Joachim Schellnhuber also chimed in, 'The global mean temperature has been at a record level for one year, and thus the probability of regional heat zones like the present one in Western Russia is rising' [95]. Yet it didn't make the headlines when seven American scientists later showed that the heatwave was a natural weather phenomenon and that no connection to global warming could be made [96].

The projected consequences of climate change stir up deep-rooted primal fears within the human psyche. The massive melting of the polar ice caps projected by the IPCC would lead to an abrupt sea level rise and flood inhabited areas. The biblical story of Noah's flood was the result of man's sinful ways. Moreover, people worry about storm surges that cause rivers to burst their banks, hurricanes and cyclones that threaten livelihoods and epic droughts. What's behind all the panic? Is civilization really on the verge of destruction?

Retreat of the ice sheets, glaciers and sea ice

Let's start with ice. There are three types of ice, namely the large ice sheets on Greenland and Antarctica, the sea ice in the Arctic and Antarctic, and the mountain glaciers all around the world. First we need to note that the current melting of ice does not come as a surprise. The temperature since the end of the Little Ice Age over the last 150 years has risen almost 1 ° C, so it is only to be expected that ice would take a hit. We have already noted the reason for the long-term warming: one part is due to the active sun (see Chapter 3), one part due to the rise in CO_2 concentration and climate gases in the atmosphere, and yet another part due to soot. Together, these factors are the primary culprits for global ice melting.

The ice sheets of Greenland and Antarctica store gigantic amounts of water in the form of ice. Hypothetically, if the Antarctic ice sheets were to melt completely, sea levels would rise about 60 metres. The melting of the Greenland ice sheets would add another

7 metres. Over the course of the earth's geological history, ice-free poles have occurred time and again, and over long phases. A complete melting of the pole caps, according to the IPCC, would in any case only take place if the warming were sustained over thousands of years. However, completely burning off all the reserves of fossil fuels on the planet and the duration of CO_2 in the atmosphere would not be enough to warm the planet for thousands of years [97].

Let's take a brief look at what has happened to the ice sheets in Greenland and in Antarctica over the last few decades. The Greenland ice sheets have really taken a hit over the past 10 years [98]. Even so, ice loss estimates there were drastically exaggerated. Recent studies have shown that the ice melt rate is only half that originally estimated and is currently about 104 gigatonnes/year [99–100]. Earlier authors had significantly underestimated the effect of the postglacial, ice unloading-related rise of both polar regions and so calculated excessive ice melt rates based on GRACE satellite gravimetry [101–103]. Furthermore, it was found that the glaciers of the Greenland ice sheet underwent a more rapid retreat in the 1930s than in the 2000s, as revealed by recently discovered air photographs taken over Greenland around that time [104].

The situation is similar in the west Antarctic and the Antarctic peninsula, which are an extension of South America and thus exhibit in part similar climate trends. The west Antarctic ice sheet is melting, but at a much slower rate than thought just a few years ago [99–100, 105]. The east Antarctic, which is cut off by the Transantarctic mountain range, represents by far the larger part of Antarctica. The climatic trends there are very different from the western part of the continent and here we find some surprises. The central part of the east Antarctic ice steadfastly refuses to take part in any of the ice melt theatrics and is even growing slowly [105]. This, by the way, can be read in the IPCC report, which explains that east Antarctica is simply too cold to melt and hence will continue to increase in mass due to snowfall. Only the periphery of east Antarctica appears to be melting.

When viewing the total Antarctic ice mass, things are quite clear: according to new ICESat laser measurements of elevation change [106], the mass gain of the total Antarctic ice sheet from snow accumulation during 2003–8 exceeded the mass loss from ice discharge by 49 gigatonnes a year. The Antarctic ice shield is actually growing!

The situation concerning the large ice sheets of Greenland and Antarctica appears to be much less dramatic than a few climate protagonists have been loudly trumpeting. Let's shed more light on the matter. Studies have shown that the bulk of the east Antarctic ice survived multiple interglacial periods with ice ages in between during the last 730,000 years, a period when at times temperatures were 6° C warmer than they are today [107]. It is thus assumed that east Antarctica will also get through the current warming with very little loss. The same applies to the Greenland ice sheets, which also did not melt away during the last interglacial 120,000 years ago [108], a time when it was up to 5° C warmer than today.

Floating ice

In the polar regions vast areas of the sea are covered by a 2–3 metre thick sheet of ice. During the summer a large part of the floating ice sheet fragments and melts, and in the winter the damage is more or less repaired and the ice cover expands. Since there have been satellites for 30 years, the annual thawing and freezing can be continuously monitored from space. From these observations it is clear that Arctic sea ice has shrunk (Figure 5.10).

Three questions are often asked. First, will this continue with summers in the North Pole eventually becoming completely ice free? Second, was the Arctic Ocean in the past always covered by vast areas of ice as we know it today or were there times when the ice melt was much greater than the present? And third, why do we hear so little about the South Pole from the media?

In answer to the first question, the IPCC is confident and forecast in its 2007 report that summer Arctic ice cover will completely disappear by the end of the twenty-first century. Over the 2000–12 period the Arctic ice indeed continuously shrank. Especially dramatic were the retreats of 2007 and 2012, which set alarm bells ringing loudly (Figure 5.10). If this trend were to continue, sea ice at the North Pole would disappear, and soon.

In the search for the cause of the unusually low 2007 and 2012 ice cover, NASA scientists found the answer. The strong ice retreat was mainly due to changes in local wind patterns [109–111]. In 2007 unusual currents carried a large part of the Arctic ice southwards and eastwards past Greenland and out into the Atlantic where it eventually melted [112–113]. The resulting exposed Arctic sea area then accumulated some additional heat because the albedo (reflectivity) of an open sea is far less than that of ice [112]. In August 2012 a powerful storm broke up the Arctic sea ice and led to a new Arctic sea ice minimum for the satellite era of the past 30 years. A few weeks later, Arctic sea ice was back at its long-term average value.

Scientists also found an explanation for the multi-year general meltdown of the Arctic ice leading up to the record 2007 year (Figure 5.10). It was probably no coincidence that the Atlantic Multidecadal Oscillation (AMO), a climate system internal fluctuation, reached its positive warm period at this time [114–115]. Such internal climate systems generally do play a prominent role in the Arctic climate. In 2004 Petr Chylek of Los Alamos National Laboratory could show that the temperature of the eastern Arctic Ocean increased by 1 ° C from 1979 to 1997 while the western Arctic cooled by 1 ° C. This demonstrates that the changes turn out to be far more complex than previously assumed, partly because of the internal climate oscillations. Notably, the NAO plays a far greater role in Arctic ice dynamics than does global warming, according to Chylek [116]. The Arctic sea ice had already undergone considerable melting in the 1930s, based on ship sightings. At that time, there were of course no satellites to record the 'damage' done by natural processes to the ice.

For this reason it is not easy to determine whether the Arctic ice will continue to shrink. Because of the natural system variability, linear trends cannot be simply projected into the future. And because temperatures may decrease due to natural factors over the coming few decades (see Chapter 7), it cannot be excluded that Arctic ice cover will once again expand in the foreseeable future. Moreover, recent models indicate that there is no danger of the Arctic ice system careening out of control and reaching a tipping point because of self-induced amplifications. The increased heat absorbed due to the reduced albedo is simply radiated into the atmosphere during the autumn and winter that follow, model calculations show [117–118].

This takes us to the next question: exactly how has Arctic sea ice been developing over the millennia? Is the current situation really unusual or even unique? As there were no satellites then, we have to use data collected from seafaring reports and coastal observations. These are less precise, but better than no data at all, and general trends can be discerned. From the historical curves we can see that the Arctic ice did not begin to retreat only a few years ago, but started in the nineteenth century, at the end of the Little Ice Age (Figures 5.11 and 3.8) [119]. The same applies to ice in the Baltic Sea [120]. The start of the Arctic sea ice retreat clearly has natural origins. It is the warming that comes with the end of a natural cold phase, the Little Ice Age.

Now let's go back a few more centuries to the Medieval Warm Period (ninth to the fourteenth centuries). Back then Arctic ice had melted so much that the Vikings were able to make expeditions to Iceland and Greenland in the ninth century, and eventually were able to settle there. Later, in around the year 1420, the Chinese were able to sail into the Arctic Ocean and found hardly any ice there [121]. The Northwest Passage around that time was ice-free [122].

As we have seen, there was a series of similar warm periods over the last 10,000 years since the end of the last ice age (see chapters 3 and 4). Arctic sea ice shrank multiple times in sync with millennial cycles, and so there were ice extents that were often considerably less than today's [123–126]. There were phases when the central Arctic

Ocean was for the most part practically ice-free during the summer [127]. Our current Arctic ice shrinkage is not unique.

The third question – why don't we hear more about the Antarctic ice from the media? – can easily be answered: the Antarctic sea ice is not shrinking at all, but according to satellite observations has been stable for the last 30 years, and is even expanding slightly [128] (Figure 5.12). Interestingly, current theoretical climate models are not able to reproduce the expanding Antarctic sea ice [129]. In 2012 the Antarctic sea ice reached a new record extent for the satellite era of the past 30 years. This, of course, limits the entertainment value of Antarctica's development for the media, and so this important information is little known to the public. Sea ice low in the Arctic and high in the Antarctic? Warming of the Arctic and cooling in Antarctica? And shrinking ice shields in Greenland while the Antarctic ice cap is growing? Scientists have found evidence for a connection, which they term the 'polar see-saw' [130]. When it gets colder at one pole, it warms up at the other. These internal climate dynamics must not be confused with changes triggered by external processes: changes in solar activity or anthropogenic influences.

To conclude, we now look briefly at how much the sea would rise if all the sea ice melted. The ice floating in the ocean displaces a certain volume of seawater. When sea ice melts, there's a density difference between fresh water and saltwater of 2.6 per cent by volume. That means there would be more melt water than displaced seawater. This in turn would lead to a sea level rise of 40 mm. Therefore, the effect in principle would be negligible [131–132].

Mountain glaciers: always on the move

All the major mountains have glaciers because temperatures in the atmosphere decline up to an altitude of at least 8–15 km on average in the Tropics. The reason for this is the lower solar energy absorption by the reduced concentration of the main and minor air components [135]. Because of the 0.8° C or more of warming since the end of the Little Ice Age, much of the global mountain glaciers have

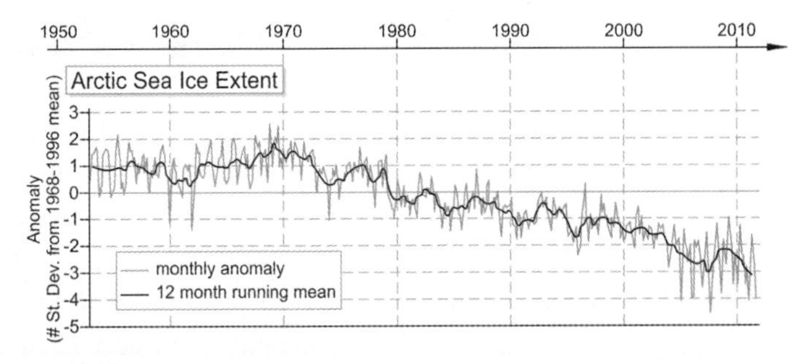

Figure 5.10 Shrinkage of Arctic sea ice [133].

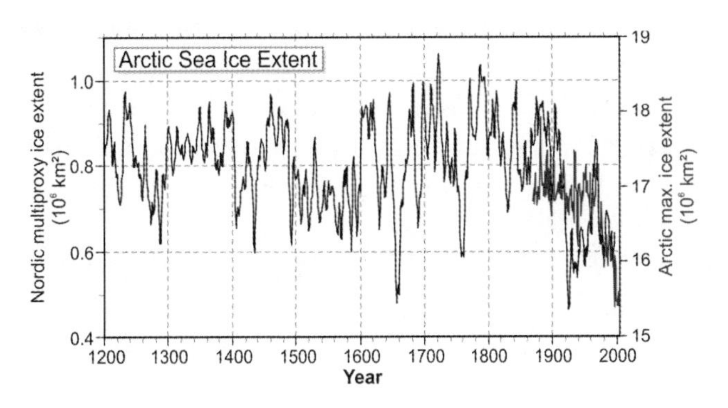

Figure 5.11 Arctic ice cover over the last 800 years. Arctic ice reached its maximum during the Little Ice Age. Since the end of the Little Ice Age, in about 1850, the sea ice cover has declined [119].

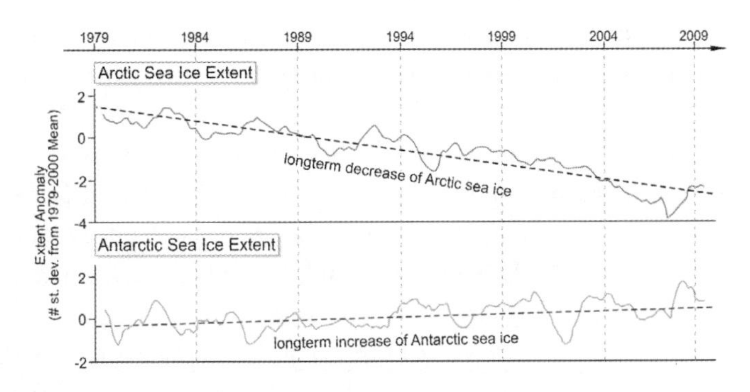

Figure 5.12 Comparison of sea ice cover for the Arctic and the Antarctic over the last 30 years. While Arctic sea ice has receded, Antarctic sea ice has expanded [134].

retreated. The exceptions are Scandinavia [136] and New Zealand [137], where many of the mountain glaciers have grown recently [138]. In the Caucasus and the Himalayas the glaciers are for the most part stable [139–141]. Some have bucked the trend, as regional effects come into play (e.g. the NAO in the case of Scandinavia) or because of increased or reduced winter precipitation [136, 142]. Don Easterbrook of Western Washington University clearly demonstrated these cyclic changes on Mt. Baker, for example. The glacial changes on the mountain are in sync with the PDO (Figure 5.13) [143].

In central Europe the historical glacier extent has been well documented since 1700. In addition, there are geological methods of reconstructing the movement of the glaciers. Therefore, it is worth taking a look at how mountain glacier movements have reacted to the climate cycles, starting with the Medieval Warm Period up to today's warm period.

In the Swiss Alps, wood originating from AD 1000 (the Medieval Warm Period) was uncovered as glaciers retreated (parts of today's glacial areas were forested during the earlier times of strong glacial retreat) [144]. Indications of massive glacier melts during this phase were also observed in Scandinavia [144]. Glaciers grew strongly across the globe during the transition to the Little Ice Age. This is documented, for example, in the Alps, Scandinavia, Alaska, Greenland, Iceland, New Zealand and Chile [144–152]. Many glaciers experienced their greatest expansion over the last 10,000 years during the Little Ice Age [149, 153]. During the transition to today's modern warming phase, the glaciers retreated in parallel with the warming (Figure 5.14) when mountain glaciers in Montana in the 1930s and 1940s melted up to six times faster than today [154].

This advance and retreat of mountain glaciers in parallel with the solar Eddy activity cycles (see Chapter 3) took place throughout the entire postglacial period [153, 155]. It is also worth mentioning that during the warmest phase of the postglacial period, the so-called Holocene Climate Optimum 4000–8000 years ago, many glaciers retreated dramatically and were considerably smaller than they are today, or had even disappeared entirely [146, 148, 156–157]. To allay any fears, a complete melt of all the mountain glaciers today would take centuries and cause sea levels to rise by about 0.5 metre [158].

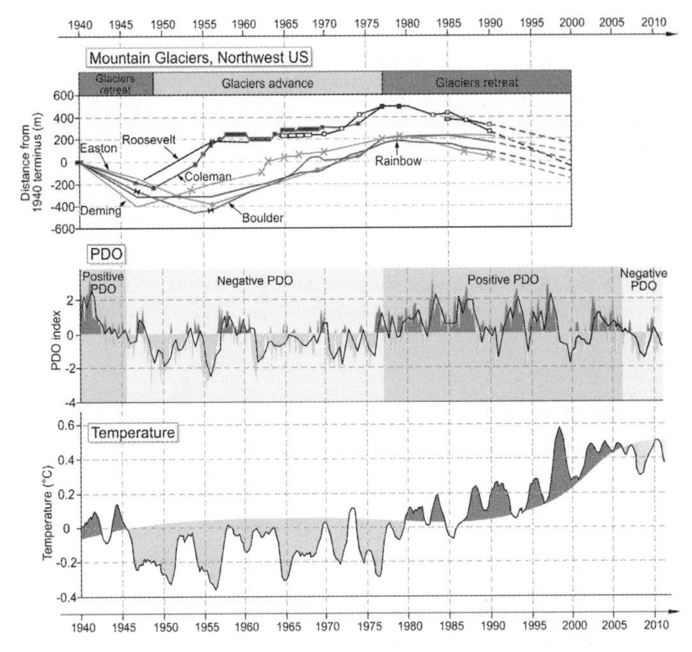

Figure 5.13 The retreat and advance of the Mt. Baker glaciers in Washington State occurred in parallel with the PDO and temperature curve over the last 70 years. Sources: glacier and PDO [8]; temperature: HadCRUT3 moving 12-month mean.

Sea-level rise

Currently the seas are rising a few millimetres a year. This is a global mean value. The sea level actually develops very differently according to location. Who would have guessed that sea levels in some regions have even fallen over the last few decades? Deeply buried in AR4 we read that the sea level over vast parts of the ocean did fall during the 1993–2003, in the eastern part of the Pacific and western Indian Ocean [159]. On the other hand, it rose in other regions. The reasons for the different sea level developments include changing ocean

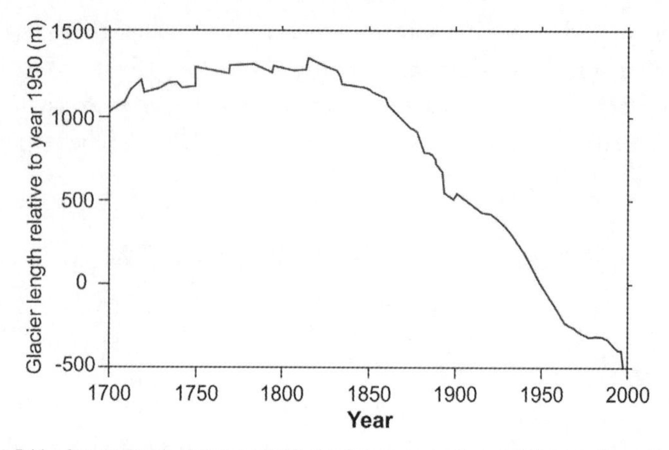

Figure 5.14 Average length of change of 169 global glaciers over the last 300 years. The melting began in around 1820, which marked the end of the Little Ice Age and initiated the transition to today's warm period [138].

currents, which are modulated by wind changes and temperature fluctuations resulting from internal climate oscillations. This is how the El Niño cycle can cause a sea level variation of about 200 mm. The surface of the sea is not flat, but characterized by slightly inclined areas caused by ocean currents. Because of the Coriolis effect, the sea surface to the right of the Gulf Stream is about 1 metre higher than to the left [160]. Meanwhile, a 60-year cycle and other natural cycles have been detected in the sea level which appear to be linked to longer-term ocean cycles, such as the PDO, the AMO and the Arctic Oscillation [161–163].

Overall, measured in all regions between 1972 and 2008, the sea level on average rose by about 2 mm per annum [164]. This is partly due to global warming of 0.8° C which began around 1820 – close to the end of the Little Ice Age, a time when many glaciers had grown considerably. The subsequent shrinkage of mountain glaciers and polar inland ice during the transition to the Modern Warm Period released huge amounts of water, more or less uniformly distributed over decades, which in turn caused sea levels to rise – slowly at first, and later more quickly once the process was fully underway (Figure

5.15) [165–166]. The rate of increase then stabilized, beginning around 1900 [165]. At the same time the oceans expanded due to global warming-induced thermal expansion, which provided an important component to the sea level rise [164]. Between 1993 and 2003 thermal expansion accounted for about half of the sea level rise.

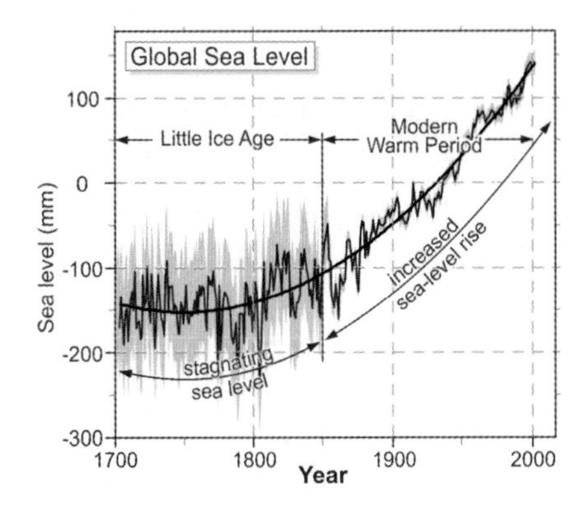

Figure 5.15 After a period of stagnation during the Little Ice Age, the sea level began to rise more quickly during the transition to the Modern Warm Period. The rate of rise then stabilized during the twentieth century [165].

Independent of global warming, there is another component that cannot be underestimated. Global pumping of inland groundwater and its subsequent discharge into the seas accounts for about 0.3–0.8 mm/year of sea level rise, or 13–25 per cent [164, 167–169]. And because groundwater extraction during the course of the twentieth century increased sharply, thus making the groundwater effect increasingly important over the course of time, the direct contribution from climatic factors again needs to be revised downwards.

The current rise in sea level has to be viewed as a part of a multi-year trend since the end of the last ice age, when vast areas of northern Europe and North America were buried under sheets of ice a kilometre or more thick. They steadily began to retreat 12,000 years ago. The gigantic quantities of melt water flowed into the oceans and caused the sea level to rise 120 metres to the present day. The rise began at a rapid 14 mm a year but then slowed over time [170]. Today the sea level is rising at no more than about 2–3 mm a year. The natural millennial climate cycles strengthened or weakened the long-term rate of rise via glacial ice gain or loss. In the German North Sea, the sea level fell repeatedly during the cold phases in the last 3000 years [171]. There is in fact evidence that the global sea level during the Medieval Warm Period was higher than it is today [172].

AR4 is very cautious in its statements on sea level increase. It reaches the conclusion that there is only a limited understanding of the causes of sea level increase and thus the report does not offer any probability, or see itself in a position to provide an estimate for an upper limit [166]. For the various greenhouse gas emission scenarios, a range of 180–590 mm of sea level rise by the year 2100 is projected. That is considerably less than the 90–880 mm given in TAR published in 2001 (Figure 5.16).

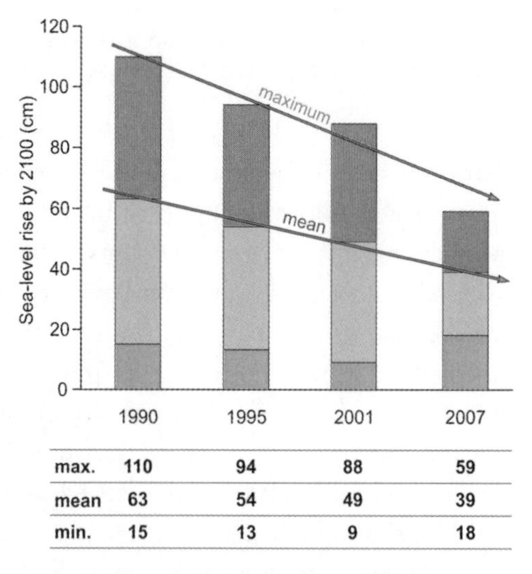

	1990	1995	2001	2007
max.	**110**	**94**	**88**	**59**
mean	**63**	**54**	**49**	**39**
min.	**15**	**13**	**9**	**18**

Figure 5.16 With the publication of each successive report, IPCC's prognoses for sea level rise by 2100 were reduced [173].

For some IPCC representatives, this development was not dramatic enough. Rahmstorf claims in *The Climate Crisis* that the

IPCC understates sea level rise [174]: 'It is obvious that a sea level rise of over one metre cannot be excluded by the end of the century.' A few sentences later he adds: 'the sea level could rise by as much as 1.4 metres.' James Hansen, one of the staunchest IPCC supporters, even fears that global sea levels could rise 2 metres by the year 2100 [174]. In 2011 Hansen increased that estimate to 5 metres [175]. With such projections one certainly can engender fear.

In order to reach such magnitudes, the authors assume that sea level rise will greatly intensify in the future [176]. In the meantime these calculations are turning out to be gross exaggerations, as a number of scientists have determined that there has been no increase in the rate of sea level rise since 1930 [177–179]. James Houston and Robert Dean recently showed that there has even been a general slowdown in sea level rise since 1930 [180]. In doing so, the effect of water exchange between land and sea through groundwater extraction or damming at reservoirs was taken into account. The authors wrote, somewhat sarcastically, that in order to reach the multi-metre levels projected for 2100 by Vermeer and Rahmstorf [176] large positive accelerations of one to two orders of magnitude greater than those yet observed in sea level data are required [180].

Yet Rahmstorf remained undaunted. Based on a study of data from a handful of measurement stations off the east coast of the United States [181], Rahmstorf and the Potsdam Institute announced in a June 2011 press release, 'The sea level is rising faster than at any time in the last 2000 years' [182]. A science showdown then took place in the July issue of the *Journal of Coastal Research* where Rahmstorf and Vermeer tried to refute Houston and Dean by claiming that the slight slowdown they had found since 1930 was due to the 1940–70 cooling and that only data from the Northern Hemisphere had been used [183]. But Rahmstorf and Vermeer once again embarrassed themselves, as in the very same issue Houston and Dean pointed out huge errors, namely that the measurements Rahmstorf and Vermeer used covered less than 60 years and are thus are subject to decadal oscillations that can generate strong sea level fluctuations up and down [184]. Then Houston and Dean delivered the knock-out blow: The most recent studies of sea level off the Australian coast show the same pattern: a slight slowdown in sea level rise, and certainly no augmentation [185]. It therefore comes as no surprise that in New Zealand too no accelerated sea level rise has been found. Over the past 100 years the sea level has been rising fairly steadily by 1.7 mm a year [186]. Many other areas worldwide do not show the alleged acceleration in sea level rise over the past few decades [187–189]. In some areas (e.g. Tasmania and parts of the North Sea), sea level rise has even slowed down significantly [190–191].

German experts doubt the unusual Rahmstorf results obtained off the east coast of the United States. Jens Schröter of the Alfred Wegener Institute, Bremerhaven suspects that the Rahmstorf team were misled by vertical coastal movements [192]. According to him, the accelerated flooding of the coast under study has little to do with global sea level development, but was the result of the coastal area sinking. Others agree [193–194]. From this dispute, we can gather that the warming over the last few decades has led to a linear increase in sea level, but that no acceleration in its rise is detectable.

With his questionable estimation, Rahmstorf concurs with the IPCC, which wrote in its 2007 report: 'Global sea level rose steadily over the twentieth century and is rising at an increasing rate' [195]. But here the IPCC overlooks one important aspect: sea level is measured by tide gauges, and since 1993 by satellites, which detect changes with respect to the earth's centre of gravity (Topex, Poseidon, Jason). The IPCC is now comparing satellite data (1993–2003) with tide gauge measurements (1961–2003). The time-point at which the alleged acceleration began is a cause for scepticism. It is 'improbable that the sea level rise began precisely in the very same year that satellites began operation,' British sea level expert Simon Holgate commented dryly [196].

As one would expect, sea level rise has slowed down somewhat since warming ceased in 2000 (Figure 5.17) as the sea temperature has not risen for over 10 years. As a result sea level rise has decreased to 2 mm a year since 2005 [197]. Small fluctuations are superimposed over the linear trend of the twentieth century and thus rates always have to be determined over periods of several decades and cannot be simply extrapolated [198].

Despite this development and actual findings of a highly distinguished scientific research group in November 2012, which showed the 'approximate constancy of the rate of the global sea level rise during the twentieth century, which shows small or no acceleration, despite the increasing anthropogenic forcing' [199], the draft AR5 asserts in its Summary for Policymakers, 'It is very likely that the global mean sea level during the twenty-first century will exceed the rate observed during 1971–2010'. In Chapter 13 the report projects a sea level rise of 0.29 m up to 0.96 m until 2100.

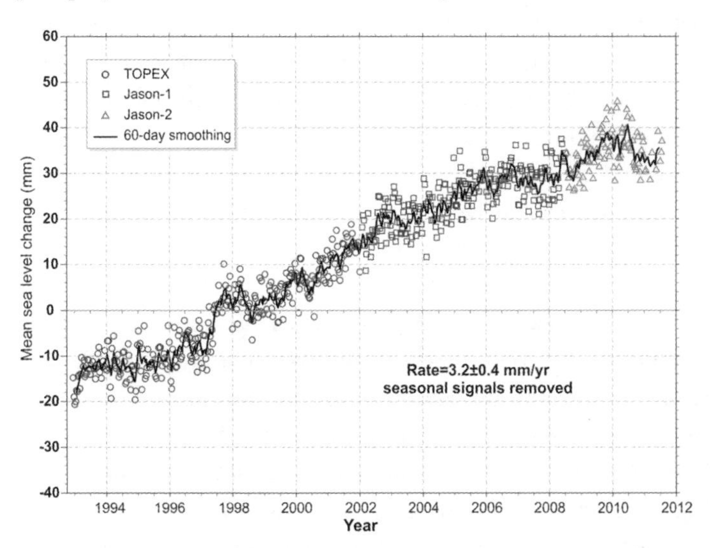

Figure 5.17 Sea level development over the last 18 years [200].

Needless to say, sea level is of great importance to the 146 million inhabitants in areas that are 1 metre or less above sea level. However, most ocean scientists do not envisage an accelerating sea level rise over the twenty-first century. Without such an acceleration, sea level would rise only 150–270 mm by 2100 [180, 196]. This does not take into account that a temperature increase of 2–4° C in all likelihood will not occur, which the IPCC assumes (see Chapter 7). Publicity stunts such as the one performed by the President of the Maldives, Mohamed Nasheed, will hopefully soon become things of the past. On 17 October 2009, he assembled his Cabinet members in scuba gear for an underwater session and passed a declaration dubbed 'SOS from the frontlines' [201]. It now appears that this Mayday distress call was premature. Scientists from Kiel recently discovered that the sea level in parts of the Indian Ocean have sunk by as much as 50 mm since the mid twentieth century [159].

Another curious sea level story took place in the South Seas a few years ago. Since 1997 on some of the islands of the Vanuatu archipelago, located east of Australia, the residents of coastal villages had been regularly getting wet feet whenever a storm system drove the sea into their homes [202]. Their valuable coconut plantations had been repeatedly flooded. The situation got so bad that in 2002 the island of Tegua began relocating the village of Lataw several hundred metres further inland. In a 2005 press release the UN Environmental Programme (UNEP) described the residents as the first 'climate-change refugees' [203]. Then UNEP Executive Director Klaus Töpfer warned in a report that Vanuatu marked the start of an ominous development where rising temperatures, melting ice and rising sea levels would lead to extensive damage worldwide [203]. As the hype approached fever pitch, some scientists took a closer look. What they found was surprising: the sea's encroachment had less to do with climate change and modest sea level rises, and more to do with the fact that the islands themselves are sinking [202]. Vanuatu is located on the boundary of two tectonic plates. In this earthquake-prone region, the land may rise and sink abruptly as well as over the long term. These movements may greatly exceed sea level dynamics. This is why between 1997 and 2009 some of the largest subsidence rates in the world were measured on the Vanuatu islands [202]. UNEP's climate refugees of Vanuatu would therefore be much more appropriately called 'plate tectonics refugees', victims of deep forces in the earth's interior which have been moving whole continents for billions of years.

Moreover, it is often forgotten that the atolls are living formations. Coral reefs in principle are able to compensate for sea level rise by growing and thus remaining close to the surface. They appear to be fully capable of keeping up with rapid sea level rises, as they did during the early phase after the last ice age, which the majority of atolls apparently survived. A recent study showed that most atolls of the central Pacific have demonstrated considerable growth over recent decades and have remained stable despite the current sea level rise in that region [204]. Another country frequently cited as already suffering from global sea level rise is Bangladesh. Here scientists from the Dhaka-based Centre for Environment and Geographic Information Services (CEGIS) have produced a big surprise. After studying 32 years of satellite images they found that Bangladesh's landmass has increased by 20 km^2 annually.

Even though the sea level debate is abating somewhat everywhere, there are still some stalwarts. In early 2011 Hansen boosted his personal horror scenario, warning that sea level would rise a full 5 metres by the year 2100 [175]. But the scam is starting to fall apart. Climatologists now shake their heads at such views and journalists are beginning to doubt the credibility of such apocalyptic shenanigans, which have as their sole purpose generating fear for political effect (see Chapter 8). Back in 1988 Hansen declared that the sea level would rise 3 metres in 40 years and that parts of New York City would be submerged [205]. Twenty-three years later the sea level around New York has risen a mere 60 mm.

In fact, we could have turned our attention to real coastal problems and used the billions of dollars spent over decades to make regions like Bangladesh, Pakistan and Southeast Asia, which are repeatedly flooded, much safer. Hans von Storch, Director of the Helmholtz Centre for Material and Coastal Research, Geesthacht, puts it in a nutshell: 'Should we allocate huge sums of money for a CO_2 reduction so that the water level in Bangladesh rises 10 cm less in 100 years – or should we use it to help the people there more effectively by financing protective barriers?' [206].

Cyclonic times

Another spectacular side effect of global warming, according to IPCC scientists, is the alleged increase in storms and high wind events. The IPCC's 2007 Summary for Policymakers states that there are indications of increasing tropical hurricane intensity in the north Atlantic since 1970. Hurricanes (also called typhoons or cyclones) form over the oceans in the trade wind zones where the sea surface temperatures are more than 26.5° C. A uniform temperature gradient is required over a great altitude to enable evaporated water to rise and form large thunderheads. The typical giant swirl of a hurricane occurs through the Coriolis effect.

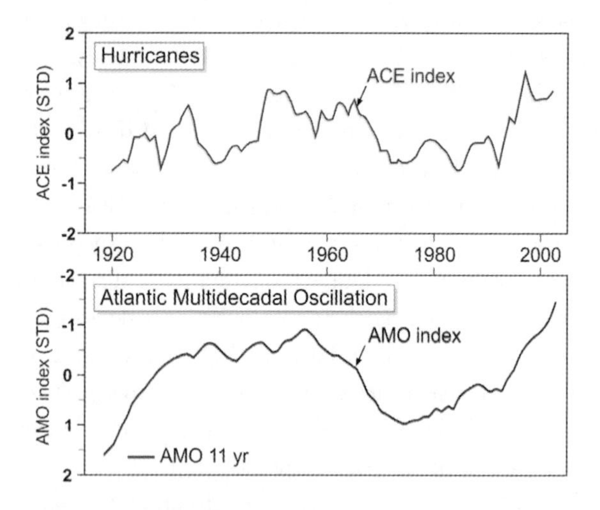

Figure 5.18 Atlantic hurricane activity (accumulated cyclone energy – ACE) over the last 60 years runs parallel to the development of the Atlantic Multidecadal Oscillation (AMO). Based on 5-year (top) and 11-year (bottom) moving averages [207].

Indeed the NOAA clearly shows there was a significant increase in the sum of hurricane strengths from 1970 to 2005 (Figure 5.18)

[208–209]. Here the NOAA adds up the energy of all hurricanes occurring in one year into the ACE (accumulated cyclone energy) index. But one notices that the cyclicality is not in line with the IPCC's climate model. Although the temperature of the sea surface between 1950 and 1970 was markedly cooler, there was strong hurricane activity during this period. Global warming alone therefore cannot be responsible. Surely the oceanic cycles (e.g. the AMO and PDO, which oscillate back and forth between warm and cold phases about every 30 years) must have something to do with it. The large synchronicity between hurricane activity and the AMO or PDO cycles is impressive [207, 210–213] (Figure 5.18). Hurricane activity has diminished considerably since 2005 and is now at levels last seen in the 1970s [214]. This has not changed since 'Hurricane Sandy', which devastated parts of the Caribbean, the mid-Atlantic and north-eastern United States during late October 2012. 'Sandy' was only at category 1 when it struck New York. There may be an additional effect from global warming, but this is not yet visible because of the impact of the internal oceanic cycles. Despite the warming, no long-term increase in tropical cyclone activity can be detected in the historical data series over the last 130 years [215–219]. Meanwhile, even climate models show a decreasing tropical cyclone trend [220].

In the past, increases were falsely postulated because historical data gaps were misinterpreted as low storm frequency, while today's complete, satellite-based data gave the impression that there is a higher storm frequency [215]. Cyclones began to be routinely observed by satellite only in 1970. It looks like even the IPCC was fooled by this false interpretation. Therefore, currently no anthropogenic influence can be discerned from the historical cyclone data. The frequency, intensity and rainfall associated with cyclones all remain within the range of natural variability [215, 221–223]. William Gray and Philip Klotzbach explained that in a warming world the temperature of the upper atmosphere develops uniformly, along with the temperature of the sea surface. If the atmospheric temperature gradient does not change, then there is 'no plausible physical reasons for believing that

Atlantic hurricane frequency or intensity will significantly change if global or Atlantic Ocean temperatures were to rise by 1–2° C' [224].

Other studies suggest that global warming even tends to curb the development of tropical storms and cyclones because the warming favours wind shear which forms in the atmosphere owing to the different wind speeds and directions. The wind shear causes rising winds to incline, and thus causes the chimney effect to collapse [225–227]. NOAA scientist Gabriel Vecchi postulates that increasing wind shear could reduce the risk of hurricanes in the Atlantic and east Pacific. Mojib Latif agrees, yet points out that future developments depend on whether 'the tropical Pacific warms up more than the tropical Atlantic. If it does, then stronger wind shear would hinder the formation of hurricanes' [228]. As a parallel process, warming also leads to water vapour content increase, which results in a slowdown in atmospheric circulation and hence the obstruction of tropical extreme wind development [229]. Meanwhile, the IPCC has back-pedalled in its draft 2013 report and has finally admitted that there are no significant observed trends in global tropical cyclone frequency.

The complicated wind

So what's the situation with the other catastrophic wind events beyond the Tropics? Today there's hardly a tornado in the United States that isn't blamed on anthropogenic climate change. As a tornado raged over Brooklyn, New York in August 2007, and led to flooding in the subways, American climate apostle James Hansen was immediately on site: 'It is fair to ask whether the human changes have altered the likelihood of such events. There the answer seems to be yes' [230]. However, the data tell a different story: strong tornadoes in the United States have declined since the mid 1970s (Figure 5.19). Between 1967 and 1977 there were on average fifty strong tornadoes a year. Since 2000 there have been, on average, only half as many. The first climate report from the Obama administration, which in general supported the IPCC position, determined: 'There are no significant changes in the frequency and strength of tornadoes since 1950' [231–232]. In May 2011 NOAA concluded that 'it is not in a position to connect the main causes of tornado outbreaks with global warming' [233]. After close scrutiny one can even find the AR4 backing this. There it is stated that tornadoes, hail, lightning and sandstorms cannot be connected to global warming. How reassuring. In the meantime there are indications that the fluctuating tornado activity is controlled much more by natural cycles. La Niña events especially are suspected because they carry cold continental air from the north of the American continent southwards until it collides with the warm, moist ocean air in the Gulf of Mexico region. The collision of cold dry air with warm humid air creates ideal tornado conditions [234].

What do things look like in Europe? Here, storm frequency in general seems to be partly decoupled from the temperature development [235–236]. Over the last 130 years storm frequency in northern and central Europe has fluctuated within a steady range. There have been two especially stormy phases during this period, both having a similar calibre. The first significant storm maximum developed in the late nineteenth to early twentieth century and lasted for a few decades [235, 237]. Beginning in 1920, despite continued warming, the storms waned and things calmed down for half a century, with the exception of a few minor interruptions. Then, beginning in 1980, the tables turned once more when the second storm maximum developed and persisted until about 2000 [235]. Currently it is relatively quiet again. Looking back over the past few hundred years, the stormiest phases in European history fell in cold phases such as the Little Ice Age [238–240]. Based on this historical relationship, storm activity may even decrease in a warming world.

As the causes for the historical storm activity in Europe are still largely unclear, it is not possible to provide a reliable prognosis. According to Jochem Marotzke of the Max Planck Institute for Meteorology, Hamburg, all computer models indicate that climate warming will have no effect on the frequency and intensity of extra-tropical storms [241]. Many models even show a possible reduction in storm frequency [242].

The climate models certainly have difficulty dealing with wind and storm frequency [243]. This is becoming obvious in the IPCC report. Here the IPCC looked at the pressure difference between Iceland and the Azores which is the decisive magnitude for the NAO and plays a dominant role in controlling the weather for western and central Europe. This pressure difference was calculated by the IPCC as fluctuating from 1900 to 1970 around a constant zero line. In 1970 the NAO then begins to rise as a more or less straight line. However, we know that this is false. The NAO fell until 1970, then rose until 1990, when it rapidly fell again. What should we make of these models? Why should we believe they can accurately forecast what will happen in the future? Even the British Hadley Centre, a proponent of the CO_2 hypothesis, is forced to admit that 'the climate models unfortunately do not yet give unequivocal predictions for the future of the NAO' [244]. So it still remains speculation by the late Theodor Landscheidt, an unfortunately mostly ignored solar scientist, who in the 1990s postulated a link between the 11-year solar cycle and the NAO [245].

Here research into the possible interplay would be urgently needed because a negative NAO brings colder and drier winters with it, and thus drives up the demand for heating energy in Europe on the one hand, and water management on the other. Norway, for example, had considerable problems meeting its demand for electric heating energy during the cold winters from 1970 until the beginning of the 1980s due to low water levels in its reservoirs. Then the NAO index flipped during the 1980s up to 1993, and so Norway's demand for heating dropped while the supply of hydropower jumped. Norway is Europe's largest hydroelectric power producer and benefits from wet winters (positive NAO). Southern Europe, on the other hand, receives less rain during positive NAOs, and this consequently brings problems for local agriculture, particularly for the wine and olive harvests.

The NAO has once again been on a downward trend since the start of the twenty-first century, and the consequences will be fewer westerly winds, drier summers, and colder and drier winters. The 2009–10 winter was the coldest in England in 30 years and 2010 had the least wind in over 100 years. It is obvious that the changes in the predominantly westerly winds have less to do with global warming than they do with the north Atlantic cycles. Even the devastating Russian forest fires referred to earlier in this chapter can be connected to the north Atlantic cycles. Thanks to the NAO, the westerly winds are driven further south over Europe, where the air is heated up and then get carried across to Russia. The most important trigger of the Moscow blazes was in any case of manmade origin -but of another kind. President Vladimir Putin 4 years earlier had abolished the state forestry administration, and with it the ability to locate and fight fires rapidly with its large fleet of light aircraft. Instead, poorly equipped private leaseholders and remote regional governments struggled to control the devastating fires [246, 153].

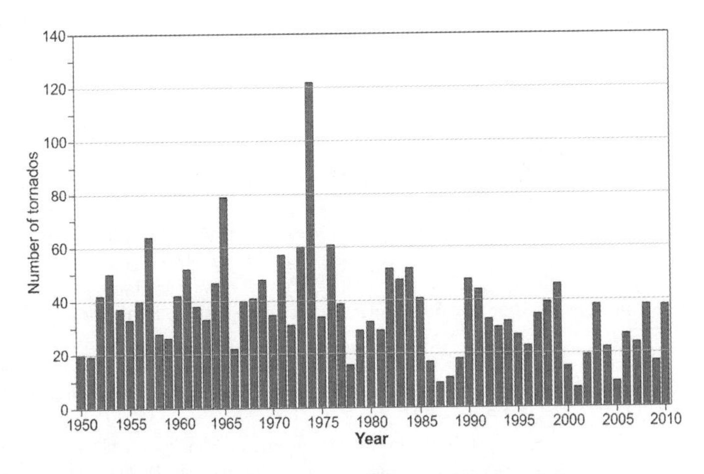

Figure 5.19 Development of US tornado activity during the last 60 years [247].

References

1. *Der Spiegel* (1974) Katastrophe auf Raten. http://www.spiegel.de/spiegel/print/d-41667249.html 33.
2. *Der Spiegel* (1986) Die Klima-Katastrophe. 33.
3. UNFCCC (2005) Kyoto Protocol. http://unfccc.int/kyoto_protocol/items/2830.php.
4. IPCC (1990) First Assessment Report. http://www.ipcc.ch/publications_and_data/publications_and_data_reports.shtml.
5. IPCC (2007) Climate Change 2007: Synthesis Report. http://www.ipcc.ch/publications_and_data/publications_and_data_reports.shtml.
6. Archer, D. and S. Rahmstorf (2010) *The Climate Crisis*, 1st edn. Cambridge University Press, Cambridge.
7. Fang, J., J. Zhu, S. Wang, C. Yue and H. Shen (2011) Global warming, human-induced carbon emissions, and their uncertainties. Science *China Earth Sciences* 54 (10), 1458–68.
8. Easterbrook, D. J. (2011) Geologic evidence of recurring climate cycles and their implications for the cause of global climate changes – the past is the key to the future. In D. J. Easterbrook (ed.), *Evidence-Based Climate Science*. Elsevier, Oxford, 3–51.
9. Klotzbach, P. J., R. A. Pielke Sr., R. A. Pielke Jr., J. R. Christy and R. T. McNider (2009) An alternative explanation for differential temperature trends at the surface and in the lower troposphere. *Journal of Geophysical Research* 114, 1–8.
10. Klotzbach, P. J., R. A. Pielke Sr., R. A. Pielke Jr., J. R. Christy and R. T. McNider (2010) Correction to 'An alternative explanation for differential temperature trends at the surface and in the lower troposphere'. *Journal of Geophysical Research* 115, 1.
11. Lomborg, B. (2008) *Cool it! Warum wir trotz Klimawandels einen kühlen Kopf bewahren sollten*. DVA, Munich.
12. Greater London Authority (2006) London's Urban Heat Island: A Summary for Decision Makers. http://static.london.gov.uk/mayor/environment/climate-change/docs/UHI_summary_report.pdf.
13. US Environmental Protection Agency (2009) Heat Island Effect. http://www.epa.gov/heatisld/about/index.htm.
14. Yang, X., Y. Hou and B. Chen (2011) Observed surface warming induced by urbanization in east China. *Journal of Geophysical Research* 116, 1–12.
15. Kim, M.-K. and S. Kim (2011) Quantitative estimates of warming by urbanization in South Korea over the past 55 years (1954–2008). *Atmospheric Environment* 45 (32), 5778–83.
16. D'Aleo, J. and A. Watts (2010) Surface Temperature Records: Policy Driven Deception? http://scienceandpublicpolicy.org/originals/policy_driven_deception.html. SPPI Original Paper, 29 January.
17. Willmot, C. J. and S. M. Robeson (1991) Influence of spatially variable instrument networks on climatic averages. *Geophysical Research Letters* 18 (2), 2249–51.
18. Long, E. R. (2010) *Contiguous U. S. Temperature Trends Using NCDC Raw and Adjusted Data for One-Per-State Rural and Urban Station Sets*. http://scienceandpublicpolicy.org/originals/temperature_trends.html. SPPI Original Paper.
19. Zhou, Y. and G. Ren (2011) Change in extreme temperature event frequency over mainland China, 1961–2008. *Climate Research* 50 (2-3), 125–39.

20. van Beelen, A. J. and A. J. van Delden (2012) Cleaner air brings better views, more sunshine and warmer summer days in the Netherlands. *Weather* 67 (1), 21–5.
21. Chaisson, E. J. (2008) Long-term global heating from energy usage. *Eos Trans.* AGU 89 (28), 253–5.
22. Mishra, V. and D. P. Lettenmaier (2011) Climatic trends in major U.S. urban areas, 1950–2009. *Geophysical Research Letters* 38, 1–8.
23. IPCC (2007) *Climate Change 2007: The Physical Science Basis: Chapter 3.2.2.2 Urban Heat Islands and Land Use Effects.* Cambridge University Press, Cambridge, and New York.
24. Fall, S., A. Watts, J. Nielsen-Gammon, E. Jones, D. Niyogi, J. Christy and R. A. Pielke Sr. (2011) Analysis of the impacts of station exposure on the U.S. Historical Climatology Network temperatures and temperature trends. *Journal of Geophysical Research* 116, 1–15.
25. Watts, A., E. Jones, S. McIntyre and J. R. Christy (in preparation) An area and distance weighted analysis of the impacts of station exposure on the U.S. Historical Climatology Network temperatures and temperature trends.
26. McKitrick, R. (2010) The Graph of Temperature vs. Number of Stations. http://www.uoguelph.ca/~rmckitri/research/nvst.html.
27. Watts, A. (2009) How Not to Measure Temperature, part 86: When in Rome, Don't Do as the Romans Do. http://wattsupwiththat.com/2009/03/28/how-not-to-measure-temperature-part-86-when-in-rome-dont-do-as-the-romans-do. Photo: Paolo Mezzasalma.
28. IPCC (2007) *Climate Change 2007: The Physical Science Basis. Kapitel 9: Understanding and Attributing Climate Change.* Cambridge University Press, Cambridge and New York.
29. IPCC (2007) *Climate Change 2007: The Physical Science Basis. Contribution of Working Group I to the Fourth Assessment Report of the Intergovernmental Panel on Climate Change.* Cambridge University Press, Cambridge and New York.
30. Lindzen, R. S. and Y.-S. Choi (2009) On the determination of climate feedbacks from ERBE data. *Geophysical Research Letters* 36, L16705.
31. Shapiro, A. I., W. Schmutz, E. Rozanov, M. Schoell, M. Haberreiter, A. V. Shapiro and S. Nyeki (2011) A new approach to long-term reconstruction of the solar irradiance leads to large historical solar forcing. *Astronomy & Astrophysics* 529, 1–8.
32. Wang, Y.-M., J. L. Lean and N. R. Sheeley Jr. (2005) Modeling the sun's magnetic field and irradiance since 1713. *The Astrophysical Journal* 625, 522–38.
33. Haigh, J. D. (2003) The effects of solar variability on the earth's climate. *Phil. Trans. R. Soc. Lond.* A 361, 95–111.
34. van Hateren, J. H. (2012) A fractal climate response function can simulate global average temperature trends of the modern era and the past millennium. *Climate Dynamics* DOI 10.1007/s00382-012-1375-3.
35. Haigh, J. D., A. R. Winning, R. Toumi and J. W. Harder (2010) An influence of solar spectral variations on radiative forcing of climate. *Nature* 467, 696–9.
36. Rawls, A. (2012) IPCC AR5 draft leaked, contains game-changing admission of enhanced solar forcing – as well as a lack of warming to match model projections, and reversal on 'extreme weather'. WUWT, 13 December. http://wattsupwiththat.com/2012/12/13/ipcc-ar5-draft-leaked-contains-game-changing-admission-of-enhanced-solar-forcing.

37. Mann, M. E., R. S. Bradley and M. K. Hughes (1998) Global-scale temperature patterns and climate forcing over the past six centuries. *Nature* 392, 779–87.
38. Mann, M. E., R. S. Bradley and M. K. Hughes (1999) Northern Hemisphere temperatures during the past millennium: inferences, uncertainties, and limitations. *Geophysical Research Letters* 26 (6), 759–62.
39. Latif, M. and N. S. Keenlyside (2011) A perspective on decadal climate variability and predictability. *Deep Sea Research II* 58 (17–18), 1880–94.
40. *Daily Mail* (2010) Glacier scientist: I knew data hadn't been verified. 24 January. http://www.dailymail.co.uk/news/article-1245636/Glacier-scientists-says-knew-data-verified.html#ixzz1N7TNzyul.
41. United States Senate Committee on Environment and Public Works (2010) 'Consensus' exposed: the CRU controversy. http://tinyurl.com/yaz8dbo.
42. IPCC (2007) Climate Change 2007: *The Physical Science Basis. FAQ 3.1 How are Temperatures on Earth Changing?* Cambridge University Press, Cambridge and New York. http://www.ipcc.ch/publications_and_data/ar4/wg1/en/faq-3-1.html.
43. Jones, P. (2010) Q&A: Professor Phil Jones 13 February. http://news.bbc.co.uk/2/hi/science/nature/8511670.stm.
44. Willson, R. C. and A. V. Mordvinov (2003) Secular total solar irradiance trend during solar cycles 21–23. *Geophysical Research Letters* 30 (5), 3/1–4.
45. Fröhlich, C. (2006) Solar irradiance variability since 1978: revision of the PMOD composite during solar cycle 21. *Space Sci. Rev.* 125, 53–65.
46. Scafetta, N. (2011) Total solar irradiance satellite composites and their phenomenological effect on climate. In D. J. Easterbrook (ed.), *Evidence-Based Climate Science*, Elsevier, Oxford, 289–316.
47. IPCC (2007) *Climate Change 2007: The Physical Science Basis. Kapitel 8.6: Climate sensitivity and Feedbacks.* Cambridge University Press, Cambridge and New York.
48. Paltridge, G., A. Arking and M. Pook (2009) Trends in middle- and upper-level tropospheric humidity from NCEP reanalysis data. *Theor Appl Climatol* 98, 351–9.
49. *Wetter Journal* (2009) Die Argumente der Klimaskeptiker I – Treibhauseffekt and Wolken. http://wetterjournal.wordpress.com/2009/05/21.
50. Trenberth, K. E., J. T. Fasullo, C. O'Dell and T. Wong (2010) Relationships between tropical sea surface temperature and top-of-atmosphere radiation. *Geophysical Research Letters* 37, 1–5.
51. Spencer, R. W. (2009) Some Comments on the Lindzen and Choi (2009) Feedback Study. http://www.drroyspencer.com/2009/11/some-comments-on-the-lindzen-and-choi-2009-feedback-study.
52. Dessler, A. E. (2011) Cloud variations and the earth's energy budget. *Geophysical Research Letters* 38, 1–3.
53. Dessler, A. E. (2010) A determination of the cloud feedback from climate variations over the past decade. *Science* 330, 1523–7.
54. Spencer, R. W. and W. D. Braswell (2011) On the misdiagnosis of surface temperature feedbacks from variations in earth's radiant energy balance. *Remote Sensing* 3, 1603–13.
55. Laken, B. A. and E. Pallé (2012) Understanding sudden changes in cloud amount: the Southern Annular Mode and South American weather fluctuations. *J. Geophys. Res.* 117 (D13), D13103.
56. Tang, Q. and G. Leng (2012) Damped summer warming accompanied with cloud cover increase over Eurasia from 1982 to 2009. *Environ. Res. Lett.* 7,

doi:10.1088/1748-9326/1087/1081/014004.

57. Sanchez-Lorenzo, A., J. Calbó and M. Wild (2012) Increasing cloud cover in the twentieth century: review and new findings in Spain. *Climate of the Past* 8, 1199–212.
58. Rahimzadeh, F., M. Pedram and M. C. Kruk (2012) An examination of the trends in sunshine hours over Iran. *Meteorological Applications*.
59. Xia, X. (2012) Significant decreasing cloud cover during 1954–2005 due to more clear-sky days and less overcast days in China and its relation to aerosol. *Annales Geophysicae* 30, 573–82.
60. Lindzen, R. S. and Y.-S. Choi (2011) On the observational determination of climate sensitivity and its implications. *Asia-Pacific J. Atmos.* Sci. 47 (4), 377–90.
61. Idso, S. B. (1998) CO_2-induced global warming: a skeptic's view of potential climate change. *Climate Research* 10, 69–82.
62. IPCC (2007) *Climate Change 2007: The Physical Science Basis*. Cambridge University Press, Cambridge and New York, Chapter 8.
63. Kirkby, J. (2010) The CLOUD Project: climate research with accelerators. *Proceedings of IPAC'10*, Kyoto, Japan. http://accelconf.web.cern.ch/AccelConf/IPAC10/talks/frymh02_talk.pdf, 4774-4778.
64. climate4you.com Climate and Clouds. http://www.climate4you.com/ClimateAndClouds.htm, data by ISCCP & HadCRUT 1983–2009.
65. US Climate Change Science Program (2006) *Temperature Trends in the Lower Atmosphere: Steps for Understanding and Reconciling Differences*, Chapter 5. http://www.climatescience.gov/Library/sap/sap1-1/finalreport/sap1-1-final-chap5.pdf.
66. Link, R. (2011) Warum die Klimamodelle des IPCC fundamental falsch sind! 25. March. http://rlrational.files.wordpress.cpm/2011/03/warum-die-klimamodelle-des-ipcc-falsch-sind.pdf.
67. Fu, Q., S. Manabe and C. M. Johanson (2011) On the warming in the tropical upper troposphere: models versus observations. *Geophysical Research Letters* 38, 1–6.
68. Veizer, J. (2009) Climate change science isn't settled. *The Australian*, 24 April. http://tinyurl.com/bt24stp.
69. Ferguson, P. R. and J. Veizer (2007) Coupling of water and carbon fluxes via the terrestrial biosphere and its significance to the earth's climate system. *Journal of Geophysical Research* 112, 1–17.
70. Veizer, J. (2005) Celestial climate driver: a perspective from 4 billion years of the carbon cycle. *Geoscience Canada* 32 (1), 13–28.
71. *The Guardian* (2010) Water vapour caused one-third of global warming in 1990s, study reveals. 29 January. http://www.guardian.co.uk/environment/2010/jan/29/water-vapour-climate-change.
72. Solomon, S., K. H. Rosenlof, R. W. Portmann, J. S. Daniel, S. M. Davis, T. J. Sanford and G.-K. Plattner (2010) Contributions of stratospheric water vapour to decadal changes in the rate of global warming. *Science* 327, 1219–23.
73. Hurst, D. F., S. J. Oltmans, H. Vömel, K. H. Rosenlof, S. M. Davis, E. A. Ray, E. G. Hall and A. F. Jordan (2011) Stratospheric water vapour trends over Boulder, Colorado: analysis of the 30-year Boulder record. *Journal of Geophysical Research* 116, 1–12.
74. Forster, P. M. d. F. and K. P. Shine (1999) Stratospheric water vapour changes as a possible contributor to observed stratospheric cooling. *Geophysical Research Letters* 26 (21), 3309–12.

75. US Climate Change Science Program (2009) Atmospheric aerosol properties and climate impacts. http://downloads.climatescience.gov/sap/sap2-3/sap2-3-final-report-all.pdf, 1–128.
76. Dentener, F., S. Kinne, T. Bond, P. Boucher, J. Cofala, S. Generoso, P. Ginoux, S. Gong, J. J. Hoelzemann, A. Ito, L. Marelli, J. E. Penner, J.-P. Putaud, C. Textor, M. Schulz, G. R. van der Werf and J. Wilson (2006) Emission of primary aerosol and precursor gases in the years 2000 and 1750 prescribed data-sets for AerCom. *Atmos. Chem. Phys.* 6, 4321–44.
77. Archer, D. and S. Rahmstorf (2010) *The Climate Crisis*, 43, 1st edn. Cambridge University Press, Cambridge.
78. Smith, S. J., R. Andres, E. Conception and J. Lurz (2004) Sulphur dioxide emissions 1850-2000. *Joint Global Change Research Institute Report PNNL-14537*, 1–13.
79. Ruckstuhl, C. and J. R. Norris (2010) Does the IPCC AR5 aerosol emission history reproduce observed trends in surface solar radiation over Europe and East Asia? *J. Geophys. Res. – Atmos.* http://meteora.ucsd.edu/~jnorris/reprints/aerosol_histories_03.pdf.
80. Kiehl, J. T. (2007) Twentieth century climate model response and climate sensitivity. *Geophysical Research Letters* 34.
81. US Climate Change Science Program (2009) Atmospheric aerosol properties and climate impacts. http://downloads.climatescience.gov/sap/sap2-3/sap2-3-final-report-all.pdf, 1–128.
82. Kaufmann, R. K., H. Kauppi, M. L. Mann and J. H. Stock (2011) Reconciling anthropogenic climate change with observed temperature 1998–2008. *PNAS* 108 (29), 11790–3.
83. Vernier, J.-P., L. W. Thomason, J.-P. Pommereau, A. Bourassa, J. Pelon, A. Garnier, A. Hauchecorne, L. Blanot, C. Trepte, D. Degenstein and F. Vargas (2011) Major influence of tropical volcanic eruptions on the stratospheric aerosol layer during the last decade. *Geophysical Research Letters* 38, 1–8.
84. Solomon, S., J. S. Daniel, R. R. Neely III, J. P. Vernier, E. G. Dutton and L. W. Thomason (2011) The persistently variable 'background' stratospheric aerosol layer and global climate change. *Science* 333 (6044), 866–70.
85. Penner, J. E., L. Xua and M. Wang (2011) Satellite methods underestimate indirect climate forcing by aerosols. *PNAS* 108 (33), 13404–8.
86. Ramanathan, V. and G. Carmichael (2008) Global and regional climate changes due to black carbon. *Nature Geoscience* 1, 221–7.
87. Greenpeace. http://www.greenpeace.de/themen/klima/nachrichten/artikel/sauber_kochen.
88. Holland, M. M., C. M. Bitz and B. Tremblay (2006) Future abrupt reductions in the summer Arctic sea ice. *Geophys. Res. Lett.* 33.
89. Jacobson, M. Z. (2006) Effects of externally-through-internally-mixed soot inclusions within clouds and precipitation on global climate. *J. Phys. Chem.* A 110, 6860–73.
90. Chakrabarty, R. K., M. A. Garro, E. M. Wilcox and H. Moosmüller (2012) Strong radiative heating due to wintertime black carbon aerosols in the Brahmaputra river valley. *Geophys. Res. Lett.* 39 (9), L09804.
91. Kondo, Y., N. Oshima, M. Kajino, R. Mikami, N. Moteki, N. Takegawa, R. L. Verma, Y. Kajii, S. Kato and A. Takami (2011) Emissions of black carbon in East Asia estimated from observations at a remote site in the East China Sea. *Journal of Geophysical Research* 116, 1–14.

92. Jacobson, M. Z. (2007) Testimony for the Hearing on Black Carbon and Global Warming. House Committee on Oversight and Government Reform, United States House of Representatives, The Honorable Henry A. Waxman, Chair, 18 October. http://www.stanford.edu/group/efmh/jacobson/0710LetHouseBC per cent201.pdf.
93. US Climate Change Science Program (2009) Atmospheric aerosol properties and climate impacts. Vol. 81. http://downloads.climatescience.gov/sap/sap2-3/sap2-3-final-report-all.pdf, 1–128.
94. Romm, J. (2010) Russian President Medvedev: 'What is happening now in our central regions is evidence of this global climate change, because we have never in our history faced such weather conditions in the past.' http://thinkprogress.org/romm/2010/08/05/206554/russia-medvedev-global-climate-change-drought-heat-wave-grain-harvest.
95. *Der Spiegel* (2010) Tritt in den Hintern. 16 August. http://www.spiegel.de/spiegel/print/d-73290108.html.
96. Dole, R., M. Hoerling, J. Perlwitz, J. Eischeid, P. Pegion, T. Zhang, X.-W. Quan, T. Xu and D. Murray (2011) Was there a basis for anticipating the 2010 Russian heat wave? *Geophysical Research Letters* 38, 1–5.
97. Kikuchi, R. (2010) Refereed papers: external forces acting on the earth's climate: an approach to understanding the complexity of climate change. *Energy & Environment* 21 (8), 953–68.
98. Hanna, E., P. Huybrechts, J. Cappelen, K. Steffen, R. C. Bales, E. Burgess, J. R. McConnell, J. Peder Steffensen, M. Van den Broeke, L. Wake, G. Bigg, M. Griffiths and D. Savas (2011) Greenland ice sheet surface mass balance 1870 to 2010 based on twentieth century reanalysis, and links with global climate forcing. *J. Geophys. Res.* 116 (D24), D24121.
99. Wu, X., M. B. Heflin, H. Schotman, B. L. A. Vermeersen, D. Dong, R. S. Gross, E. R. Ivins, A. W. Moore and S. E. Owen (2010) Simultaneous estimation of global present-day water transport and glacial isostatic adjustment. *Nature Geoscience* 3, 642–6.
100. Bromwich, D. H. and J. P. Nicolas (2010) Ice-sheet uncertainty. *Nature Geoscience* 3, 596–7.
101. van den Broeke, M., J. Bamber, J. Ettema, E. Rignot, E. Schrama, W. J. van de Berg, E. van Meijgaard, I. Velicogna and B. Wouters (2009) Partitioning recent Greenland mass loss. *Science* 326, 984–6.
102. Velicogna, I. (2009) Increasing rates of ice mass loss from the Greenland and Antarctic ice sheets revealed by GRACE. *Geophysical Research Letters* 36, 1–4.
103. Rinne, E. J., A. Shepherd, S. Palmer, M. R. van den Broeke, A. Muir, J. Ettema and D. Wingham (2011) On the recent elevation changes at the Flade Isblink ice cap, northern Greenland. *J. Geophys. Res.* 116 (F3), F03024.
104. Bjørk, A. A., K. H. Kjær, N. J. Korsgaard, S. A. Khan, K. K. Kjeldsen, C. S. Andresen, J. E. Box, N. K. Larsen and S. Funder (2012) An aerial view of 80 years of climate-related glacier fluctuations in southeast Greenland. *Nature Geoscience* 5, 427–32.
105. Zwally, H. J. and M. B. Giovinetto (2011) Overview and assessment of antarctic ice-sheet mass balance estimates: 1992–2009. *Surv Geophys*, 1–26.
106. Zwally, H. J., J. Li, J. Robbins, J. L. Saba, D. Yi, A. Brenner and D. Bromwich (2012) Mass gains of the Antarctic ice sheet exceed losses. Abstract of ISMASS (Ice-Sheet

Mass Balance and Sea Level) Workshop of SCAR Scientific Committee on Antarctic Research; 14 July, Portland, OR. http://ntrs.nasa.gov/search.jsp?R=20120013495.

107. Sime, L. C., E. W. Wolff, K. I. C. Oliver and J. C. Tindall (2009) Evidence for warmer interglacials in East Antarctic ice cores. *Nature* 462, 342–6.
108. Willerslev, E., E. Cappellini, W. Boomsma, R. Nielsen, M. B. Hebsgaard, T. B. Brand, M. Hofreiter, M. Bunce, H. N. Poinar, D. Dahl-Jensen, S. Johnsen, J. P. Steffensen, O. Bennike, J.-L. Schwenninger, R. Nathan, S. Armitage, C.-J. d. Hoog, V. Alfimov, M. Christl, J. Beer, R. Muscheler, J. Barker, M. Sharp, K. E. H. Penkman, J. Haile, P. Taberlet, M. T. P. Gilbert, A. Casoli, E. Campani and M. J. Collins (2007) Ancient biomolecules from deep ice cores reveal a forested Southern Greenland. *Science* 317, 111–14.
109. Screen, J. A., I. Simmonds and K. Keay (2011) Dramatic interannual changes of perennial Arctic sea ice linked to abnormal summer storm activity. *Journal of Geophysical Research* 116, 1–10.
110. Smedsrud, L. H., A. Sirevaag, K. Kloster, A. Sorteberg and S. Sandven (2011) Recent wind driven high sea ice export in the Fram Strait contributes to Arctic sea ice decline. *The Cryosphere Discuss.* 5, 1311–34.
111. NASA (2012) Arctic Cyclone Breaks Up Sea Ice. http://www.nasa.gov/multimedia/videogallery/index.html?media_id=152489941.
112. Zhang, J., R. Lindsay, M. Steele and A. Schweiger (2008) What drove the dramatic retreat of Arctic sea ice during summer 2007? *Geophysical Research Letters* 35, 1–5.
113. NASA (2007) NASA Examines Arctic Sea Ice Changes Leading to Record Low in 2007. http://www.nasa.gov/vision/earth/lookingatearth/quikscat-20071001.html.
114. Della-Marta, P. M., J. Luterbacher, H. v. Weissenfluh, E. Xoplaki, M. Brunet and H. Wanner (2007) Summer heat waves over western Europe 1880–2003, their relationship to large-scale forcings and predictability. *Clim Dyn* 29, 251–75.
115. Titz, S. (2008) Im Wechselbad des Klimas. Spektrum der Wissenschaft, http://www.spektrum.de/artikel/960476 8, 54-60.
116. Chylek, P., M. K. Dubey and G. Lesins (2006) Greenland warming of 1920–1930 and 1995–2005. *Geophysical Research Letters* 33, 1–5.
117. Tietsche, S., D. Notz, J. H. Jungclaus and J. Marotzke (2011) Recovery mechanisms of Arctic summer sea ice. *Geophysical Research Letters* 38, 1–4.
118. Armour, K. C., I. Eisenman, E. Blanchard-Wrigglesworth, K. E. McCusker and C. M. Bitz (2011) The reversibility of sea ice loss in a state-of-the-art climate model. *Geophys. Res. Lett.* 38 (16), L16705.
119. Polyak, L., R. B. Alley, J. T. Andrews, J. Brigham-Grette, T. M. Cronin, D. A. Darby, A. S. Dyke, J. J. Fitzpatrick, S. Funder, M. Holland, A. E. Jennings, G. H. Miller, M. O'Regan, J. Savelle, M. Serreze, K. St. John, J. W. C. White and E. Wolff (2010) History of sea ice in the Arctic. *Quaternary Science Reviews* 29, 1757–78.
120. Omstedt, A., C. Pettersen, J. Rodhe and P. Winsor (2004) Baltic Sea climate: 200 years of data on air temperature, sea level variation, ice cover, and atmospheric circulation. *Climate Research* 25, 205–16.
121. Menzies, G. (2004) *1421: Als China die Welt entdeckte.* Knaur Taschenbuch.
122. Müller, J., K. Werner, R. Stein, K. Fahl, M. Moros and E. Jansen (2012) Holocene cooling culminates in sea ice oscillations in Fram Strait. *Quaternary Science Reviews* 47, 1–14.

123. Jakobsson, M., A. Long, Ó. Ingólfsson, K. H. Kjær and R. F. Spielhagen (2010) New insights on Arctic quaternary climate variability from palaeo-records and numerical modelling. *Quaternary Science Reviews* 29, 334–58.
124. Geological Survey of Norway (2008) Less ice in the Arctic Ocean 6000–7000 years ago. http://www.ngu.no/en-gb/Aktuelt/2008/Less-ice-in-the-Arctic-Ocean-6000-7000-years-ago.
125. McKay, J. L., A. de Vernal, C. Hillaire-Marcel, C. Not, L. Polyak and D. Darby (2008) Holocene fluctuations in Arctic sea-ice cover: dinocyst-based reconstructions for the eastern Chukchi Sea. *Can. J. Earth Sci.* 45, 1377–97.
126. Funder, S., H. Goosse, H. Jepsen, E. Kaas, K. H. Kjær, N. J. Korsgaard, N. K. Larsen, H. Linderson, A. Lyså, P. Möller, J. Olsen and E. Willerslev (2011) A 10,000-year record of Arctic Ocean sea-ice variability – view from the beach. *Science* 333, 747–50.
127. Gard, G. (1993) Late quaternary coccoliths at the North Pole: evidence of ice-free conditions and rapid sedimentation in the central Arctic Ocean. *Geology* 21 (3), 227–30.
128. Turner, J., J. C. Comiso, G. J. Marshall, T. A. Lachlan-Cope, T. Bracegirdle, T. Maksym, M. P. Meredith, Z. Wang and A. Orr (2009) Non-annular atmospheric circulation change induced by stratospheric ozone depletion and its role in the recent increase of Antarctic sea ice extent. *Geophysical Research Letters* 36, 1–5.
129. Turner, J., T. Bracegirdle, T. Phillips, G. J. Marshall and J. S. Hosking (2012) An initial assessment of Antarctic sea ice extent in the CMIP5 models. *Journal of Climate.*
130. Chylek, P., C. K. Folland, G. Lesins and M. K. Dubey (2010) Twentieth century bipolar seesaw of the Arctic and Antarctic surface air temperatures. *Geophysical Research Letters* 37, 1–4.
131. Noerdlinger, P. D. and K. R. Brower (2007) The melting of floating ice raises the ocean level. *Geophys. J. Int.* 170 (1), 145–50.
132. Jenkins, A. and D. Holland (2007) Melting of floating ice and sea level rise. *Geophysical Research Letters* 34, 1–5.
133. National Snow and Ice Data Center http://nsidc.org.
134. National Snow and Ice Data Center State of the Cryosphere. http://nsidc.org/sotc/sea_ice.html.
135. Graßl, H. (2007) *Klimawandel: Was stimmt? Die wichtigsten Antworten.* (Herder).
136. Imhof, P., A. Nesje and S. U. Nussbaumer (2012) Climate and glacier fluctuations at Jostedalsbreen and Folgefonna, southwestern Norway and in the western Alps from the 'Little Ice Age' until the present: the influence of the North Atlantic Oscillation. *The Holocene* 22 (2), 235–47.
137. Chinn, T., S. Winkler, M. J. Salinger and N. Haakensen (2005) Recent glacier advances in Norway and New Zealand: a comparison of their glaciological and meteorological causes. *Geogr. Ann.* 87A (1), 141–57.
138. Oerlemans, J. (2005) Extracting a climate signal from 169 glacier records. *Science* 308, 675–7.
139. Jacob, T., J. Wahr, W. T. Pfeffer and S. Swenson (2012) Recent contributions of glaciers and ice caps to sea level rise. *Nature* 482, 514–18.
140. Hewitt, K. (2011) Glacier change, concentration, and elevation effects in the Karakoram Himalaya, Upper Indus Basin. *Mountain Research and Development* 31 (3), 188–200.

141. Bolch, T., A. Kulkarni, A. Kääb, C. Huggel, F. Paul, J. G. Cogley, H. Frey, J. S. Kargel, K. Fujita, M. Scheel, S. Bajracharya and M. Stoffel (2012) The state and fate of Himalayan glaciers. *Science* 336 (6079), 310–14.
142. IPCC (2007) *Climate Change 2007: The Physical Science Basis.* Cambridge University Press, Cambridge and New York, Chapter 4.5.2.
143. Easterbrook, D. J. The Looming Threat of Global Cooling. http://myweb.wwu.edu/dbunny/research/global/looming-threat-of-global-cooling.pdf.
144. Grove, J. M. and R. Switsur (1994) Glacial geological evidence for the medieval warm period. *Climatic Change* 26, 143–69.
145. Lowell, T. V. (2000) As climate changes, so do glaciers. *PNAS* 97 (4), 1351–4.
146. Geirsdóttir, Á., G. H. Miller, Y. Axford and S. Ólafsdóttir (2009) Holocene and latest Pleistocene climate and glacier fluctuations in Iceland. *Quaternary Science Reviews* 28, 2107–18.
147. Ivy-Ochs, S., H. Kerschner, M. Maisch, M. Christl, P. W. Kubik and C. Schlüchter (2009) Latest Pleistocene and Holocene glacier variations in the European Alps. *Quaternary Science Reviews* 28, 2137–49.
148. Goehring, B. M., J. M. Schaefer, C. Schluechter, N. A. Lifton, R. C. Finkel, A. J. T. Jull, N. Akçar and R. B. Alley (2011) The Rhone glacier was smaller than today for most of the Holocene. *Geology* 39 (7), 679–82.
149. Larsen, D. J., G. H. Miller, Á. Geirsdóttir and T. Thordarson (2011) A 3000-year varved record of glacier activity and climate change from the proglacial lake Hvítárvatn, Iceland. *Quaternary Science Reviews* 30, 2715–31.
150. Briner, J. P., N. E. Young, E. K. Thomas, H. A. M. Stewart, S. Losee and S. Truex (2011) Varve and radiocarbon dating support the rapid advance of Jakobshavn Isbræ during the Little Ice Age. *Quaternary Science Reviews* 30, 2476–86.
151. Brook, M. S., V. E. Neall, R. B. Stewart, R. C. Dykes and D. L. Birks (2011) Recognition and paleoclimatic implications of late-Holocene glaciation on Mt Taranaki, North Island, New Zealand. *The Holocene*, 1–8.
152. Rivera, A., M. Koppes, C. Bravo and J. C. Aravena (2012) Little Ice Age advance and retreat of Glaciar Jorge Montt, Chilean Patagonia, recorded in maps, air photographs and dendrochronology. *Clim. Past* 8, 403–14.
153. Davis, P. T., B. Menounos and G. Osborn (2009) Holocene and latest Pleistocene alpine glacier fluctuations: a global perspective. *Quaternary Science Reviews* 28, 2021–33.
154. Munroe, J. S., T. A. Crocker, A. M. Giesche, L. E. Rahlson, L. T. Duran, M. F. Bigl and B. J. C. Laabs (2012) A lacustrine-based neoglacial record for Glacier National Park, Montana, USA. *Quaternary Science Reviews* 53, 39–54.
155. Koch, J. and J. J. Clague (2006) Are insolation and sunspot activity the primary drivers of Holocene glacier fluctuations? *PAGES News* 14 (3), 20–1.
156. Berner, U. and H. Streif (2004) *Klimafakten: Der Rückblick – Ein Schlüssel für die Zukunft*, 4th edn. Schweizerbart, Stuttgart.
157. Young, N. E., J. P. Briner, H. A. M. Stewart, Y. Axford, B. Csatho, D. H. Rood and R. C. Finkel (2011) Response of Jakobshavn Isbræ, Greenland, to Holocene climate change. *Geology* 39 (2), 131–4.
158. Latif, M. (2008) *Bringen wir das Klima aus dem Takt? Hintergründe and Prognosen.* Fischer Taschenbuch Verlag, Frankfurt a.M.
159. Schwarzkopf, F. U. and C. W. Böning (2011) Contribution of Pacific wind stress to multi-decadal variations in upper-ocean heat content and sea level in the tropical south Indian Ocean. *Geophysical Research Letters* 38, 1–6.

160. Archer, D. and S. Rahmstorf (2010) *The Climate Crisis*. 1st edn. Cambridge University Press, Cambridge.
161. Chambers, D. P., M. A. Merrifield and R. S. Nerem (2012) Is there a 60-year oscillation in global mean sea level? *Geophys. Res. Lett.* 39 (18), L18607.
162. Henry, O., P. Prandi, W. Llovel, A. Cazenave, S. Jevrejeva, D. Stammer, B. Meyssignac and N. Koldunov (2012) Tide gauge-based sea level variations since 1950 along the Norwegian and Russian coasts of the Arctic Ocean: contribution of the steric and mass components. *J. Geophys. Res.* 117 (C6), C06023.
163. Meyssignac, B., D. Salas y Melia, M. Becker, W. Llovel and A. Cazenave (2012) Tropical Pacific spatial trend patterns in observed sea level: internal variability and/or anthropogenic signature? *Climate of the Past* 8, 787–802.
164. Church, J. A., N. J. White, L. F. Konikow, C. M. Domingues, J. G. Cogley, E. Rignot, J. M. Gregory, M. R. van den Broeke, A. J. Monaghan and I. Velicogna (2011) Revisiting the earth's sea-level and energy budgets from 1961 to 2008. *Geophysical Research Letters* 38, 1–8.
165. Jevrejeva, S., J. C. Moore, A. Grinsted and P. L. Woodworth (2008) Recent global sea level acceleration started over 200 years ago? *Geophysical Research Letters* 35, 1–4.
166. Gehrels, W. R., B. P. Horton, A. C. Kemp and D. Sivan (2011) Two millennia of sea level data: the key to predicting change. *Eos* 92 (35), 289–90.
167. Wada, Y., L. P. H. van Beek, C. M. van Kempen, J. W. T. M. Reckman, S. Vasak and M. F. P. Bierkens (2010) Global depletion of groundwater resources. *Geophysical Research Letters* 37, 1–5.
168. Konikow, L. F. (2011) Contribution of global groundwater depletion since 1900 to sea-level rise. *Geophysical Research Letters* 38, 1–5.
169. Pokhrel, Y. N., N. Hanasaki, P. J.-F. Yeh, T. J. Yamada, S. Kanae and T. Oki (2012) Model estimates of sea-level change due to anthropogenic impacts on terrestrial water storage. *Nature Geoscience* doi:10.1038/ngeo1476.
170. Engelhart, S. E., W. R. Peltier and B. P. Horton (2011) Holocene relative sea-level changes and glacial isostatic adjustment of the U.S. Atlantic coast. *Geology* 39 (8), 751–4.
171. Behre, K.-E. (2004) Die Schwankungen des mittleren Tidehochwassers an der deutschen Nordseeküste in den letzten 3000 Jahren nach archäologischen Daten. In G. Schernewski and T. Dolch (eds.), *Geographie der Meere and Küsten – Coastline Reports* 1, 1–7.
172. Grinsted, A., J. C. Moore and S. Jevrejeva (2010) Reconstructing sea level from paleo and projected temperatures 200 to 2100 AD. *Climate Dynamics* 34 (4), 461–72.
173. Herold, J. and K. E. Puls pers. comm.
174. Archer, D. and S. Rahmstorf (2010) *The Climate Crisis*. 1st edn. Cambridge University Press, Cambridge.
175. Hansen, J. E. and M. Sato (2011) paleoclimate implications for human-made climate change. In A. Berger, F. Mesinger and D. Šijači (eds.), *Climate Change at the Eve of the Second Decade of the Century: Inferences from Paleoclimate and Regional Aspects*. Proceedings of Milutin Milankovitch 130th Anniversary Symposium, Springer, New York.
176. Vermeer, M. and S. Rahmstorf (2009) Global sea level linked to global temperature. *PNAS* 106 (51), 21527–32.
177. Woodworth, P., N. J. White, S. Jevrejeva, S. J. Holgate, J. A. Church and W. R. Gehrels (2009) Evidence for the accelerations of sea level on multi-decade and century timescales. Int. J. *Climatology*, 29, 777–89.

178. Khandekar, M. L. (2009) New perspective on global warming and sea level rise: modest future rise with reduced threat. *Energy & Environment* 20 (7), 1067–74.
179. Wenzel, M. and J. Schröter (2010) Reconstruction of regional mean sea level anomalies from tide gauges using neural networks. *Journal of Geophysical Research* 115, 1–15.
180. Houston, J. R. and R. G. Dean (2011) Sea-level acceleration based on U.S. tide gauges and extensions of previous global-gauge analyses. *Journal of Coastal Research* 27 (3), 409–17.
181. Kemp, A. C., B. P. Horton, J. P. Donnelly, M. E. Mann, M. Vermeer and S. Rahmstorf (2011) Climate related sea-level variations over the past two millennia. *PNAS* 108 (27), 11017–22.
182. Potsdam-Institut für Klimafolgenforschung (2011) Meeresspiegel steigt heute schneller als je zuvor in den letzten 2000 Jahren. *Pressemitteilung,* 20 June. http://idw-online.de/pages/de/news428920.
183. Rahmstorf, S. and M. Vermeer (2011) Discussion of Houston, J. R. and Dean, R. G., 2011. Sea-Level acceleration based on U.S. tide gauges and extensions of previous global-gauge analyses. *Journal of Coastal Research*, 27(3), 409–17. *Journal of Coastal Research* 27 (4), 784–7.
184. Houston, J. R. and R. G. Dean (2011) Reply to Rahmstorf, S. and Vermeer, M., 2011. Discussion of Houston, J.R. and Dean, R.G., 2011. Sea-level acceleration based on U.S. tide gauges and extensions of previous global-gauge analyses. *Journal of Coastal Research*, 27(3), 409–417. *Journal of Coastal Research* 27 (4), 788–90.
185. Watson, P. J. (2011) Is there evidence yet of acceleration in mean sea level rise around mainland Australia? *Journal of Coastal Research* 27, 368–77.
186. Hannah, J. and R. G. Bell (2012) Regional sea level trends in New Zealand. *J. Geophys. Res.* 117 (C1), C01004.
187. Wöppelmann, G., C. Letetrel, A. Santamaria, M. N. Bouin, X. Collilieux, Z. Altamimi, S. D. P. Williams and B. M. Miguez (2009) Rates of sea-level change over the past century in a geocentric reference frame. *Geophys. Res. Lett.* 36 (12), L12607.
188. Leorri, E., A. Cearreta and G. Milne (2012) Field observations and modelling of Holocene sea-level changes in the southern Bay of Biscay: implication for understanding current rates of relative sea-level change and vertical land motion along the Atlantic coast of SW Europe. *Quaternary Science Reviews* 42, 59–73.
189. Donner, S. (2012) Sea level rise and the ongoing Battle of Tarawa. *Eos Trans. AGU* 93 (17).
190. Gehrels, W. R., S. L. Callard, P. T. Moss, W. A. Marshall, M. Blaauw, J. Hunter, J. A. Milton and M. H. Garnett (2012) Nineteenth and twentieth century sea-level changes in Tasmania and New Zealand. *Earth and Planetary Science Letters* 315–316, 94–102.
191. Albrecht, F., T. Wahl, J. Jensen and R. Weisse (2011) Determining sea level change in the German Bight. *Ocean Dynamics* 61 (12), 2037–50.
192. *Der Spiegel* (2011) Meeresspiegel-Studie entzweit Forschergemeinde. 21 June. http://www.spiegel.de/wissenschaft/natur/0,1518,769424,00.html.
193. Grinsted, A., S. Jevrejeva and J. C. Moore (2011) Comment on the subsidence adjustment applied to the Kemp et al. proxy of North Carolina relative sea level. *PNAS* 108 (40), E781–782.
194. Grinsted, A. (2011) Comment on the subsidence adjustment applied to the Kemp et al. proxy of North Carolina relative sea level. http://tinyurl.com/624sllt.

195. IPCC (2007) *Climate Change 2007: The Physical Science Basis*. Abschnitt 5.5., S. 409. Cambridge University Press and New York.
196. *Der Spiegel* (2011) Klimarat feilscht um Daten zum Meeresspiegel-Anstieg. 14 July. http://www.spiegel.de/wissenschaft/natur/0,1518,774312,00.html.
197. Ablain, M., A. Cazenave, G. Valladeau and S. Guinehut (2009) A new assessment of the error budget of global mean sea level rate estimated by satellite altimetry over 1993–2008. *Ocean Sci.* 5, 193–201.
198. Church, J. A. and N. J. White (2006) A twentieth century acceleration in global sea-level rise. *Geophysical Research Letters* 33, 1–4.
199. Gregory, J. M., N. J. White, J. A. Church, M. F. P. Bierkens, J. E. Box, M. R. van den Broeke, J. G. Cogley, X. Fettweis, E. Hanna, P. Huybrechts, L. F. Konikow, P. W. Leclercq, B. Marzeion, J. Oerlemans, M. E. Tamisiea, Y. Wada, L. M. Wake and R. S. W. van de Wal (2012) Twentieth-century global-mean sea-level rise: is the whole greater than the sum of the parts? *Journal of Climate*.
200. CU Sea Level Research Group (2011) Global Mean Sea Level Time Series (seasonal signals removed). http://sealevel.colorado.edu/content/global-mean-sea-level-time-series-seasonal-signals-removed.
201. *Der Spiegel* (2009) Der Unterwasser-Obama. 26 October. http://www.spiegel.de/spiegel/print/d-67510041.html.
202. Ballu, V., M.-N. Bouin, P. Siméoni, W. C. Crawford, S. Calmant, J.-M. Boré, T. Kanas and B. Pelletier (2011) Comparing the role of absolute sea-level rise and vertical tectonic motions in coastal flooding, Torres Islands (Vanuatu). *PNAS*, 1–4.
203. United Nations Environment Programme (2005) Pacific Island Villagers First Climate Change 'Refugees'. http://tinyurl.com/3v276ee.
204. Webb, A. P. and P. S. Kench (2010) The dynamic response of reef islands to sea-level rise: Evidence from multi-decadal analysis of island change in the Central Pacific. *Global and Planetary Change* 72, 234–46.
205. Hansen, J. (2011) Singing in the Rain. http://www.columbia.edu/~jeh1/mailings/2011/20110126_SingingInTheRain.pdf.
206. *Der Spiegel* (2003) Wir werden das wuppen. 18 August. http://www.spiegel.de/spiegel/print/d-28325115.html.
207. Hetzinger, S., M. Pfeiffer, W.-C. Dullo, N. Keenlyside, M. Latif and J. Zinke (2008) Caribbean coral tracks Atlantic Multidecadal Oscillation and past hurricane activity. *Geology* 36 (1), 11–14.
208. NOAA Climate Prediction Center The North Atlantic Hurricane Season http://www.cpc.ncep.noaa.gov/products/outlooks/background_information.shtml.
209. Vecchi, G. A. and T. R. Knutson (2008) On estimates of historical North Atlantic tropical cyclone activity. *Journal of Climate* 21, 3580–600.
210. Poore, R. Z., T. Quinn, J. Richey and J. L. Smith (2007) Cycles of hurricane landfalls on the eastern United States linked to changes in Atlantic sea-surface temperatures. In G. S. Farris, G. J. Smith, M. P. Crane, C. R. Demas, L. L. Robbins and D. L. Lavoie (eds.), *Science and the Storm – The USGS Response to the Hurricanes of 2005*. U.S. Geological Survey Circular 1306, 7–11.
211. Maue, R. N. (2011) Recent historically low global tropical cyclone activity. *Geophysical Research Letters* 38, 1–6.
212. Maue, R. N. (2009) Northern Hemisphere tropical cyclone activity. *Geophysical Research Letters* 36, 1–5.
213. Kubota, H. and J. C. L. Chan (2009) Interdecadal variability of tropical cyclone

landfall in the Philippines from 1902 to 2005. *Geophys. Res. Lett.* 36 (12), L12802.
214. Gray, W. M. (2011) Gross errors in the IPCC-AR4 report regarding past and future changes in global tropical cyclone activity – (A Nobel disgrace). SPPI Original Paper, http://scienceandpublicpolicy.org/images/stories/papers/originals/gross_errors_ipcc.pdf, 1–22.
215. Knutson, T. R., J. L. McBride, J. Chan, K. Emanuel, G. Holland, C. Landsea, I. Held, J. P. Kossin, A. K. Srivastava and M. Sugi (2010) Tropical cyclones and climate change. *Nature Geoscience* 3, 157–63.
216. Villarini, G., G. A. Vecchi, T. R. Knutson and J. A. Smith (2011) Is the recorded increase in short-duration North Atlantic tropical storms spurious? *Journal of Geophysical Research* 116, 1–11.
217. Ying, M., Y. Yang, B. Chen and W. Zhang (2011) Climatic variation of tropical cyclones affecting China during the past 50 years. *Science China Earth Sciences* 54 (8), 1226–37.
218. Chan, J. C. L. and M. Xu (2009) Inter-annual and inter-decadal variations of landfalling tropical cyclones in East Asia. Part I: time series analysis. *International Journal of Climatology* 29 (9), 1285–93.
219. Kuleshov, Y., R. Fawcett, L. Qi, B. Trewin, D. Jones, J. McBride and H. Ramsay (2010) Trends in tropical cyclones in the South Indian Ocean and the South Pacific Ocean. *J. Geophys. Res.* 115 (D1), D01101.
220. Sugi, M. and J. Yoshimura (2012) Decreasing trend of tropical cyclone frequency in 228-year high-resolution AGCM simulations. *Geophys. Res. Lett.* 39 (19), L19805.
221. Nott, J. (2011) Tropical cyclones, global climate change and the role of quaternary studies. *Journal of Quaternary Science* 26, 468-473.
222. Callaghan, J. and S. B. Power (2010) Variability and decline in the number of severe tropical cyclones making land-fall over eastern Australia since the late nineteenth century. *Climate Dynamics* 37, 647–62.
223. Mumby, P. J., R. Vitolo and D. B. Stephenson (2011) Temporal clustering of tropical cyclones and its ecosystem impacts. *PNAS* 108 (43), 17626–30.
224. Gray, W. M. and P. J. Klotzbach (2011) Have increases in CO_2 contributed to the recent large upswing in Atlantic basin major hurricanes since 1995? In D. Easterbrook (ed.), *Evidence-Based Climate Science: Data Opposing CO_2 Emissions as the Primary Source of Global Warming*, Elsevier, Oxford, 223–49.
225. Vecchi, G. A. and B. J. Soden (2007) Increased tropical Atlantic wind shear in model projections of global warming. *Geophysical Research Letters* 34, 1–5.
226. Latif, M., N. Keenlyside and J. Bader (2007) Tropical sea surface temperature, vertical wind shear, and hurricane development. *Geophysical Research Letters* 34, 1–4.
227. Villarini, G., G. A. Vecchi, T. R. Knutson, M. Zhao and J. A. Smith (2011) North Atlantic tropical storm frequency response to anthropogenic forcing: projections and sources of uncertainty. *Journal of Climate* 24, 3224–38.
228. *Der Spiegel* (2007) Klimawandel als Hurrikan-Bremse. 18 April. http://www.spiegel.de/wissenschaft/natur/0,1518,477915,00.html.
229. Gastineau, G. and B. J. Soden (2011) Evidence for a weakening of tropical surface wind extremes in response to atmospheric warming. *Geophysical Research Letters* 38, 1–5.
230. Caruso, D. B. (2007) Did global warming cause NYC tornado? *Associated Press*, 9 August. http://tinyurl.com/3k6kodv.

231. US Global Change Research Program (2009) *Global Climate Change Impacts in the United States*. Vol. 38. Cambridge University Press, New York. www.globalchange.gov/usimpacts.
232. *Die Zeit* (2011) Das extreme Tornado-Jahr 2011 ist nur ein Ausreißer. 25 May. http://www.zeit.de/wissen/umwelt/2011-05/tornados-usa-interview.
233. NOAA (2011) Preliminary Assessment of Climate Factors Contributing to the Extreme 2011 Tornadoes. http://www.esrl.noaa.gov/psd/csi/events/2011/tornadoes/climatechange.html.
234. Reuters (2011) La Niña weather pattern may be factor in more tornadoes. 23 May http://tinyurl.com/3r9m3ev.
235. Matulla, C., W. Schöner, H. Alexandersson, H. von Storch and X. L. Wang (2007) European storminess: late nineteenth century to present. *Clim Dyn* 31 (2–3), 125–30.
236. Esteves, L. S., J. J.Williams and J. M. Brown (2011) Looking for evidence of climate change impacts in the eastern Irish Sea. *Nat. Hazards Earth Syst. Sci.* 11, 1641–56.
237. Donat, M. G., D. Renggli, S. Wild, L. V. Alexander, G. C. Leckebusch and U. Ulbrich (2011) Reanalysis suggests long-term upward trends in European storminess since 1871. *Geophysical Research Letters* 38, 1–6.
238. Cunningham, A. C., M. A. J. Bakker, S. van Heteren, B. van der Valk, A. J. F. van der Spek, D. R. Schaart and J. Wallinga (2011) Extracting storm-surge data from coastal dunes for improved assessment of flood risk. *Geology* 39 (11), 1063–6.
239. Sorrel, P., M. Debret, I. Billeaud, S. L. Jaccard, J. F. McManus and B. Tessier (2012) Persistent non-solar forcing of Holocene storm dynamics in coastal sedimentary archives. *Nature Geosci* 5 (12), 892–6.
240. Sabatier, P., L. Dezileau, C. Colin, L. Briqueu, F. Bouchette, P. Martinez, G. Siani, O. Raynal and U. Von Grafenstein (2012) 7000 years of paleostorm activity in the NW Mediterranean Sea in response to Holocene climate events. *Quaternary Research* 77 (1), 1–11.
241. *Der Spiegel* (2010) Die Wolkenschieber. 29 March. http://www.spiegel.de/spiegel/0,1518,686437,00.html.
242. Sienz, F., A. Schneidereit, R. Blender, K. Fraedrich and F. Lunkeit (2010) Extreme value statistics for North Atlantic cyclones. *Tellus*, 1–14.
243. Rodgers, K. B., S. E. Mikaloff-Fletcher, D. Bianchi, C. Beaulieu, E. D. Galbraith, A. Gnanadesikan, A. G. Hogg, D. Iudicone, B. R. Lintner, T. Naegler, P. J. Reimer, J. L. Sarmiento and R. D. Slater (2011) Interhemispheric gradient of atmospheric radiocarbon reveals natural variability of Southern Ocean winds. *Clim. Past* 7, 1123–38.
244. Osborn, T. (2000) North Atlantic Oscillation. http://www.cru.uea.ac.uk/cru/info/nao.
245. Landscheidt, T. Solar Eruptions Linked to North Atlantic Oscillation. http://www.john-daly.com/theodor/solarnao.htm.
246. *Frankfurter Allgemeine Zeitung* (2010) Für viele ist Putin ein Brandstifter. 4 August. http://www.faz.net/artikel/C30721/waldbraende-in-russland-fuer-viele-ist-putin-ein-brandstifter-30295382.html.
247. National Climatic Data Center (2011) US Tornado Climatology. http://lwf.ncdc.noaa.gov/oa/climate/severeweather/tornadoes.html.

Cosmic rays and clouds: experiments and observations

Henrik Svensmark
DANISH NATIONAL SPACE INSTITUTE, TECHNICAL UNIVERSITY OF DENMARK

The most profound questions with the most surprising answers are often the simplest to frame. One is: why is the climate endlessly changing? Historical and archaeological evidence of global warming and cooling that occurred long before the Industrial Revolution, and the geological traces of far greater variations before human beings even existed, require natural explanations. Regrettably, the global sense of urgency about understanding current climate change does not extend to profound investigations of natural influences. As a result, a small group of physicists has enjoyed a near monopoly in one unfashionable but highly productive line of research for more than 15 years.

Here the surprising answer about those never-ending natural changes of climate is that galactic cosmic rays, atomic particles coming from the supernova remnants left by exploded stars, appear to play a major part. Although unseen, unfelt, generally harmless and therefore often ignored, cosmic rays are as much a part of our environment as sunlight or rainfall. About two particles go through our heads every second, but luckily the air is a good shield. When highly energetic protons and other atomic nuclei hit the earth's outer atmosphere, travelling close to the speed of light, they react with nuclei in the gas molecules to make all sorts of atomic particles, most of which are stopped high overhead. Survivors reaching the ground include neutrons, commonly counted to gauge the varying influx due

Figure 5.A.1 Cosmic ray muons activate an art show called *Cosmic Revelation* installed at the KASCADE project of the Forschungszentrum Karlsruhe. Spread across a field of 4.4 hectares, 252 white cabins contain particle detectors, and 1500-watt strobe lights connected to groups of sixteen cabins flash unpredictably as the cosmic rays descend. Artist: Tim Otto Roth. Photo M. Breig, Forschungszentrum Karlsruhe.

to solar activity or lack of it, and charged muons, which are mainly responsible for ionizing the air at low altitudes.

This ionization of the air by penetrating cosmic rays helps to form the aerosols, cloud condensation nuclei, required for water droplets to condense and create low-altitude clouds. As these clouds exert a strong cooling effect, increases or decreases in the cosmic ray influx and in cloud cover can significantly lower or raise the world's mean temperature. This is our central hypothesis and it provides a ready explanation for the frequent cooling and warming experienced by our ancestors over many thousands of years – most recently in the Medieval Warm Period, the Little Ice Age and the Modern Warm Period of the twentieth century.

Over 11,500 years, as revealed in the seabed record [1], repeated cold spells coincided with times of high cosmic ray influx, recorded by high counts of radioisotopes ^{14}C and ^{10}Be, made when cosmic rays react with terrestrial atoms. Variations in solar irradiance measured by satellites suggest that they would be too small to explain the cooling and warming. An amplifier is needed, and the hypothesis is that the cosmic rays are not just a proxy for solar activity but a causal agent. The cold spells occurred whenever a magnetically weak sun failed to repel as many of the cosmic rays coming from the galaxy as it does when it is most active.

Experimental results

Such explorations through time would be less meaningful if there was no clear evidence from experiments and observations that cosmic rays really do affect the clouds and the climate, here and now. Our experimental work has revealed the existence of a microphysical and chemical mechanism by which cosmic rays help to produce aerosols. It is a matter of ion-induced nucleation of micro-aerosols, small clusters of sulphuric acid and water molecules, which can then grow into the cloud condensation nuclei on which water droplets form.

In 2005, our team completed the world's first laboratory experiment on the role of cosmic rays in aerosol formation in the basement of the Rockefeller Complex, Juliane Maries Vej, Copenhagen [2]. We called the experiment SKY (cloud in Danish) and used natural cosmic ray muons coming through the ceiling, supplemented by gamma ray sources when we wanted to check the effect of increased ionization of the air. A 7 m^3 reaction chamber contained purified air and the trace gases that occur naturally in unpolluted air over the ocean. UV lamps mimicked the sun's rays and instruments traced the chemical events. We were surprised by how quickly the ionizing radiation worked – in a split second – suggesting that electrons released in the air act as efficient catalysts for building the micro-aerosols (Figure 5.A.2).

A small-scale (50 litre) version of the experimental chamber confirmed the SKY results at sea level, before going 1.1 km underground in the deep Boulby mine in north Yorkshire. Called SKY@Boulby, its aim was to investigate aerosol production in the absence of cosmic ray muons, which were blocked by the overlying rocks. With the equipment encased in lead, radioactivity was

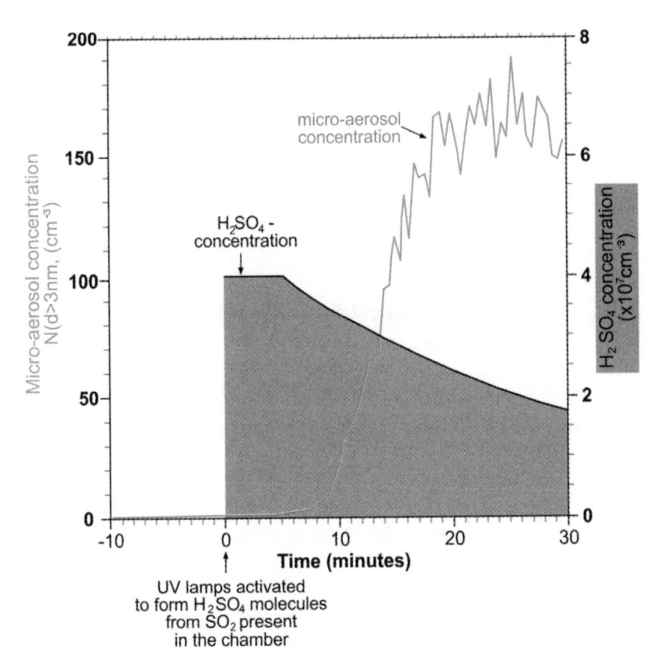

Figure 5.A.2 Formation of micro-aerosols (clusters of sulphuric acid and water molecules) in a SKY experimental chamber that is ionized by gamma rays simulating cosmic rays. The ionization of the air by gamma rays promotes the formation of micro-aerosols. Micro-aerosol formation peaks 20–30 minutes after their source material (sulphuric acid, H_2SO_4) has been provided through activating UV lamps that form sulphuric acid from sulphur dioxide (SO_2). This mimics the natural formation of sulphuric acid from sulphur compounds in the real atmosphere. Sulphuric acid concentration reaches a maximum before decaying.

effectively excluded from the experiment as well, allowing the team to investigate aerosol formation both in the absence of ionization and with controlled ionization using radioactive sources, while experimenting with various concentrations of vapours.

We then adapted the SKY@Boulby chamber for exposure to a beam of energetic electrons from the ASTRID accelerator at Aarhus University, simulating cosmic rays. Again we detected the production of micro-aerosols and were able to show that the results were no different from what we obtained using gamma rays to ionize the air [3]. Shortly after publication of the Aarhus results, independent support for our findings came in the summer of 2011.

A large international collaboration at the European Organization for Nuclear Research, CERN, announced findings of the CLOUD experiment, which also uses a beam of accelerated particles [4]. These results were very satisfactory for our hypothesis, and confirmed and extended our experimental results by showing an amplification of the formation of small aerosols by ions by a factor of 10 or more.

Although the results of both SKY and CLOUD point to the existence of a micro-physical mechanism that links ions and aerosol formation, they do not reveal how important the mechanism is in the real atmosphere. An argument advanced against an important influence of cosmic rays on clouds is that any effect may be drowned out in the real atmosphere by other sources of aerosols. So an important question is whether a significant fraction of the small of ion-induced aerosols grows to cloud-seeding size. Following our own experiments in 2005, we looked to the real atmosphere for enlightenment.

Cosmic rays, aerosols and clouds in the real atmosphere

It turns out that the question of aerosol growth and effects on low-level clouds can be answered by what one can call 'natural experiments' where solar explosions (coronal mass ejections) send out plasma clouds that shield the earth and cause the cosmic ray flux to decrease within half a day. The largest of these Forbush decreases, as they are called, cut the cosmic ray ionization by about 10–15 per cent. Obviously, one should look for an impact on the earth's aerosols and cloud cover following such events.

Our results published in 2009 [5] show clear effects in cloud properties following the sudden decrease in cosmic ray ionization. In fact, the whole chain from solar activity to cosmic rays ionization to aerosol impact to cloud properties could be observed for the stronger events. The effect is quite large – nearly 7 per cent loss in cloud liquid water averaged over the five strongest events. That translates into the loss of about 3 billion tonnes of liquid water from the sky.

The water remains there in vapour form, but unlike cloud droplets it does not get in the way of sunlight warming the ocean.

An important preliminary step was to calculate changes in ionization in the lower atmosphere due to each Forbush decrease considered, using responses in about 130 neutron monitors worldwide and in the Nagoya muon detector to compute the changes in the primary cosmic ray spectrum in the earth's vicinity. To see the effect of cosmic ray loss on aerosols in the atmosphere, we turned to the data from the solar photometers of the AERONET programme, widely scattered around the world, which record changes in the intensity of sunlight at different frequencies. With fewer fine aerosols in the lower atmosphere to scatter it, more violet light reaches the photometers at the earth's surface. As expected, in the aftermath of major Forbush decreases the AERONET data revealed falls in the relative abundance of fine aerosol particles which, in normal circumstances, could have evolved into cloud condensation nuclei.

The effects on low-altitude liquid water clouds were traced using three independent sources of satellite data. The Special Sounder Microwave Imager (SSM/I) observes changes in the cloud liquid water content (CWC) over the world's oceans. The Moderate Resolution Imaging Spectroradiometer (MODIS) on NASA's *Terra* and *Aqua* satellites gives the liquid water cloud fraction (LWCF). The International Satellite Cloud Climate Project (ISCCP) provides data on infrared detection of low clouds (below 3.2 km) over the oceans. The main results are summarized in Figure 5.A.3.

There have been a few papers on this subject that dispute our 2009 results, but on close analysis there is no contradiction. We have always emphasized that meteorological 'noise' due to the natural daily variability of the weather tends to mask the effects of relatively weak Forbush decreases and the critics have relied on weak events to deny any effect of cosmic ray variations. Indeed, the fact that the detectable impact is proportional to the magnitude of the decreases supports our hypothesis rather than refutes.

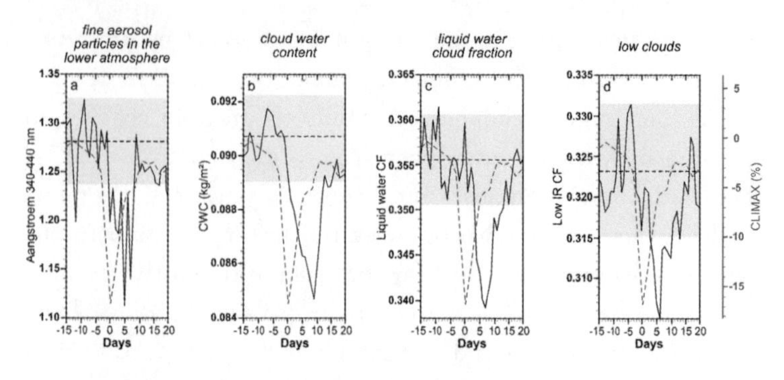

Figure 5.A.3 Short-term solar explosions typically lead to a significant decrease of the cosmic rays arriving in the earth atmosphere (Forbush decreases). Satellite data indicate that a few days after the cosmic ray decrease a reduction occurs in the aerosol concentration (AERONET), cloud water content (SSM/I), liquid water cloud fraction (MODIS) and the low cloud coverage (ISCCP). The illustrated data are averaged for the five strongest Forbush decreases that the datasets have in common between 2000 and 2005. The broken red line is the average of cosmic ray neutron counts at the Climax (Colorado) station.

Recent support for our results comes from a study that uses routinely observed changes of the diurnal temperature range (DTR) as a proxy for cloud variations [6]. Reductions in cloud cover tend to have a warming effect by day and cooling at night, thereby increasing the DTR. The advantages of this approach are first, that it is not confined to the satellite era for data on cloudiness; and second, it extends the analysis to continents (where there are most weather stations) in contrast to the primarily oceanic data used in our 2009 report. When Aleksandar Dragi and his colleagues at the Institute of Physics, Belgrade selected the thirteen strongest Forbush decreases between 1954 and 1995 (with more than 10 per cent reduction in cosmic rays) and looked at meteorological data from 184 stations, they found that in the days following the fall in cosmic rays, the DTR in Europe increased on average by 0.6° C – an impressive result.

There is, therefore, strong support for the existence and importance of the experimentally found link between cosmic ray ionization, aerosols and clouds, in response to variations in solar activity. It continues to offer the simplest explanation of why the frequent cooling and warming during the last 11,500 years have followed solar-induced variations in the cosmic ray flux, through to the Medieval Warm Period and the Little Ice Age. Even in the Modern Warm Period of the twentieth century, cosmic rays continued to have a significant impact (see Shaviv in Chapter 3).

Discussion of the solar influence on climate via cosmic rays is complicated by lesser solar effects due to changes in total irradiance or ultraviolet intensities. An unambiguous perspective comes from a totally different cause of much larger changes in the cosmic ray flux, namely the variation over millions of years in the rate of supernova explosions in the vicinity of the solar system. Strong evidence links the changing flux to the alternations between 'hothouse' and 'icehouse' conditions over geological time (see Chapter 6).

References

1. Bond, G., B. Kromer, J. Beer, R. Muscheler, M. N. Evans, W. Showers, S. Hoffmann, R. Lotti-Bond, I. Hajdas and G. Bonani (2001) Persistent solar influence on North Atlantic climate during the Holocene. *Science* 294, 2130–6.
2. Svensmark, H., J. O. P. Pedersen, N. D. Marsh and M. B. Enghoff (2007) Experimental evidence for the role of ions in particle nucleation under atmospheric conditions. *Proc. R. Soc.* A 463, 385–96.
3. Enghoff, M. B., J. O. P. Pedersen, U. I. Uggerhøj, S. M. Paling and H. Svensmark (2011) Aerosol nucleation induced by a high energy particle beam. *Geophysical Research Letters* 38, 1–4.
4. Kirkby, J., J. Curtius, J. Almeida, E. Dunne, J. Duplissy, S. Ehrhart, A. Franchin, S. Gagné, L. Ickes, A. Kürten, A. Kupc, A. Metzger, F. Riccobono, L. Rondo, S. Schobesberger, G. Tsagkogeorgas, DanielaWimmer, A. Amorim, F. Bianchi, M. Breitenlechner, A. David, J. Dommen, A. Downard, M. Ehn, R. C. Flagan, S. Haider, A. Hansel, D. Hauser, W. Jud, H. Junninen, Fabian Kreissl, A. Kvashin, A. Laaksonen, K. Lehtipalo, J. Lima, E. R. Lovejoy, V. Makhmutov, S. Mathot, J. Mikkilä, P. Minginette, S. Mogo, T. Nieminen, A. Onnela, P. Pereira, T. Petäjä, R. Schnitzhofer, J. H. Seinfeld, M. Sipilä, Y. Stozhkov, F. Stratmann, A. Tomé, J. Vanhanen, Y. Viisanen, A. Vrtala, P. E. Wagner, H. Walther, E. Weingartner, H. Wex, P. M.Winkler, K. S. Carslaw, D. R.Worsnop, U. Baltensperger and M. Kulmala (2011) Role of sulphuric acid, ammonia and galactic cosmic rays in atmospheric aerosol nucleation. *Nature* 476, 429–35.
5. Svensmark, H., T. Bondo and J. Svensmark (2009) Cosmic ray decreases affect atmospheric aerosols and clouds. *Geophysical Research Letters* 36, 1–4.
6. Dragic, A., I. Anicin, R. Banjanac, V. Udovicic, D. Jokovic, D. Maletic and J. Puzovic (2011) Forbush decreases – clouds' relation in the neutron monitor era. *Astrophys. Space Sci. Trans.* 7, 315–18.

6. The misunderstood climate amplifiers

Small causes, huge effects

At first glance the major climate drivers do not appear to change much. The energy emitted by the sun fluctuates a lowly 0.1 per cent over the course of an entire decade, and the annual anthropogenic contribution to atmospheric CO_2 concentrations is a measly 2 ppm (0.0002 per cent). Yet both climate factors have the potential to exert a significant impact on the climate. The reason is that nature has multiple amplification processes that can magnify small signal changes into major impacts. It is similar to how an electronic microphone functions. The human voice produces sound waves, which in turn produce vibrations that are converted into electrical signals in the microphone. These signals are measured in millivolts which the microphone amplifier multiplies a thousand times to a few volts. Now the signal can be processed by mixers, recorders and small speakers. Further amplification makes the speaker's voice audible in large stadiums.

In general, amplification is nothing more than a small output signal modulating and controlling another much more powerful process. Here the term positive feedback is often used. (The opposite is negative feedback, indicating a powerful signal is weakened.) In the earth's climate system, both feedback types are found and play a crucial role. Research into climate feedback systems today, however, is still very much in its infancy, and this means quantitative relationships are poorly understood and defined. This leads to great difficulties in producing reliable climate models. Despite inadequate knowledge, feedback processes are very important elements in IPCC models. And because of the highly selective and non-transparent choice and dimensioning of the feedback elements in the IPCC models, the models often come under attack from critics.

While the IPCC models take the possible feedback effects that amplify CO_2 into account, they ignore the feedback and amplification effects that magnify the sun's influence – cosmic rays or UV. Who would have guessed that, without its poorly determinable water vapour amplifier, CO_2 is only about 1° C of warming for a doubling of CO_2 concentration? The climate impact of CO_2 only skyrockets to the worrisome levels of up to 4.5° C when a substantial water vapour amplifier effect is added to the models [1]. If it turns out that the water vapour amplification is grossly exaggerated, then society will be able to breathe a sigh of relief and focus on other more pressing problems, such as hunger, disease or water shortages. The amplifier effects are therefore the key question in the scientific and political dispute over whether we are drifting towards a dramatic manmade climate catastrophe or whether we have to prepare for a natural climate change that is only influenced by man in a minor way. In this chapter we consider the most important candidates for natural climate amplification processes and discuss their potential magnitudes in the poorly understood climate equation.

The ominous water vapour amplifier

Because it absorbs a broad spectrum of long-wave radiation, water vapour is the most important natural greenhouse gas in the atmosphere [1–2] (Figure 6.1). The other greenhouse gases, among them CO_2, play only a subordinate role because water vapour absorbs a large part of the solar irradiative energy: 66–85 per cent of the natural greenhouse effect can be traced back to water vapour and small droplets in clouds.

The IPCC assumes that a small, manmade warming leads to an increased water vapour concentration in the atmosphere [1] because the capacity of air to absorb water vapour rises 7 per cent with each degree Celsius of temperature increase. In other words, warmer air holds more water vapour – and there is no shortage of water vapour because the evaporation rate also rises at higher temperatures, so more water vapour is mobilized and supplied to the atmosphere.

Temperature increases of every type could therefore be amplified multiple times due to the potential coupling with the water vapour concentration and the powerful greenhouse effect of water vapour. Because the water vapour amplifier is indifferent to what originally led to the warming, the process is not limited to warming from greenhouse gases like CO_2. Clearly, it has to apply to all climate drivers. It makes no difference whether a temperature increase is due to CO_2 or to the sun; the water vapour amplifier has the same effect for both.

The real question is whether water vapour amplification exists at all in its described manner and simplicity. Recall that the earth is characterized by complex feedback processes. Some act to dampen, while others act to amplify. In its models, the IPCC considers only the positive water vapour feedback processes with varying degrees of magnitude. The other possibilities are missing from the IPCC approach. Some experts find that it is entirely possible to have negative water vapour feedback in the interplay among climate factors [3]. According to them, more sunlight-reflecting clouds are produced as water vapour concentrations rise; thus increased water vapour could act to dampen temperatures.

A vast array of processes complicates the calculation of water vapour feedbacks:

1. It is true that the maximum possible water vapour saturation rises as temperatures rise. But if insufficient water vapour is produced – for example, over arid regions – then the water vapour concentration in the atmosphere cannot increase in these areas.
2. Warmer, moisture-saturated air masses rise, then cool, and thus lose water vapour in the form of rain and cloud formation. When the air sinks back down and heats up again, it is no longer saturated with water vapour. The strength of the water vapour feedback thus also depends on the movement of air masses [4].

3. Higher temperatures could also lead to an increase in low cloud cover via enhanced water vapour formation. Low clouds act as an effective parasol and reflect about half of the solar energy back into space. Additional earth-cooling low clouds stemming from rising temperatures could then suppress the CO_2 greenhouse warming [5].
4. Indications of an additional dampening effect were found by a research group led by former IPCC author Richard Lindzen [6]. In the western tropical Pacific they observed that high atmospheric cloud cover systematically decreased as the sea surface warmed. Clouds in the high atmosphere act as a cap for heat radiating from the earth. Fewer high clouds allow more heat to escape into space. Like a safety valve, a part of the sea surface warming could be offset by the increased thermal radiation escaping into space because the higher temperatures may lead to higher precipitation discharge of high atmosphere clouds. Because high atmosphere clouds hardly prevent the sun's radiation from penetrating, the amount of energy radiating from the sun also hardly changes. All in all, cloud and water vapour feedbacks are still poorly understood and represent a weakness in the current climate models [7–10].

CO_2 climate forcing, with and without amplification

It cannot be emphasized enough that without assuming a CO_2 amplification mechanism, only an unspectacular 1.1 ° C of warming would theoretically occur for a doubling of the atmospheric CO_2 concentration [11]. This is recognized and accepted for the most part by both sides of the debate. CO_2's alleged climatic threat comes from the IPCC assuming a massive amplification process for the climate gas. Water vapour and clouds, and snow and ice cover to a lesser extent, are alleged to be the amplification factors that act to compound CO_2's greenhouse effect. Although the feedback mechanisms are still poorly understood, the IPCC uses the amplification process with little hesitation [4]. This is the only way the IPCC is able to make

CO_2 look dangerous. In AR4 (2007) IPCC assumes that a doubling of CO_2 leads to a global temperature increase of 2–4.5° C. The 2013 draft report sticks to that range [12]. Depending on the scenario, about 50–80 per cent of this temperature rise would supposedly be caused by the amplification process alone and only 20–50 per cent would be caused by the unamplified CO_2 original temperature signal [13]. The threat from the climate scenarios comes from arbitrarily assigned massive positive feedbacks. The dependence of the climatic CO_2 effect on this amplification process is little known to the public and is not actively communicated by IPCC scientists.

Possible negative feedback scenarios such as the one postulated by Lindzen, which would dampen the warming rather than enhance it, have not been taken into account in IPCC climate models so far. According to Lindzen, a doubling of atmospheric CO_2 concentration (a rise of 100 per cent) and taking all feedbacks into account would lead to a warming of only 0.5–0.7° C [3]. That means natural dampening processes may absorb up to 0.6° C of the 1.1° C of warming for each doubling of atmospheric CO_2 concentration. In that case CO_2 could not have been the main cause of the warming that has occurred so far. Rather, CO_2 could have been responsible for less than half, perhaps a third, of the 0.8° C temperature increase we have witnessed over the last 150 years.

Further evidence for negative feedback processes have come from satellite-based measurements of the Earth Radiation Budget Experiment (ERBE) [3, 14]. The data has shown again that energy radiated by the earth increased noticeably in the Tropics whenever sea surface temperatures rose. That means a part of the warming energy escaped into outer space and thus to some extent put the brakes on the earth's temperature rise. At higher latitudes the dampening effect may be less distinct, but a strong amplification does not occur there, according to Lindzen's group (see Chapter 5).

It had to have been a political decision to keep the non-alarmist side of the CO_2–climate sensitivity spectrum out of the IPCC models. Frustrated by the IPCC's stubborn adherence to its alarmist scenarios, Lindzen resigned from the group in 2001. Is this how the IPCC consensus opinion is reached? This scientific deficiency has been a sad but recurrent theme in all the IPCC's works: inconvenient scientific results and models that show manmade greenhouse gases play a small role are either dismissed or suppressed.

Slippery and hyperactive: the water vapour development

It would be very interesting to compare the global water vapour development of the last 100 years to the temperature curve. Has water vapour concentration risen parallel to the long-term warming? If so, by how much? The results of such an analysis could indeed shed some light on the ominous water vapour amplifier. Unfortunately, that is easier said than done. Water vapour is not an easy matter to deal with [15]. It is very unevenly distributed throughout the atmosphere and involves profound differences, even in limited spaces, over both area and altitude. In addition, water vapour is in a constant state of flux. The wind blows it up mountains, where the air mass cools and a part of the water vapour is lost as clouds and rain. By the time it descends back down into the valley, the water vapour concentration has already changed dramatically.

How can one determine a sensible mean value for water vapour with so many small-scale and short-term changes, and those on a global basis? Surface stations measure useful time-series, but only for a fixed point and for a certain altitude. Weather balloons give detailed vertical altitude profiles, but only at one point and at fixed times. Satellites, on the other hand, are able to cover and measure vast regions on a regular basis, yet data points that measure many square kilometres are much too crude for small-scale, highly dynamic water vapour events. Unlike weather balloons, satellites cannot record detailed vertical profiles and form an average over wide altitude intervals. Due to the numerous problems involved, we should anticipate that lower atmospheric water vapour will remain difficult to measure in the future. Within the NASA Water Vapour Project a data set is currently being compiled which it is hoped will shed light on the water vapour development in the atmosphere. Initial results indicate that the water vapour content of the total atmosphere has slightly decreased over the past 10 years [16].

Fortunately, water vapour conditions in the stratosphere are much simpler, so reliable measurements are available. Stratospheric water vapour concentration increased significantly during 1980–2000. Then, beginning in 2000, water vapour concentration in this middle atmospheric layer decreased and remained at a widely fluctuating plateau value [17–19] (Figure 6.1). The stratospheric water vapour development shows a striking similarity to the temperature development of the lower atmosphere (troposphere) over the same period. Also, the surface temperatures increased considerably between 1980 and 2000 before levelling out, beginning in 2000. The synchronicity of the curves indicates that the climate processes of the stratosphere and troposphere may be coupled although the exact relationship remains unexplained, particularly the fundamental question of whether the warming causes the stratospheric water vapour increase effect [17]. Nevertheless, the synchronicity tells us that there are processes that connect these two atmospheric layers. This finding is especially interesting for the solar UV amplifier discussion where increased UV radiation during a solar activity maximum significantly heats up the stratospheric ozone layer. However, the temperature increase transfer to the underlying troposphere is still under investigation (see below).

Overall, there are strong indications that solar activity has an impact on the water vapour concentration of the high atmosphere. Interestingly, the specific water vapour concentration at 10 km in the upper troposphere pulsed in close synchronicity with solar activity during the past 60 years (Figure 6.1) [20].

The sun and its helpers

The solar-pulsed water vapour trend in the stratosphere is another sign that the sun plays a significant role in a host of climatic subsystems, and does so while the variation in total irradiance is only

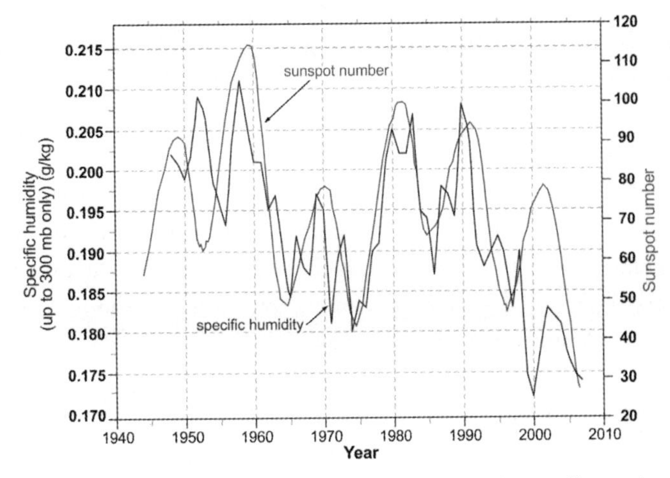

Figure 6.1 The specific water vapour concentration at 10 km fluctuated over a wide range in parallel with solar activity [20].

0.1 per cent over an 11-year solar cycle. How can that be? And why has solar activity been able to leave such a significant mark on the climate development of the last 10,000 years [21] even though the IPCC avers the sun has little influence today? One may be surprised to learn that even a research group led by one of the co-authors of the infamous hockey stick curve, Raymond Bradley, explicitly emphasizes the importance of the sun for climate development today. In 2006, Bradley's PhD student Anne Waple wrote in her dissertation that she was able to find clear indications that relatively small changes in solar radiation significantly influence the climate, over both 100-year and decadal scales [22]. So just how could a connection between the sun and the earth's climate, which would explain the observed historical synchronicity of both systems, function in detail?

First, scientific findings postulate a significantly stronger fluctuation in solar energy output over the last 1000 years [23] (see Chapter 3). This could raise the solar base signal by a factor of 6 compared to the IPCC's assumptions. Independent of this, there are also strong indications that two pre-amplification mechanisms are at work in drastically increasing the original solar signal. The processes are for the most part independent of each other and may operate in parallel. These are the UV effect in the ozone layer or stratosphere and a chained mechanism by which the strength of the solar magnetic field indirectly influences the earth's cloud cover via cosmic rays. Once the work is done by the two amplifiers, the water vapour master amplifier must be added and thus provides the final temperature result of the feedback chain. As seen with the example of stratospheric water vapour content, the sun has the ability to influence water vapour concentration.

The UV amplifier

We briefly considered the UV amplifier in Chapter 3. The UV part fluctuates within the eleven-solar cycle with a magnitude that is far greater than that of total irradiation, by a few percentage points [24–27] compared to the 0.1 per cent change for total solar irradiance. The raised UV radiation during the solar activity maximum spurs the formation of ozone at altitudes of 50–15 km. A large number of oxygen molecules (O_2) are converted into ozone (O_3) via the added UV energy input. A higher ozone concentration in turn catches more UV radiation and converts the energy into heat, which then leads to warming of the ozone layer and the stratosphere. Satellite measurements over the last few years have documented corresponding changes in ozone concentration and temperature in the stratosphere, and even to a certain extent in the ionosphere above (Figure 6.2; see Chapter 3). This part of the UV amplifier is becoming clearer. UV warms in the stratosphere. Infrared and the visible spectrum of light, on the other hand, warm the lower layers of the atmosphere.

Scientists are now looking for a process that connects the powerful stratospheric fluctuations to the tropospheric climate under 15 km [29]. Here they have identified two possible mechanisms that may be working in tandem. First, it appears that the UV warming of the ozone layer generates anomalies in the atmospheric temperature gradient, which in turn cause changes in the tropical circulation

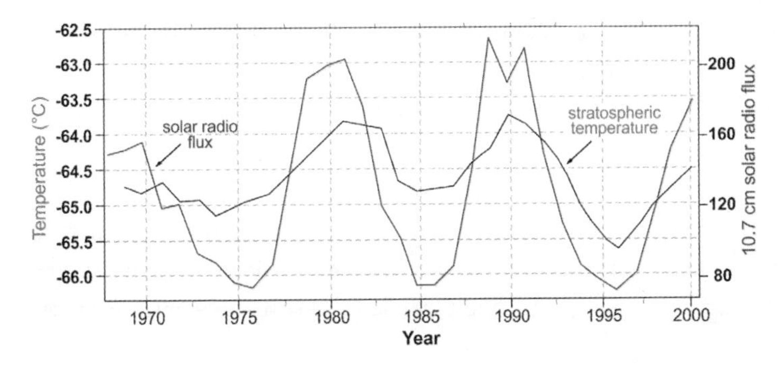

Figure 6.2 The temperature of the stratosphere fluctuates in sync with the 11-year solar cycle [28].

systems of the lower atmosphere and shifts in precipitation zones [25, 30–33]. The increased solar radiation warms the ocean water in parallel which, in connection with powerful trade winds, leads to fewer clouds in the Tropics. This in turn allows more solar radiation to reach the surface and warm the ocean [25, 34].

While the first process manufactures the connection between the stratosphere and troposphere, the second delivers an additional amplification contribution to the total effect of solar changes. Recent studies have corroborated the existence of a link between the stratosphere and climate in the troposphere. Ozone changes due to solar activity variations have been shown to affect the winds south of Greenland and in parts the Southern Hemisphere [35–36].

UV amplification effects have not been included in the IPCC models so far. Moreover, changes in the individual spectral classes of solar radiation are insufficiently differentiated by the IPCC, and hence the UV effects cannot be adequately taken into account [26].

The cosmic ray amplifier

In addition to the UV amplifier, there is another very remarkable amplification process triggered by changes in solar activity that is now the focus of research. One of the studies is taking place at the CERN facility in Geneva, where a comprehensive series of experiments is being conducted as part of the CLOUD Project [37]. The cosmic ray amplifier is based on the linkage of multiple intermediate steps. In principle the average cloud coverage of the earth changes in sync with solar activity so that the clouds act as a parasol that cools the earth, to varying degrees.

Everything starts with the sun. The sun's magnetic field fluctuates with solar activity. The solar magnetic field encompasses the solar system and thus the earth, thereby shielding it from cosmic rays from outer space. The stronger the sun's magnetic field, the more it shields the earth from the flux of cosmic rays. It is precisely these cosmic rays that appear to be the decisive switch for the earth's climate for it is suspected that cosmic rays provide part of the cloud condensation nuclei needed for forming low-level clouds for the first 3 km of the atmosphere above the earth's surface. The charged cosmic ray particles trigger the condensation of atmospheric water vapour, similar to what happens inside a condensation chamber.[11]

In a nutshell: the weaker the solar activity, the weaker is the sun's magnetic field, and so the weaker is the shield protecting the earth from cosmic rays. This means more cosmic condensation nuclei are able to penetrate the earth's atmosphere, which in turn leads to more condensation and thus to an increase in the formation of cooling clouds (Figure 6.3). It's an interwoven structure of dependent processes, yet it seems a very effective one. The 11-year solar cycle's weakly fluctuating solar irradiation is in this way overshadowed by the much more powerful fluctuating solar magnetic field. The argument that looks only at the minimal 0.1 per cent solar irradiation fluctuation is simply a stratagem to divert attention from the powerful impacts of the solar magnetic field.

The solar radiation amplifier model was developed by the Danish physicist Henrik Svensmark in cooperation with Eigil Friis-Christensen beginning in the late 1990s [38–41]. Svensmark's

[11] Here it is particularly the highest energy part of the secondary cosmic radiation, the so-called muons. They are the only particles that can penetrate to the lowest levels of the atmosphere. Clouds of the middle and upper atmosphere, on the other hand, are not influenced by cosmic rays as here there is always sufficient cosmic radiation.

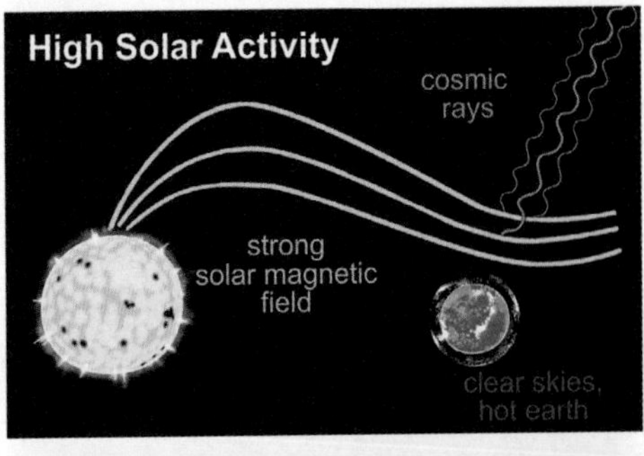

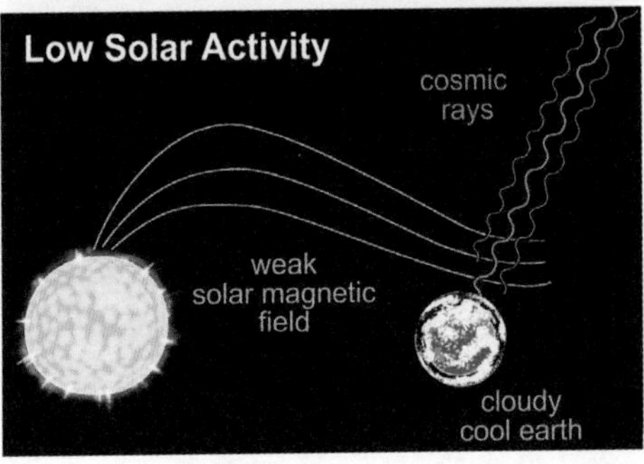

Figure 6.3 This is how Svensmark's solar amplifier model functions. Fluctuating solar activity changes the solar magnetic field, which regulates the cosmic rays reaching the earth's surface. The cosmic particles are suspected of acting as condensation nuclei for forming low clouds that cool the earth. Through this chain, the sun gains its climatic importance, as evidenced by the high level of synchronicity between temperatures and solar activity [42].

model was met by stout resistance – criticism that was less factual and professional than personal. What on earth was he thinking of, playing down dangerous CO_2? Indeed, if Svensmark's model is confirmed by further research, it would mean nothing less than CO_2 having been given a far too prominent role as a climate factor for years. Even more embarrassing, it would mean that policy makers had prematurely used the IPCC's CO_2-centric model as the basis for major decisions.

No one need wonder why Svensmark's model has not found its place in any of the IPCC climate models. It is a major threat to the IPCC's overall thrust. The mechanism was mentioned briefly in AR4, but was promptly dismissed as improbable. It's yet another example of the IPCC limiting the scientific spectrum of possibilities at an inconvenient place, despite mounting scientific evidence. The Svensmark amplification model may be a decisive key that explains how the sun has been able to impact the earth's climate so profoundly in the past, even though the primary solar irradiance variability appears at first glance to be small. We shall now explain the individual elements of the cosmic ray amplifier in more detail and discuss the chain of evidence.

The sun's magnetic field

The starting point of Svensmark's amplification process is the sun's magnetic field. As for the dynamo principle, the movement of ionized particles in the sun's interior generates electric currents that lead to the formation of a powerful magnetic field [43–44]. The magnetic field on the sun's surface is about twice as strong as the earth's magnetic field. The strength of the sun's magnetic field is therefore closely coupled to the sun's activity. During maximum solar activity, the sun's magnetic field strengthens, and when solar activity subsides, the sun's magnetic field weakens. Not only do solar irradiation and the number of sunspots change during the 11-year cycle, but so does the sun's magnetic field (Figure 6.3).

The magnetic field of a quiet sun more or less corresponds to a dipole field, which reverses polarity about every 11 years. After approximately 22 years, the original alignment is restored. This is also the cause of the 11- and 22-year solar cycles (the Schwabe and Hale cycles). During turbulent phases, the solar magnetic field is strongly distorted and takes on a very complex configuration. The gigantic solar magnetic field expands well beyond the orbit of Pluto, past the solar system's outermost planets some 7 billion km away.

The sun rotates about its own axis. And because it consists exclusively of mobile gases and plasma, the sun needs only twenty-five days for one revolution at the equator, while the polar areas need ten days longer to complete a full revolution. Because of the sun's rotation, the magnetic field lines extend through interplanetary space as a giant spiral, the so-called Parker spiral. Charged cosmic ray particles normally follow the solar magnetic field lines as they travel through the solar system.

During an 11-year solar cycle, the strength of the sun's magnetic field varies by a factor of two [45–46]. But even more decisive is the observation that the mean strength of the solar magnetic field more than doubled between 1901 and 1995 [47–48]. Over this period cosmic rays striking the atmosphere became ever more effectively batted away from the earth, and so fewer cooling-type clouds were produced. This magnetic trend coincides with the main phase of the modern climate warming. This is little known among the public and is certainly not brought into the discussion by the proponents of the IPCC CO_2-centric models, unless demanded. So it comes down to one question: How much does each factor contribute to warming? We have two possible climatic control factors that have significantly increased over the last 100 years or so: CO_2 concentration and the solar magnetic field (which is a measure of solar activity) (Figure 3.8). Looking at the overall trend, either one alone could be responsible for global warming. However, it is more likely that it is a combination of the two, which now can be narrowed down using scientific methods. The IPCC currently assumes that the trigger for the warming is about 95 per cent manmade CO_2 and 5 per cent sun. The historical synchronicity between solar activity and the climate development during pre-industrial times, however, strongly suggests that the sun plays a much greater role in the climate equation.

An increase in solar activity can be observed in a number of ways, and hence it can be determined by measuring several parameters. In addition to a strengthening solar magnetic field, the number of sunspots also increases and the radiation striking the top of the earth's atmosphere intensifies. All this can be measured by satellites. Moreover, historical cosmic ray intensity can be measured in ice and sediment cores, and they show a decrease during phases of strong solar activity. All these solar indicators for the most part run parallel to each other. Nevertheless, there are some interesting deviations which indicate that each of these magnitudes reacts slightly differently to solar activity fluctuations. Therefore, reconstructing solar activity using sunspot observations, solar magnetic field measurements or other parameters yields slightly different results at detailed levels, and thus they have to be taken into account when using different methods and when comparing data [28, 49].

Notable here is the discrepancy surrounding solar cycle 19 around 1960, when sunspots reached their highest value of the twentieth century. However, this was not the case for the solar magnetic field, which reached its maximum during 1980–2000 (Figure 6.4).

This is especially interesting because we see that the main 1980–2000 warming period occurred precisely during the solar magnetic field maximum. Perhaps this is an indication that the solar magnetic field plays a far greater role than changes in the primary electromagnetic solar irradiation output. CO_2-based climate models are unable to reproduce the cooling period from 1945 to 1975 properly. Up to now the strong solar cycle 19 has been used as evidence to show that higher solar irradiation contradicts the cooling of this period. But if the IPCC had considered the impact of the sun, it would have noticed the significance of the solar magnetic field. But then it would not have been able to load extreme CO_2 impacts into the models. Ultimately, policy makers would have been forced to address how to protect the planet from the vagaries of a sometimes 'hot' and sometimes 'cold' sun.

Due to its possible climatic significance, it is clear that more precise data on the development of the solar magnetic field are necessary. Nevertheless, direct measurement of the solar magnetic field at the surface of the sun is complicated. The joint *Ulysses* probe of the European Space Organization and NASA orbited the sun over the poles and measured the magnetic field in the vicinity of the sun from 1994 to 2009. However, *Ulysses* did not have any instruments to record directly the magnetic field at the sun's surface. This is one

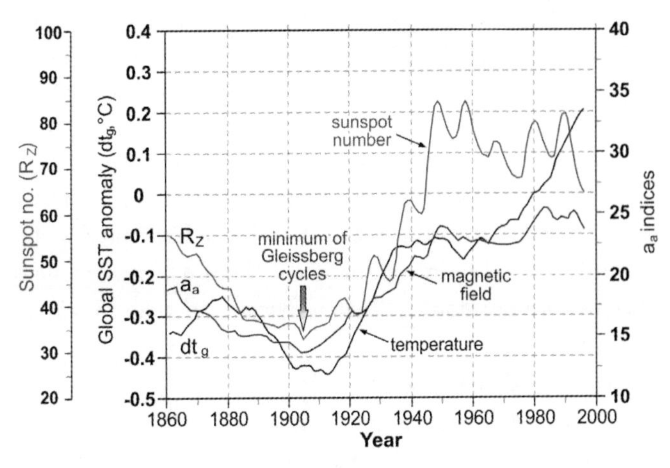

Figure 6.4 The temperature curve follows solar activity over the last 150 years with remarkable synchronicity. The sunspot record, however, shows a discrepancy in the 1950s and 1960s. But if the solar magnetic field is used as a measure for solar activity, then once again a good agreement with temperature is observed, possibly indicating the importance of the solar magnetic field (represented here by the aa proxy) in this process. The 11-year solar cycle is calculated using a 23-year mean smoothing [50].

reason why the European Space Organization plans to launch the 200 million euro *Solar Orbiter* space probe in 2017 to measure the magnetic field of the sun's surface at close quarters and in detail. Measurements taken by the probe may improve our understanding of the relationship between the sun and climate.

The earth's magnetic field as a junior partner

Like the solar magnetic field, the earth's magnetic field acts as a shield against cosmic rays [51]. A change in the earth's magnetic field, therefore, has the potential to influence the climate. A study by a French team appears to confirm this [52]. Over the course of thousands of years, several 100-year phases occurred where the dipole axis of the earth's magnetic field tilted. Characteristic climate changes in Europe (i.e. glacial advances in the Swiss Alps) occurred precisely at these times.

It is very little known that the magnetic poles are in constant movement. A thousand years ago the magnetic pole was located over the Siberian archipelago of Severnaya Zemlya; today it is located in northern Greenland [53]. And the magnetic pole continues to wander – every day it moves another 90 metres. Because charged cosmic particles first reach the earth at the magnetic poles, the shifting of the magnetic pole should lead to local changes in incoming cosmic rays in the Arctic and Antarctica, as well as at the middle latitudes. These effects need to be taken into account and deducted when it comes to global climate observation and solar activity fluctuations [53–57].

Cosmic rays as a climate factor

The vast expanses of space are constantly flooded by electrically charged particles that have their origins in rotating neutron stars, supernovas and black holes [58]. This is what we call cosmic rays or cosmic radiation. Only a tiny part of the particle stream is contributed by the sun [54]. Excluding the sun's share, the remainder is known as galactic cosmic rays. The energy that is transported by cosmic rays to the earth's atmosphere is negligible; however, cosmic rays are the main source of ionization in the troposphere and the lower stratosphere, and thus play an important role in the formation of clouds that cool the earth [59].

Even though the solar magnetic field provides a protective shield as they journey through space, some cosmic rays collide with the earth. However, the stronger the sun's magnetic field, the fewer the number of cosmic particles that succeed in penetrating deep into the earth's atmosphere. Because of the solar magnetic field's great influence on the number of cosmic rays reaching the earth, the latter is often used as a measure of the sun's activity.

The atmosphere forms the earth's second line of defence against cosmic rays. Here the cosmic particles crash into atmospheric particles and are thus transformed [54, 60]. The greater the density of the atmosphere, or atmospheric pressure, the greater the number of particle collisions that take place, and so the more the cosmic rays are weakened.

Cosmic rays in sync with the sun's activity

The strength of the cosmic rays reaching the earth has been measured globally by a series of neutron monitor stations since the International Geophysical Year in 1957 [61]. The development of cosmic rays is characterized by the 11-year solar cycle (the Schwabe cycle) [58]. Nevertheless, the development of cosmic radiation lags behind solar activity by 3–10 months [58], resulting in a slight delay mechanism.

Using neutron monitors in addition to the 11-year cycle, we can recognize another long-term trend in the cosmic ray curves. Using the Kiel station as an example, we see that the cosmic rays during the cold phase of the 1970s was extraordinarily strong and that it dropped successively from one Schwabe cycle to the next during the subsequent 1980–2000 warming period [62]. The minimum of each successive cycle was weaker than the one before (Figure 6.5). Did the decline in cosmic rays during this period lead to fewer clouds and hence warmer temperatures? Was it pure coincidence or an indication that there is something behind the Svensmark Mechanism which has attracted so much criticism in the past? During the first decade of the twenty-first century cosmic rays in Kiel increased again. And around the same time the global mean temperature stopped rising and has been static at a plateau ever since. More cosmic rays means more low clouds, which means less solar radiation reaching the earth's surface, which leads to lower temperatures. This development fits well with Svensmark's theory. The trends of the neutron station at Kiel are not just a flash in the pan. Similar cosmic radiation developments have been observed worldwide, from Magadan in Siberia to Oulu in Finland.

The first measurements of cosmic rays were taken in 1957. Before then the reconstruction of cosmic rays was done by measuring cosmogenic nuclides in ice cores, ice-rafted debris, dripstones and tree rings. As described earlier, cosmogenic nuclides are generated when converted cosmic rays collide in the atmosphere and find their way to the earth's surface, and hence into the natural climate archives. Especially useful are the ^{14}C of carbon and ^{10}Be of beryllium isotopes,

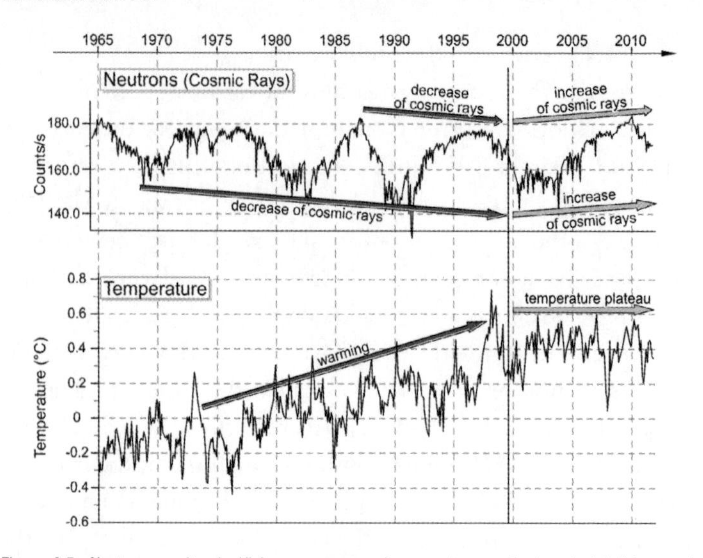

Figure 6.5 Neutron monitor in Kiel, as a measure for cosmic rays (top) and global temperature development since 1965. Cosmic rays declined during the 1977–2000 warming period. When the warming ended in 2000 cosmic radiation increased. Sources: neutron monitor Kiel [63]; temperature, HadCRUT3 data series.

which are produced by reactions between the secondary cosmic rays and nitrogen and oxygen [54, 64]. In practice the cosmic ray curve reconstructed from 10Be lags behind solar activity by about 2 years [54]. That fits rather well with the time lag of the modern cosmic ray measurements, whereby cosmogenic nuclides finally reach the earth's surface after wandering in the atmosphere for about a year.

The cosmogenic nuclides form a well-accepted proxy for the sun's activity [65]. The calculated solar activity curve for the most part runs surprisingly well with the temperature development of the last 10,000 years (see Chapter 3). During the main warming phase of the past 150 years, cosmic ray intensity declined over the long term by about 20 per cent (Figure 6.6) [48, 62, 66–68].

The good agreement between the development of cosmic rays and temperature can have only two causes. On the one hand, cosmic rays could be a passive indicator of the power of solar irradiation. In

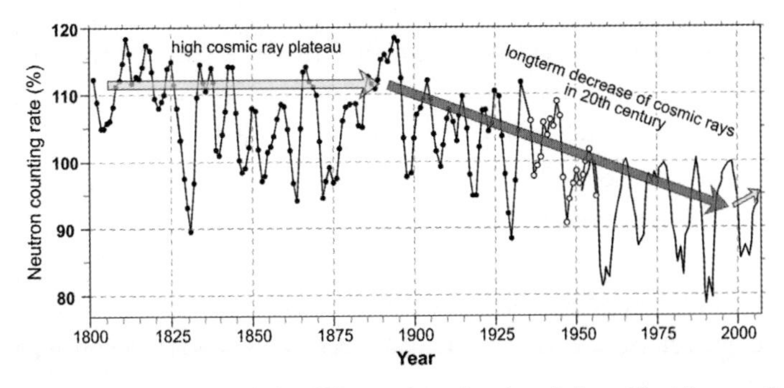

Figure 6.6 Cosmic rays over the last 200 years show a long-term decline, while at the same time global temperatures climb [48].

this case climate change would be traceable only to solar irradiance changes, without the involvement of any cosmic ray amplification mechanism. On the other, the cosmic rays may be a decisive factor controlling the temperature signal largely by influencing low cloud cover formation. The relatively small changes in the sun's electromagnetic irradiance output would then no longer play a major role. Our view is that the amplification effects of cosmic rays and other mechanisms are the most important climate changers. Determining the precise extent of each contribution remains the task of scientific research in the years ahead [54, 69].

An important piece of the puzzle is found deep in the earth's history. Over shorter timescales, the intensity of cosmic radiation is for the most part regulated by solar activity. But over timescales of hundreds of millions of years, the dosage of cosmic rays coming from space varies. Depending on whether or not the solar system traverses a spiral arm of the Milky Way, or if it is traveling through a relatively empty place between the spiral arms, the available primary cosmic ray intensity also changes. The crossing of a Milky Way spiral arm occurs about every 135 million years. Nir Shaviv and Jan Veizer found changes in cosmic radiation following this rhythm as well as largely synchronous long-term climate changes [45, 70–71]. Whenever our solar system traversed a spiral arm and thus was exposed to large doses of cosmic radiation, dramatic cooling occurred on earth.

This is more evidence indicating that the climatic impact of cosmic rays has been grossly underestimated in the past [52, 72]. It is becoming increasingly apparent that the IPCC should have included climate models based on cosmic radiation in its spectrum of possibilities, even if all processes are not yet well understood.

Svensmark's model comes under scrutiny

The model developed by Henrik Svensmark and his colleagues describing climate amplification via cosmic rays began, like most models, in a simple version. Following criticism, scientific discussion and subsequent research the model was then incrementally transformed, refined and adjusted. That's how science is supposed to work. A good example of an important development stage in this respect was the discussion of the so-called Laschamp Event, a short-term weak phase of the earth's magnetic field that occurred 40,000 years ago when cosmic rays were able to penetrate the earth's atmosphere with greater intensity for a few thousand years [73–74]. This led to an abrupt increase in cosmogenic ^{10}Be nuclides showing up in the natural climate archives. More cosmic rays? According to Svensmark's model, it should have led to more clouds and more cooling. However, a temperature decline in connection with the Laschamp Event did not occur, which appeared to cast serious doubt on the Svensmark effect [75]. The simple basic model obviously needed revision. The key to the puzzle had to have something to do with the earth's magnetic field because this was the only thing that changed during the Laschamp Event and not the much stronger solar magnetic field, which is normally the main driver of the cosmic ray amplifier. It was therefore time to take a closer look at the composition of cosmic rays and to determine which particles are responsible for forming low cloud cover.

And Svensmark found it [47, 76]. Foremost are the highest energy cosmic particles that have enough power to penetrate through to the lower layers of the atmosphere and to deliver condensation nuclei for forming low clouds. Because of their high energy, these fast particles cannot be halted by the earth's weak magnetic field and can only be influenced by the more powerful solar magnetic field. Typical cosmogenic nuclides such as ^{10}Be, which are usually used for the reconstruction of cosmic ray intensity, are however created mainly by the low-energy component of cosmic rays. Because of their lower penetration power, the low-energy cosmic particles are stopped in the upper atmospheric layers, which is where most ^{10}Be production plays out. The crux is that the earth's magnetic field is able to influence the low-energy cosmic radiation, which explains the ^{10}Be anomaly in the climate archives. However, because the process plays out in the upper atmospheric layers, the low-energy cosmic particles play an insignificant role in the low atmosphere climate-impacting cloud formation. This explains why temperatures during the Laschamp Event did not change. The activity of the solar magnetic field is thus vital because that is what decides whether the climatically relevant high-energy cosmic ray particles are shielded from the earth or allowed to pass through the atmosphere.

To develop Svensmark's amplifier model further, new high-resolution measurement data sets of cosmic radiation and atmospheric aerosol concentration are urgently needed. These data are neither easy nor inexpensive to obtain. A satellite costs about 300 million euros (excluding start-up costs). NASA invested this amount in its *Glory* satellite, which was intended to measure solar activity and aerosols over a period of years. However, the mission ended 5 minutes 17 seconds after launch when the rocket carrying the satellite developed a malfunction and crashed, taking the satellite down with it into the Pacific Ocean. We really could have used the data.

However, things improved a few months later with the start of the AMS-02 magnet-spectrometer. This sophisticated 1.4 billion euro particle detector was delivered without a hitch by the US shuttle Endeavour to its destination on the ISS International Space Station in May 2011. The instruments will measure the strength and composition of cosmic rays for the next 18 years.

A parasol of clouds for the earth

The last link in the Svensmark solar amplifier are clouds, which are the real climate regulator. Whatever controls the clouds, rules the climate. The objective, therefore, is to determine if cloud cover changed with the upstream-connected steps of the Svensmark chain, in other words, whether solar activity, cosmic rays and cloud cover oscillated in parallel in the past. If the relationship were to be confirmed, then the next step would be to compare it to temperature. If a synchronicity does exist, this would be followed by determining how to build the effect into the climate models. That's the general roadmap. Note that nobody is proposing that the CO_2 model be replaced. What is suggested is that the CO_2 model be supplemented with the missing climate drivers.

The impact of clouds as a climate factor is not disputed and is relatively easy to imagine. As soon as a cloud blocks out the sun, it becomes noticeably cooler. Clouds form a giant parasol and block out 30 watts of irradiative energy from the earth per square metre. That's hardly small change. Even a few percentage variation in cloud cover can produce a change in the earth's energy budget that corresponds in magnitude to the IPCC's projected effect of manmade CO_2 gases emitted since 1750, which is pegged at 2.63 W/m^2 [131, 54]. Here low clouds are responsible for about half of the solar radiation that is reflected back into space when it reaches the earth's atmosphere [5]. Changes in low cloud cover thus appear to be highly climate-relevant [59, 76]. On the other hand, clouds in the upper levels of the atmosphere appear to be transparent for the most part and so allow the sun's radiation to pass through [6]. These high-level clouds then act as insulation and trap the heat of infrared radiation that is emitted by the earth. In summary, more high clouds produce warming while low clouds produce cooling.

Cloud condensation nuclei from cosmic radiation

More than 50 years ago scientists suspected that cosmic rays could impact the weather on earth through cloud formation [77–78].

But just how could the connection between cosmic particles and clouds function in detail? What has to take place for cosmic rays to create condensation nuclei and produce cooling water droplets and cloud layers?

The primary cosmic rays arriving from space consist mainly of protons, supplemented by electrons and fully ionized atoms. Even though a huge number of particles enter the atmosphere, they are much too small to act as effective condensation nuclei. But as we have seen, this is only the start of a complex particle cascade mechanism. Through multiple collisions with atmospheric components, particles of the so-called secondary cosmic rays are created; these are known as cosmogenic nuclides. Are larger particles suitable for condensation nuclei perhaps also created in this process? We shall now explore this in more depth.

The relationship between airborne particles (aerosols) and cloud formation is one of the unknowns in current climate models [79]. What appears to be very simple in cloud chamber experiments is far more complicated under atmospheric conditions. The over-saturation of water vapour in a laboratory experiment is much higher than what we find in the natural environment. In a cloud chamber traces of vapour form along the cosmic ray paths, but in nature an entire series of fundamental requirements has to be fulfilled before a similar effect takes place [78].

Ion clusters and charged aerosols as cloud magnets

Currently two main mechanisms are being investigated that may explain the interplay between cosmic rays and cloud formation [80]. The first involves the agglomeration of many small charged cosmic particles from which a larger particle grows. These resulting ion clusters form ideal condensation nuclei for clouds [40, 78, 80–84]. In the second mechanism, cosmic rays assist in the electrical charging of other aerosols, which as a result become useful condensation nuclei for cloud formation. Cloud droplets preferentially form around charged particles and hence clouds are more electrically charged than their surroundings. The charge differences at cloud edges with respect to their vicinity lead to a charging of aerosols located in the immediate vicinity of the cloud, which in turn leads to the formation of new condensation nuclei for more cloud formation. The formation of condensation nuclei progresses more intensely the greater the charge difference to the cloud edge. Here cosmic rays take effect because they influence electrical conductivity and the electric current regime in the atmosphere. Strong cosmic rays lead to greater charge differences and hence to more condensation nuclei and more cooling clouds [59, 78, 80–81, 85–86]. Currently, it is unclear which of the two mechanisms is leading the race [80]. It is even possible that both processes are active and reinforce each other, according to the ambient conditions. Other condensation nuclei scenarios are under research [87–89].

The cloudless alternative

Cosmic rays and other aerosols are not only of interest for the formation of condensation nuclei for clouds. They can also absorb and diffuse solar radiation. Through this shielding effect, the amount of solar energy reaching the earth's surface is reduced and hence results in a cooling effect. Just how much cooling is caused by this process is uncertain [90–91]. Their role is unknown in current IPCC models. Moreover, cosmic rays too may play a part. Werner Weber of Dortmund suspects that cosmic rays form oxygen molecule ions to which water droplets can attach (see Weber below). These water droplets then diffuse sunlight and weaken it at the earth's surface [92]. Weber's cosmic ray amplifier does not need a cloud effect and makes up another variant in climate influence via solar-dependent changes to cosmic radiation. Another study group has recently found that reactions activated by light can lead to the growth of particles in the atmosphere, a process that has not been included in climate models [93]. Yet another study suggests that cosmic rays may affect the ozone concentration in the stratosphere, which by way of water vapour affects climate in the lower atmosphere and on the earth's surface [94].

The inconvenient CLOUD Project

There is no shortage of suggestions when it comes to the climate impact of cosmic rays, as we have seen. More research is necessary in order to reproduce the observed synchronicity between cosmic rays and climate values in the climate models [95]. At the end of the 1990s, inspired by Svensmark, Jasper Kirkby of the European Organization for Nuclear Research (CERN), Geneva suggested a series of extensive experiments in order to cast light on the subject [96]. Within the framework of the CLOUD Project, he wanted to build a modern version of the cloud chamber. A particle stream from the CERN Proton Synchrotron would reproduce cosmic rays, which in turn would be shot into a 3-metre diameter cylindrical chamber containing various mixes of atmospheric gases. After firing the particles into the chamber, it would be investigated whether aerosols had formed and if they could serve as condensation nuclei for clouds. The project proposal received strong support from the scientific community. At last, the gigantic CERN apparatus would be used for something terrestrial and practical, and not only for investigating abstract subatomic particles that have a lifetime of fractions of a second.

However, the appointed project reviewers were less enthusiastic. One believed it was not possible to replicate the atmosphere inside a small cylindrical chamber; the other had very different worries. The CERN Nuclear Research Centre is very expensive to operate and is funded in full by its twenty member states. With the CLOUD Project, CERN would be researching a physical effect that could eventually marginalize the anthropogenic contribution to climate change. Kirkby had made this clear when he submitted his application. At this point many of the governments funding CERN were already fully behind the IPCC's CO_2 stance. Inconvenient scientific doubts were no longer welcome in the politically charged debate surrounding climate change, and it was feared that CERN's reputation and flow of funding might be jeopardized [35].

Hence the project was rejected – a real defeat for scientific freedom and open-mindedness, and a victory for research politics. But Kirkby didn't give up. Instead, he stepped up his efforts to convince his peers, and eventually succeeded. Delayed by years, he finally was able to convince CERN's leadership of the scientific importance of the CLOUD Project in 2005. In 2006, in a pilot experiment, the formation of particles in the experimentation chamber could be shown [97] and thus CLOUD finally began in 2009 [98–99]. Today a team of physicists from eighteen institutes in nine countries is attempting to solve the big puzzle behind cloud formation through cosmic rays. The first report of partial aspects of the project appeared in the journal Nature in mid 2011. The experiments showed that up to ten times more aerosol particles formed inside a chamber fired at with simulated cosmic rays than inside a neutral, uninfluenced chamber [100–101]. For the next step, scientists will attempt to discover whether the small particles can form larger nuclei that serve as condensation nuclei for cloud formation. The results are due in 2013–14.

Measuring clouds

So far we have been able to confirm the first stages of the Svensmark amplifier:

1. Changes in solar activity cause strong fluctuations in the solar magnetic field, which are much more powerful than changes in solar irradiance.
2. The cosmic radiation reaching the earth is for the most part regulated by the solar magnetic field; that is generally accepted by scientists. Secondary products from cosmic rays – so-called cosmogenic isotopes – are generally employed to approximate solar activity.
3. Two fundamental physical models are being investigated which may explain how the low cloud cover is regulated via cosmic rays. Comprehensive laboratory experiments within the scope of the CLOUD Project are currently testing the fundamental theoretical assumptions. Now everything comes down to the last

stage of the Svensmark amplifier, namely whether cloud cover really does change in sync with solar activity. Can historical measurement data answer this?

While there are numerous measurements and historical reconstructions of solar activity, when it comes to cloud cover the situation is less than desirable. It wasn't until the satellite age that it became possible to record cloud cover over vast areas of the globe. Before that, the records were limited to a handful of recording stations. Clouds can be categorized under different altitudes, which in turn exert different impacts on the climate. The cooling clouds of the Svensmark amplifier are found at the lower atmospheric levels below the 3.2 km altitude. Despite modern satellite technology, it remains something of a challenge to decipher the extent of cloud cover for the various atmospheric altitudes. If layers of clouds are stacked one on top of another, then the lower layers are hidden from the satellite sensors. Likewise, upper cloud layers cannot be detected by sensors on the earth's surface when low cloud cover hides them [4, 102]. Despite this, various data sets for clouds are available for the last 25 years and hence cover more than two 11-year (Schwabe) solar cycles.

Before we take a closer look at the data concerning cloud cover, we should first discuss our expectations. We know that cosmic rays do not strike the earth with the same intensity everywhere. The intensity increases as you move away from the equator towards the poles. Also the distribution of aerosols in the atmosphere is anything but homogeneous across the globe. In regions where there are already sufficient cloud condensation nuclei in the atmosphere, cosmic rays can only have a minimal influence on cloud cover. However, in areas that are under-saturated with nuclei, newly arriving cosmic particles may play a larger role. From a theoretical point of view we would expect that the influence of cosmic radiation on low cloud formation occurs with varying strength over different parts of the globe. The Svensmark effect should thus be of importance mostly in areas where the atmospheric fundamental conditions are right [80]. When these regions in total make up a significant part of the entire earth's surface, then the global total climate could be regulated [53]. For this reason different regions should always be compared in research studies [103].

Clouds in sync with the solar 11-year beat

Now let's look at the cloud records and try to find out if there is a relationship with cosmic radiation over the course of the 11-year solar activity cycle. For the period 1983–94 the development of the coverage extent of low clouds was in almost perfect harmony with cosmic radiation [104] (Figure 6.7). That in itself is a minor sensation. The correlation that was so clearly presented by Svensmark and his colleague Nigel Marsh in 2000 electrified the climate science community. A heated controversy developed, as nothing less than the life of the IPCC-postulated CO_2 domination as a climate-regulating factor was at stake.

Because of the potential explosiveness, the result had to be checked as quickly as possible by independent parties. Confirmation came swiftly. Other scientists were able in principle to confirm the synchronicity of global low cloud cover and cosmic rays, and even expanded on it over the subsequent years [81, 105–108] (Figure 6.7).

Yet during some periods inconsistencies were found [109] which forced scientists to check the approximate global approach and further refine the model. Were there regions where both parameters correlated especially well with each other, while in others they did not? More comprehensive studies showed that low clouds reacted to changes in cosmic rays foremost at the middle latitudes (Figure 6.7) [53, 81, 110]. The coupling of both processes worsened near the equator and the poles, on which an entire series of previously observed discrepancies were based.

So why did the interaction function worse at the poles and equatorial regions than it did at the middle latitudes? Especially at high latitudes, cosmic rays are intensive. However, because of the Arctic

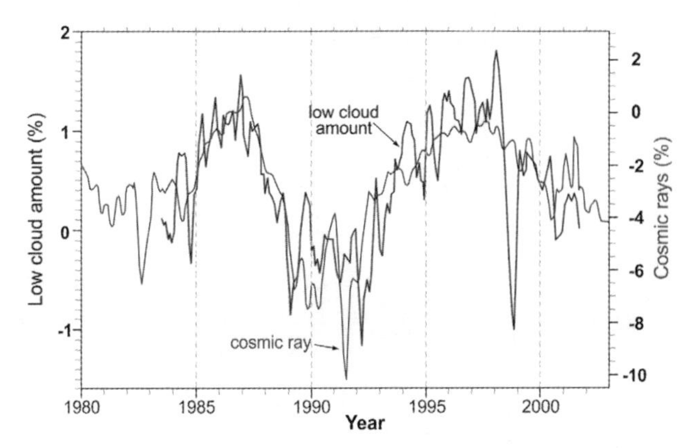

Figure 6.7 There is a very good correlation between cosmic rays and low cloud cover over the last 30 years, since the start of systematic cloud data determination [76].

cold, the proportion of ice crystals to cloud droplets in polar low clouds is higher. Despite the extremely clean air and thus the related minimal overall aerosol concentration, the cosmic condensation nuclei apparently are limited in their ability to stimulate cloud formation in the usual way [110]. The process appears somehow to 'freeze' in the cold latitudes. Moreover, due to observational limitations, cloud coverage recordings in the polar regions are much more incomplete compared to the data from lower latitudes [80, 110]. In addition, ice algae in the Arctic produce marine micro-gels [111] which seem to compete with the cosmic rays when it comes to condensation nuclei formation during the summer.

At the middle latitudes, on the other hand, there are sufficient stratified clouds, liquid cloud droplets and suitable quantities of cosmic rays, so cloud formation via cosmic rays functions very well there. Thanks to reliable satellite data for low cloud cover, it can all be documented [80].

If one differentiates between the various cloud levels in the atmosphere and the different geographic regions, then a distinct and consistent relationship between cosmic rays and low cloud cover appears and is most apparent at the middle latitudes. This was known when AR4 was written [81, 104, 109].

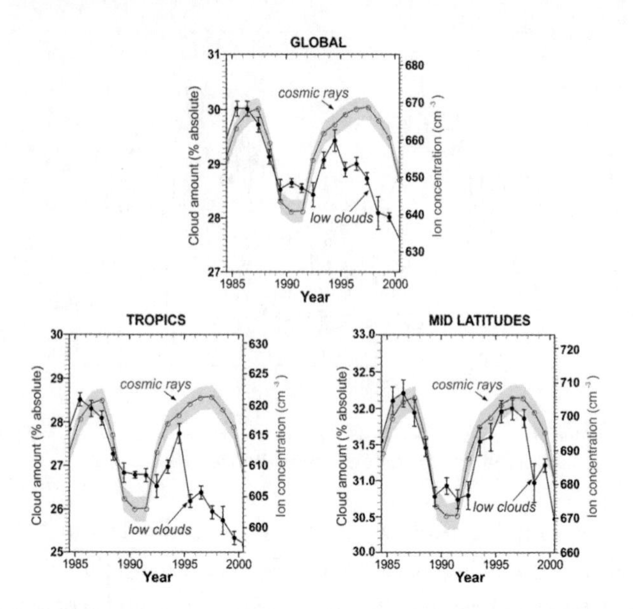

Figure 6.8 Comparison of cosmic rays and low cloud cover during 1984–2000 for the entire earth, Tropics and mid-geographical latitudes. Synchronicity between low clouds and cosmic rays is foremost in the mid-geographical latitudes where the fluctuations are pulsed by the 11-year solar activity cycle [81].

While the Svensmark effect is not even touched on in AR4 in its Summary for Policymakers, it is mentioned in the draft of AR5. However, even though the draft concludes that cosmic rays enhance aerosol nucleation and cloud condensation nuclei production in the free troposphere, it is judged to be 'too weak to have any significant climatic influence during a solar cycle' [12].

Long-term cloud trends

The 11-year solar cycle and its coupling with cloud cover and cosmic rays help us to understand the physical process behind it. But to understand how the climate developed over the last century, we need to identify long-term trends in cloud cover that extend over multiple decades and overlie the oscillating 11-year rhythm. Here the Finnish solar physicist Ilya Usoskin was successful [81]. He examined changes in low cloud cover for 1984–2000 and focused in a part study exclusively on the Tropics. The results were astounding. Cloud cover for the period under investigation declined steadily (Figure 6.7). While low clouds in the middle latitudes expanded and shrank according to the 11-year cycle, a long-term trend could be observed in the Tropics [81]. Fewer low clouds mean more solar radiation reaching the earth and hence more warming. That is precisely what the global temperature data show. The decline of tropical low cloud cover for 1984–2000 coincides precisely with the main warming phase of the late twentieth century. A coincidence? It is notable that the reduction of clouds occurs in parallel with a decrease in cosmic rays when the longer-term trend over multiple 11-year solar cycles up to 2000 is considered (Figure 6.5). Furthermore, the PDO, which increased markedly around the same time, may play a role here (Figure 4.7).

A team of researchers from California reconstructed the development of the earth's albedo [112] for the same period and were able to show a steady decline during 1984–2000. Less solar radiation reflected into space means more solar energy warming the earth's surface. Simple model calculations show that the observed decline in albedo equals 2–6 W/m^2 of radiation change and thus would imply a significant climatic impact [113]. Interestingly enough, the data for 2001–3 show that the albedo of the earth's atmosphere again increased. That too appears to fit well with the global temperature curve, which shows that the temperature rose during 1975–2000, then stopped.

Shortly after that two British scientists added another piece to the puzzle. They had studied the statistics for clouds over Great Britain since 1947 and were able to show that overcast weather statistically occurred 20 per cent more often during times of high cosmic rays [78].

Cosmic ray experiments in an open-air laboratory: cloud effects on a daily scale

It would be useful if we could test the cloud formation initiated by cosmic rays in an open-air experiment using the real atmosphere and not only in the CLOUD laboratory. We're in luck. Nature is very cooperative in this respect and is ready to participate in a trial. This is possible because of so-called Forbush events, which last a few days. During Forbush events, after a powerful solar outbreak, a sudden decline in cosmic radiation occurs because these solar winds keep the high energy cosmic particles away from the earth particularly effectively. The changes in cosmic radiation during a Forbush event correspond to the magnitude of the 11-year solar cycle [114].

A British group led by the physicist Ben Laken looked at a number of these Forbush events and studied how cloud cover developed during these periods of low cosmic ray activity [80]. As it does during the course of the 11-year cycle [81], the middle latitude cloud cover decreased considerably during Forbush events. Furthermore, an increase in air temperature at the surface of the earth was detected. This confirms the relationship between cosmic rays and clouds. Other studies produced similar results [59, 78, 115–117] and so the lead to the Svensmark amplifier keeps getting hotter. It also underscores the need to study in detail the atmospheric basic requirements for low-altitude cloud regulation by cosmic rays. A more precise regional and time prognosis on the effectiveness of the process is necessary in order to understand the effect.

Antarctica plausibility check

Low clouds are possibly also the key to a temperature paradox that has been occurring for decades in east Antarctica and has mostly gone unnoticed. While the rest of the world warmed during this time, east Antarctica cooled [118]. The explanation here could be that low clouds in Antarctica provide a warming rather than a cooling effect [119]. When cloud cover is missing, the brilliant white Antarctic ice sheet reflects a huge amount of solar energy back into space and thus cools, contrary to what is happening over the rest of the planet. But when there are clouds, they absorb part of the reflected energy and radiate it back to the surface in the form of warmth [47]. Because of the high level of solar activity recently, the associated decrease in cosmic rays could have led to a decline in low cloud cover, thus reducing the heat insulation effect over east Antarctica.

Anti-cyclic temperature developments in Antarctica with respect to the rest of the world have been nothing unusual over the last 10,000 years [47, 118, 120]. The hypothesis that climate gases are mostly responsible for climate change could not explain Antarctica's special role. The solar influence on clouds and climate, however, may explain the anomalous temperature development in that region.

What's the use of climate models that don't include solar amplifiers?

None of the current IPCC climate models take solar amplification mechanisms into account, whether via UV or cosmic rays [1, 4, 121]. IPCC scientists justify this on the grounds that the physical processes are not understood. Without solar amplifiers, the solar output signal in the models is so small that the postulated water vapour amplifier cannot properly support the sun. The IPCC, therefore, calculates an extremely low radiative forcing of 0.12 W/m^2 since 1750 for the sun, compared to a relatively hefty 1.66 W/m^2 for CO_2 (Figure 5.2) [1].

In this chapter we have observed how the IPCC continues to ignore the sun despite an overwhelming body of evidence pointing to solar amplification processes. The UV amplifier is supported by evidence showing that solar UV radiation fluctuates much more intensively than previously assumed. This causes temperature fluctuations in the stratosphere of over several degrees Celsius as well as measurable chemical changes [24, 26]. There is also strong support for the Svensmark amplifier. The solar magnetic field and cosmic rays vary in sync with solar activity, as does low cloud cover – the latter at least over parts of the globe. Evidence for the effectiveness of this amplification mechanism is overwhelming. Cosmic rays and the earth's climate seem to be very tightly linked [59].

Opponents of the Svensmark model have tried time and again to refute the solar amplifier [122–129]. However, their arguments have never held water because they use single events of non-synchronicity as a blanket argument and imprecise mean values instead of high resolution data. Furthermore, possible phase shifts in the solar–climate interaction have been ignored [130–131]. Svensmark's opponents do not differentiate between low and high clouds or between the various latitudes. Moreover, they do not properly process contradictory cloud cover data from different data sources [132]. They also mistake short-term effects for long-term effects [133–136].

The exact physical processes of the solar amplifier are a long way from being understood in detail and are currently the subject of intensive research [137]. The cutting edge of research continues to progress and produces new results that add one piece of the puzzle after another almost every month. Therefore, it cannot make any sense to take climate models that do not include solar amplification processes and call them mature and reliable. IPCC-calculated climate scenarios that do not include the solar factors must therefore be taken with a serious dose of doubt, and certainly cannot claim to indicate what the future of the climate will be like.

On the way to a realistic climate mix: how much sun, how much carbon dioxide?

One of the most important tasks for climate science in the years ahead will be to assign in a realistic way the importance of each climate control factor. To achieve this, it will first be necessary to do away with the dominance by CO_2 and replace it with a quantitative mix of control factors that includes the sun's variability as a crucial factor [138]. This is what scientists working independently of the IPCC believe. They propose that the sun accounts for 40–70 per cent of the observed climate warming of the last few decades [66, 139–147]. Nir Shaviv calculates from oceanic fluctuations that, via the 11-year solar activity cycle, there has to be a feedback mechanism that amplifies the output solar signal by at least a factor of 5–7 in order to explain the observed changes in the oceans [144].

Here one must acknowledge that CO_2 is by no means firmly in the driving seat. Less than half of the temperature increase due to CO_2 postulated by the IPCC climate models has actually occurred [148–149]. The reason for this discrepancy could be that the climate sensitivity of CO_2 was overestimated and/or because of the cooling effect of aerosols [148]. The IPCC vehemently favours the latter and as a result introduced a series of poorly or not at all understood factors that acted to counter the warming of CO_2 over the last 150 years. This is the only way the IPCC models are able to simulate the past. In the same models, however, it is assumed that these cooling aerosol factors will steadily disappear [149].

Another fundamental problem with the models is that the CO_2 climate sensitivity can only be poorly estimated using the climate data of the last 100 years because, parallel to CO_2, the atmospheric concentration of climatic 'short-lived' substances has risen [150]. Some of these substances (e.g. methane, tropospheric ozone and soot) have a strong warming effect. Others (e.g. sulphates, nitrates, organic aerosols) have a cooling effect. The exact, individual, quantitative share of the short-lived substances and CO_2 on warming is relatively poorly understood and is currently the subject of much study.

Some authors attempt to reconstruct the climate forcing of CO_2 using case studies from the distant geological past. Such comparisons are somewhat limited, however, because the intensity of the various amplification mechanisms can vary, depending on the overall climate at the time [151–152].

In conclusion, we are now familiar with a very convincing sun and so the following scenario probably arises: the primary solar variability is possibly much more powerful than first thought [23]. The sun's signal is greatly increased by pre-amplifiers such as UV processes and cosmic rays, and is then handed over to a moderate water vapour final amplifier, just like the one the IPCC assumes for CO_2. This water vapour amplifier in turn drives up the temperature signal further. In this way one can imagine that the sun plays in the same league as CO_2, if not higher, in terms of its impact. Our climate prognoses in the next chapter will take this more realistic distribution of forcing into account. This area of research is currently being intensively investigated by numerous scientific groups [153].

References

1. IPCC (2007) *Climate Change 2007: The Physical Science Basis. Contribution of Working Group I to the Fourth Assessment Report of the Intergovernmental Panel on Climate Change*. Cambridge University Press, Cambridge and New York.
2. Willett, K. M., N. P. Gillett, P. D. Jones and P. W. Thorne (2007) Attribution of observed surface humidity changes to human influence. *Nature* 449, 710–13.
3. Lindzen, R. S. and Y.-S. Choi (2009) On the determination of climate feedbacks from ERBE data. *Geophysical Research Letters* 36, L16705.
4. Archer, D. and S. Rahmstorf (2010) *The Climate Crisis*, 1st edn. Cambridge University Press, Cambridge.
5. L'Ecuyer, T. S. and J. H. Jiang (2010) Touring the atmosphere aboard the A-train. *Physics Today* July, 36–41.
6. Lindzen, R. S., M.-D. Chou and A. Y. Hou (2001) Does the earth have an adaptive infrared iris? *Bulletin of the American Meteorological Society* 82 (3), 417–32.
7. de Szoeke, S. P., S. Yuter, D. Mechem, C. W. Fairall, C. Burleyson and P. Zuidema (2012) Observations of stratocumulus clouds and their effect on the eastern Pacific surface heat budget along 20°S. *Journal of Climate*.
8. Miller, M. A., V. P. Ghate and R. K. Zahn (2012) The radiation budget of the West African Sahel and its controls: a perspective from observations and global climate models. *Journal of Climate* 25 (17), 5976–96.
9. Davies, R. and M. Molloy (2012) Global cloud height fluctuations measured by MISR on Terra from 2000 to 2010. *Geophys. Res. Lett.* 39 (3), L03701.
10. Cho, H., C.-H. Ho and Y.-S. Choi (2012) The observed variation in cloud-induced longwave radiation in response to sea surface temperature over the Pacific warm pool from MTSAT-1R imagery. *Geophys. Res. Lett.* 39 (18), L18802.
11. Curry, J. (2010) CO_2 No-Feedback Sensitivity. http://judithcurry.com/2010/12/11/co2-no-feedback-sensitivity.
12. IPCC (2012) Climate Change 2013: The Physical Science Basis. Summary for Policymakers. IPCC WGI Fifth Assessment Report, First Order Draft. http://www.stopgreensuicide.com/SummaryForPolicymakers_WG1AR5-SPM_FOD_Final.pdf.
13. Lacis, A. A., G. A. Schmidt, D. Rind and R. A. Ruedy (2010) Atmospheric CO_2: principal control knob governing earth's temperature. *Science* 330, 356–9.
14. Spencer, R. W., W. D. Braswell, J. R. Christy and J. Hnilo (2007) Cloud and radiation budget changes associated with tropical intraseasonal oscillations. *Geophysical Research Letters* 34.
15. Iassamen, A., H. Sauvageot, N. Jeannin and S. Ameur (2009) Distribution of tropospheric water vapour in clear and cloudy conditions from microwave radiometric profiling. *Journal of Applied Meteorology and Climatology* 48, 600–15.

16. Vonder Haar, T. H., J. Bytheway and J. M. Forsythe (in press) Weather and climate analyses using improved global water vapour observations. *Geophysical Research Letters.*
17. Solomon, S., K. H. Rosenlof, R. W. Portmann, J. S. Daniel, S. M. Davis, T. J. Sanford and G.-K. Plattner (2010) Contributions of stratospheric water vapour to decadal changes in the rate of global warming. *Science* 327, 1219–23.
18. Fujiwara, M., H. Vömel, F. Hasebe, M. Shiotani, S.-Y. Ogino, S. Iwasaki, N. Nishi, T. Shibata, K. Shimizu, E. Nishimoto, J. M. Valverde Canossa, H. B. Selkirk and S. J. Oltmans (2010) Seasonal to decadal variations of water vapour in the tropical lower stratosphere observed with balloon-borne cryogenic frost point hygrometers. *Journal of Geophysical Research* 115, 1–15.
19. Hurst, D. F., S. J. Oltmans, H. Vömel, K. H. Rosenlof, S. M. Davis, E. A. Ray, E. G. Hall and A. F. Jordan (2011) Stratospheric water vapour trends over Boulder, Colorado: analysis of the 30-year Boulder record. *Journal of Geophysical Research* 116, 1–12.
20. Tallbloke (2010) Interesting Correlation: Sunspots vs. Specific Humidity. http://tallbloke.wordpress.com/2010/08/08/interesting-correlation-sunspots-vs-specific-humidity; sunspot numbers smoothed with an 8-year moving average.
21. Marsh, N., H. Svensmark and F. Christiansen (2005) Climate Variability Correlated with Solar Activity. http://www.space.dtu.dk/upload/institutter/space/forskning/06_projekter/isac/wp_103.pdf.
22. Waple, A. M. (2006) The Identification of a Solar Signal in Climate Records of the Last 500 Years Using Proxy and Model-Based Analysis and the Implications for Natural Climate Variability. PhD thesis. Graduate School of the University of Massachusetts, Amherst.
23. Shapiro, A. I., W. Schmutz, E. Rozanov, M. Schoell, M. Haberreiter, A. V. Shapiro and S. Nyeki (2011) A new approach to long-term reconstruction of the solar irradiance leads to large historical solar forcing. *Astronomy & Astrophysics* 529, 1–8.
24. Bard, E. and M. Frank (2006) Climate change and solar variability: what's new under the sun? *Earth and Planetary Science Letters* 248, 1–14.
25. Meehl, G. A., J. M. Arblaster, K. Matthes, F. Sassi and H. v. Loon (2009) Amplifying the Pacific climate system response to a small 11-year solar cycle forcing. *Science* 325, 1114–18.
26. Haigh, J. D., A. R. Winning, R. Toumi and J. W. Harder (2010) An influence of solar spectral variations on radiative forcing of climate. *Nature* 467, 696–9.
27. Gray, L. J., J. Beer, M. Geller, J. D. Haigh, M. Lockwood, K. Matthes, U. Cubasch, D. Fleitmann, G. Harrison, L. Hood, J. Luterbacher, G. A. Meehl, D. Shindell, B. van Geel and W. White (2010) Solar influences on climate. *Reviews of Geophysics* 48, 1–53.
28. Almasi, P. F. and G. Bond (2009) Sun-climate connections. In V. Gornitz (ed.), *Encyclopedia of Paleoclimatology and ancient environments.* Springer, New York, 929–35.
29. Niranjankumar, K., T. K. Ramkumar and M. Krishnaiah (2011) Vertical and lateral propagation characteristics of intraseasonal oscillation from the tropical lower troposphere to upper mesosphere. *Journal of Geophysical Research* 116, 1–10.
30. Kang, S. M., L. M. Polvani, J. C. Fyfe and M. Sigmond (2011) Impact of polar ozone depletion on subtropical precipitation. *Science* 332, 951–4.

31. Kodera, K. (2006) The role of dynmics in solar forcing. *Space Science Reviews* 125, 319–30.
32. Bal, S., S. Schimanke, T. Spangehl and U. Cubasch (2011) On the robustness of the solar cycle signal in the Pacific region. *Geophysical Research Letters* 38, 1–5.
33. Ineson, S., A. A. Scaife, J. R. Knight, J. C. Manners, N. J. Dunstone, L. J. Gray and J. D. Haigh (2011) Solar forcing of winter climate variability in the Northern Hemisphere. *Nature Geoscience* 4, 753–7.
34. Udelhofen, P. M. and R. D. Cess (2001) Cloud cover variations over the United States: an influence of cosmic rays or solar variability? *Geophys. Res. Lett.* 28 (13), 2617–20.
35. Varma, V., M. Prange, T. Spangehl, F. Lamy, U. Cubasch and M. Schulz (2012) Impact of solar-induced stratospheric ozone decline on Southern Hemisphere westerlies during the late Maunder minimum. *Geophys. Res. Lett.* 39 (20), L20704.
36. Reichler, T., J. Kim, E. Manzini and J. Kröger (2012) A stratospheric connection to Atlantic climate variability. *Nature Geoscience* 5, 783–7.
37. Kanipe, J. (2006) A cosmic connection. *Nature* 443 (September), 141–3.
38. Svensmark, H. (2000) Cosmic rays and earth's climate. *Space Science Reviews* 93, 155–66.
39. Svensmark, H. and E. Friis-Christensen (1997) Variation of cosmic ray flux and global cloud coverage – a missing link in solar–climate relationships. *Journal of Atmospheric and Solar–Terrestrial Physics* 59 (11), 1225–32.
40. Svensmark, H., J. O. P. Pedersen, N. D. Marsh and M. B. Enghoff (2007) Experimental evidence for the role of ions in particle nucleation under atmospheric conditions. *Proc. R. Soc.* A 463, 385–96.
41. Svensmark, H. and E. Friis-Christensen (2007) Reply to Lockwood and Fröhlich – the persistent role of the sun in climate forcing. *Danish National Space Center, Scientific Report* 3.
42. JoNova (2009) The Skeptics Handbook II. http://joannenova.com.au/globalwarming/skeptics-handbook-ii/the_skeptics_handbook_II-sml.pdf.
43. Tobias, S. M. (2002) The solar dynamo. *Phil. Trans. R. Soc. Lond.* A 360, 2741–56.
44. Schüssler, M. (2005) The sun and its restless magnetic field. In K. Scherer, H. Fichtner, B. Heber and U. Mall (eds.), *Space Weather*: The Physics Behind a Slogan. Lecture Notes in Physics 656, Springer, New York, 23–49.
45. Owens, M. J., N. U. Crooker and M. Lockwood (2011) How is open solar magnetic flux lost over the solar cycle? *Journal of Geophysical Research* 116, 1–13.
46. Svalgaard, L. and E. W. Cliver (2007) A floor in the solar wind magnetic field. *The Astrophysical Journal* 661, L203–L206.
47. Svensmark, H. and N. Calder (2007) *The Chilling Stars*. Icon Books, Cambridge.
48. McCracken, K. G. (2007) Heliomagnetic field near earth, 1428–2005. *Journal of Geophysical Research* 112, 1–9.
49. Muscheler, R., F. Joos, J. Beer, S. A. Müller, M. Vonmoos and I. Snowball (2007) Solar activity during the last 1000 years inferred from radionuclide records. *Quaternary Science Reviews* 26, 82–97.
50. Mufti, S. and G. N. Shah (2011) Solar-geomagnetic activity influence on earth's climate. *Journal of Atmospheric and Solar–Terrestrial Physics* 73, 1607–15.
51. St-Onge, G., J. S. Stoner and C. Hillaire-Marcel (2003) Holocene paleomagnetic records from the St. Lawrence Estuary, eastern Canada: centennial- to millennial-

scale geomagnetic modulation of cosmogenic isotopes. *Earth and Planetary Science Letters* 209, 113–30.

52. Courtillot, V., Y. Gallet, J.-L. Le Mouël, F. Fluteau and A. Genevey (2007) Are there connections between the earth's magnetic field and climate? *Earth and Planetary Science Letters* 253, 328–39.
53. Kovaltsov, G. A. and I. G. Usoskin (2007) Regional cosmic ray induced ionization and geomagnetic field changes. *Adv. Geosci.* 13, 31–5.
54. Scherer, K., H. Fichtner, T. Borrmann, J. Beer, L. Desorgher, E. Flükiger, H.-J. Fahr, S. E. S. Ferreira, U. W. Langner, M. S. Potgieter, B. Heber, J. Masarik, N. J. Shaviv and J. Veizer (2006) Interstellar–terrestrial relations: variable cosmic environments, the dynamic heliosphere, and their imprints on terrestrial archives and climate. *Space Science Reviews* 127, 327–465.
55. Usoskin, I. G., M. Korte and G. A. Kovaltsov (2008) Role of centennial geomagnetic changes in local atmospheric ionization. *Geophysical Research Letters* 35, L05811.
56. Usoskin, I. G., S. K. Solanki and M. Korte (2006) Solar activity reconstructed over the last 7000 years: the influence of geomagnetic field changes. *Geophysical Research Letters* 33, 1–4.
57. De Santis, A., E. Qamili, G. Spada and P. Gasperini (2012) Geomagnetic South Atlantic anomaly and global sea level rise: a direct connection? *Journal of Atmospheric and Solar–Terrestrial Physics* 74, 129–35.
58. Ma, L. H., Y. B. Han and Z. Q. Yin (2009) Influence of the 11-year solar cycle on variations of cosmic ray intensity. *Solar Physics* 255, 187–91.
59. Usoskin, I. G. and G. A. Kovaltsov (2007) Link between cosmic rays and clouds on different timescales. In W.-H. Ip and M. Duldig (eds.), *Advances in Geosciences, Volume 2: Solar Terrestrial* (ST). World Scientific Co., Singapore, 321–31.
60. Calisto, M., I. Usoskin, E. Rozanov and T. Peter (2011) Influence of galactic cosmic rays on atmospheric composition and dynamics. *Atmos. Chem. Phys.* 11, 4547–56.
61. Institute for Experimental and Applied Physics – Christian-Albrechts-Universität zu Kiel Neutron Monitor Database. http://www.nmdb.eu/nest/search.php.
62. Rouillard, A. P. and M. Lockwood (2007) Centennial changes in solar activity and the response of galactic cosmic rays. *Advances in Space Research* 40, 1078–86.
63. NMBD (2011) Kiel Neutronen-Monitor. http://www.nmdb.eu/nest/search.php.
64. Mann, M., J. Beer, F. Steinhilber and M. Christl (2012) 10Be in lacustrine sediments – a record of solar activity? *Journal of Atmospheric and Solar – Terrestrial Physics* 80, 92–9.
65. Potgieter, M. S. (2010) The dynamic heliosphere, solar activity, and cosmic rays. *Advances in Space Research* 46 (4), 402–12.
66. Rao, U. R. (2011) Contribution of changing galactic cosmic ray flux to global warming. *Current Science* 100 (2), 223–5.
67. Berggren, A.-M., J. Beer, G. Possnert, A. Aldahan, P. Kubik, M. Christl, S. J. Johnsen, J. Abreu and B. M. Vinther (2009) A 600-year annual ^{10}Be record from the NGRIP ice core, Greenland. *Geophysical Research Letters* 36, 1–5.
68. McCracken, K. G. and J. Beer (2007) Long-term changes in the cosmic ray intensity at earth, 1428–2005. *Journal of Geophysical Research* 112, 1–15.
69. Fichtner, H., K. Scherer and B. Heber (2006) A criterion to discriminate between solar and cosmic ray forcing of the terrestrial climate. *Atmos. Chem. Phys.* Discuss. 6, 10811–36.

70. Shaviv, N. J. (2002) Cosmic ray diffusion from the galactic spiral arms, iron meteorites, and a possible climatic connection. *Physical Review Letters* 89 (5), 051102/051101-051104.
71. Shaviv, N. J. and J. Veizer (2003) Celestial driver of Phanerozoic climate? GSA *Today* 7, 4–10.
72. Dergachev, V. A., P. B. Dmitriev, O. M. Raspopov and B. V. Geel (2004) The effects of galactic cosmic rays, modulated by solar terrestrial magnetic fields, on the climate. *Russian Journal of Earth Sciences* 6 (5), 323–38.
73. Plenier, G., J.-P. Valet, G. Guérin, J.-C. Lefèvre, M. LeGoff and B. Carter-Stiglitz (2007) Origin and age of the directions recorded during the Laschamp event in the Chaîne des Puys (France). *Earth and Planetary Science Letters* 259, 414–31.
74. Laj, C., C. Kissel, A. Mazaud, J. E. T. Channell and J. Beer (2000) North Atlantic palaeointensity stack since 75 ka (NAPIS-75) and the duration of the Laschamp Event. *Phil. Trans. R. Soc. Lond.* A 358, 1009–25.
75. Beer, J. (2005) Kosmische Strahlung und Wolken. *EAWAG News* 58, 16–18.
76. Svensmark, H. (2007) Cosmoclimatology: a new theory emerges. Astronomy & *Geophysics* 48, 1.18–11.24.
77. Ney, E. P. (1959) Cosmic radiation and the weather. *Nature* 183, 451–2.
78. Harrison, R. G. and D. B. Stephenson (2006) Empirical evidence for a nonlinear effect of galactic cosmic rays on clouds. *Proc. R. Soc.* A 462, 1221–33.
79. Costantino, L. and F.-M. Bréon (2010) Analysis of aerosol–cloud interaction from multi-sensor satellite observations. *Geophysical Research Letters* 37, 1–5.
80. Laken, B. A., D. R. Kniveton and M. R. Frogley (2010) Cosmic rays linked to rapid mid latitude cloud changes. *Atmos. Chem. Phys.* 10, 10941–8.
81. Usoskin, I. G., N. Marsh, G. A. Kovaltsov, K. Mursula and O. G. Gladysheva (2004) Latitudinal dependence of low cloud amount on cosmic ray induced ionization. *Geophysical Research Letters* 31, 1–4.
82. Gray, L. J., J. D. Haigh and R. G. Harrison (2005) The influence of solar changes on the earth's climate. *Hadley Centre Technical Note* 62, 1–81.
83. Enghoff, M. B., J. O. P. Pedersen, U. I. Uggerhøj, S. M. Paling and H. Svensmark (2011) Aerosol nucleation induced by a high energy particle beam. *Geophysical Research Letters* 38, 1–4.
84. Enghoff, M. B. and H. Svensmark (2008) The role of atmospheric ions in aerosol nucleation – a review. *Atmos. Chem. Phys.* 8, 4911–23.
85. Nicoll, K. A. and R. G. Harrison (2010) Experimental determination of layer cloud edge charging from cosmic ray ionisation. *Geophysical Research Letters* 37.
86. Tinsley, B. A., G. B. Burns and L. Zhou (2007) The role of the global electric circuit in solar and internal forcing of clouds and climate. *Advances in Space Research* 40, 1126–39.
87. Mendoza, B. and V. Velasco (2009) High-latitude methane sulphonic acid variability and solar activity: the role of the total solar irradiance. *Journal of Atmospheric and Solar–Terrestrial Physics* 71, 33–40.
88. Vallina, S. M., R. Simó, S. Gassó, C. d. Boyer-Montegut, E. del Rio, E. Jurado and J. Dachs (2007) Analysis of a potential 'solar radiation dose–dimethylsulfide–cloud condensation nuclei' link from globally mapped seasonal correlations. *Global Biochemical Cycles* 21, 1–16.
89. Osorio, J., B. Mendoza and V. Velasco (2008) Methane sulphonic acid trend associated with Beryllium-10 and solar irradiance. In R. Caballero, J. C. D'Olivo,

G. Medina-Tanco, L. Nellen, F. A. Sánchez and J. F. Valdés-Galicia (eds.), *Proceedings of the 30th International Cosmic Ray Conference*. Universidad Nacional Autónoma de México, Mexico City, Mexico,501-4.

90. Kulmala, M., I. Riipinen, T. Nieminen, M. Hulkkonen, L. Sogacheva, H. E. Manninen, P. Paasonen, T. Petäjä, M. Dal Maso, P. P. Aalto, A. Viljanen, I. Usoskin, R. Vainio, S. Mirme, A. Mirme, A. Minikin, A. Petzold, U. Horrak, C. Plaß-Dülmer, W. Birmili and V.-M. Kerminen (2010) Atmospheric data over a solar cycle: no connection between galactic cosmic rays and new particle formation. *Atmos. Chem. Phys.* 10, 1885–98.
91. Peters, K., J. Quaas and N. Bellouin (2011) Effects of absorbing aerosols in cloudy skies: a satellite study over the Atlantic Ocean. *Atmos. Chem. Phys.* 11, 1393–404.
92. Weber, W. (2010) Strong signature of the active sun in 100 years of terrestrial insolation data. *Annalen der Physik* 522 (6), 372–81.
93. Monge, M. E., T. Rosenørn, O. Favez, M. Müller, G. Adler, A. Abo Riziq, Y. Rudich, H. Herrmann, C. George and B. D'Anna (2012) Alternative pathway for atmospheric particles growth. *Proceedings of the National Academy of Sciences* 109 (18), 6840–4.
94. Kilifarska, N. A. (2012) Climate sensitivity to the lower stratospheric ozone variations. *Journal of Atmospheric and Solar–Terrestrial Physics* 90–1, 9–14.
95. Carslaw, K. S., R. G. Harrison and J. Kirkby (2002) Cosmic rays, clouds, and climate. *Science* 298, 1732–7.
96. Kirkby, J. (2007) Cosmic rays and climate. *Surveys in Geophysics* 28, 333–75.
97. Duplissy, J., M. B. Enghoff, K. L. Aplin, F. Arnold, H. Aufmhoff, M. Avngaard, U. Baltensperger, T. Bondo, R. Bingham, K. Carslaw, J. Curtius, A. David, B. Fastrup, S. Gagné, F. Hahn, R. G. Harrison, B. Kellett, J. Kirkby, M. Kulmala, L. Laakso, A. Laaksonen, E. Lillestol, M. Lockwood, J. Mäkelä, V. Makhmutov, N. D. Marsh, T. Nieminen, A. Onnela, E. Pedersen, J. O. P. Pedersen, J. Polny, U. Reichl, J. H. Seinfeld, M. Sipilä, Y. Stozhkov, F. Stratmann, H. Svensmark, J. Svensmark, R. Veenhof, B. Verheggen, Y. Viisanen, P. E.Wagner, G.Wehrle, E. Weingartner, H. Wex, M. Wilhelmsson and P. M.Winkler (2010) Results from the CERN pilot CLOUD experiment. *Atmos. Chem. Phys.* 10, 1635–47.
98. Kirkby, J. (2010) The CLOUD Project: climate research with accelerators. *Proceedings of IPAC'10, Kyoto, Japan*, http://accelconf.web.cern.ch/AccelConf/IPAC10/talks/frymh02_talk.pdf, 4774-4778.
99. CLOUD Collaboration (2010) 2009 Progress Report on PS215/CLOUD. *CERN-SPSC-2010-013, SPSC-SR-061*, 1–29.
100. Kirkby, J., J. Curtius, J. Almeida, E. Dunne, J. Duplissy, S. Ehrhart, A. Franchin, S. Gagné, L. Ickes, A. Kürten, A. Kupc, A. Metzger, F. Riccobono, L. Rondo, S. Schobesberger, G. Tsagkogeorgas, DanielaWimmer, A. Amorim, F. Bianchi, M. Breitenlechner, A. David, J. Dommen, A. Downard, M. Ehn, R. C. Flagan, S. Haider, A. Hansel, D. Hauser, W. Jud, H. Junninen, Fabian Kreissl, A. Kvashin, A. Laaksonen, K. Lehtipalo, J. Lima, E. R. Lovejoy, V. Makhmutov, S. Mathot, J. Mikkilä, P. Minginette, S. Mogo, T. Nieminen, A. Onnela, P. Pereira, T. Petäjä, R. Schnitzhofer, J. H. Seinfeld, M. Sipilä, Y. Stozhkov, F. Stratmann, A. Tomé, J. Vanhanen, Y. Viisanen, A. Vrtala, P. E. Wagner, H. Walther, E. Weingartner, H. Wex, P. M.Winkler, K. S. Carslaw, D. R.Worsnop, U. Baltensperger and M. Kulmala (2011) Role of sulphuric acid, ammonia and galactic cosmic rays in atmospheric aerosol nucleation. *Nature* 476, 429–35.

101. *Die Welt* (2011) Sonnenwinde befördern womöglich den Klimawandel. 24 August. http://www.welt.de/wissenschaft/umwelt/article13563144/Sonnenwinde-befoerdern-womoeglich-den-Klimawandel.html.
102. Usoskin, I. G., M. Voiculescu, G. A. Kovaltsov and K. Mursula (2006) Correlation between clouds at different altitudes and solar activity: fact or artifact? *Journal of Atmospheric and Solar–Terrestrial Physics* 68, 2164–72.
103. Le Mouél, J.-L., E. Blanter, M. Shnirman and V. Courtillot (2009) Evidence for solar forcing in variability of temperatures and pressures in Europe. *Journal of Atmospheric and Solar–Terrestrial Physics* 71 (12), 1309–21.
104. Marsh, N. and H. Svensmark (2000) Cosmic rays, clouds, and climate. Space Sci. Rev. 94 (1/2), 215–30.
105. Pallé, E. and C. J. Butler (2000) The influence of cosmic rays on terrestrial clouds and global warming. Astronomy and Geophysics 41, 18–22.
106. Yu, F. (2002) Altitude variations of cosmic ray induced production of aerosols: implications for global cloudiness and climate. *Journal of Geophysical Research* 107, A7.
107. Zarrouk, N. and R. Bennaceur (2010) Link nature between low cloud amounts and cosmic rays through wavelet analysis. *Acta Astronautica* 66 (9–10), 1311–19.
108. Rohs, S., R. Spang, F. Rohrer, C. Schiller and H. Vos (2010) A correlation study of high-altitude and midaltitude clouds and galactic cosmic rays by MIPAS-Envisat. *J. Geophys. Res.* 115 (D14), D14212.
109. Marsh, N. and H. Svensmark (2003) Galactic cosmic ray and El Niño – southern oscillation trends in international satellite cloud climatology project D2 low-cloud properties. *Journal of Geophysical Research* 108, 6-1–6-11.
110. Pallé, E., C. J. Butler and K. O'Brien (2004) The possible connection between ionization in the atmosphere by cosmic rays and low level clouds. *Journal of Atmospheric and Solar–Terrestrial Physics* 66, 1779–90.
111. Orellana, M. V., P. A. Matrai, C. Leck, C. D. Rauschenberg, A. M. Lee and A. E. Coz (2011) Marine microgels as a source of cloud condensation nuclei in the high Arctic. *PNAS* 108 (33), 13612–17.
112. Pallé, E., P. R. Goode, P. Montanés-Rodriguez and S. E. Koonin (2004) Changes in earth's reflectance over the past two decades. *Science* 304, 1299–301.
113. Pallé, E., P. Montanes-Rodriguez, R. Goode, S. E. Koonin, M. Wild and S. Casadio (2005) A multi-data comparison of shortwave climate forcing changes. *Geophysical Research Letters* 32, 1–4.
114. Laken, B., D. Kniveton and A. Wolfendale (2011) Forbush decreases, solar irradiance variations, and anomalous cloud changes. *Journal of Geophysical Research* 116, 1–10.
115. Svensmark, H., T. Bondo and J. Svensmark (2009) Cosmic ray decreases affect atmospheric aerosols and clouds. *Geophysical Research Letters* 36, 1–4.
116. Harrison, R. G. and M. H. P. Ambaum (2010) Observing Forbush decreases in cloud at Shetland. *Journal of Atmospheric and Solar–Terrestrial Physics* 72, 1408–14.
117. Svensmark, J., M. B. Enghoff and H. Svensmark (2012) Effects of cosmic ray decreases on cloud microphysics. *Atmos. Chem. Phys.* Discuss. 12, 3595–617.
118. Chylek, P., C. K. Folland, G. Lesins and M. K. Dubey (2010) Twentieth-century bipolar seesaw of the Arctic and Antarctic surface air temperatures. *Geophysical Research Letters* 37, 1–4.
119. Pavolonis, M. J. and J. R. Key (2003) Antarctic cloud radiative forcing at the surface estimated from the AVHRR Polar Pathfinder and ISCCP D1 data sets, 1985–93. *J. Appl. Meteor.* 42, 827–40.

120. Severinghaus, J. P. (2009) Southern see-saw seen. *Nature* 457, 1093–4.
121. Latif, M. (2008) *Bringen wir das Klima aus dem Takt? Hintergründe und Prognosen.* Fischer Taschenbuch Verlag, Frankfurt a.M.
122. Sloan, T. and A. W. Wolfendale (2008) Testing the proposed causal link between cosmic rays and cloud cover. *Environmental Research Letters* 3, 1–6.
123. Sloan, T. and A. W. Wolfendale (2011) The contribution of cosmic rays to global warming. *Journal of Atmospheric and Solar–Terrestrial Physics* 73 (16), 2352–5.
124. Erlykin, A. D. and A. W. Wolfendale (2011) Cosmic ray effects on cloud cover and their relevance to climate change. *Journal of Atmospheric and Solar–Terrestrial Physics* 73, 1681–6.
125. Calogovic, J., C. Albert, F. Arnold, J. Beer, L. Desorgher and E. O. Flueckiger (2010) Sudden cosmic ray decreases: no change of global cloud cover. *Geophysical Research Letters* 37, 1–5.
126. Agee, E. M., K. Kiefer and E. Cornett (2011) Relationship of lower troposphere cloud cover and cosmic rays: an updated perspective. *Journal of Climate.*
127. Kazil, J., K. Zhang, P. Stier, J. Feichter, U. Lohmann and K. O'Brien (2012) The present-day decadal solar cycle modulation of earth's radiative forcing via charged H2SO4/H2O aerosol nucleation. *Geophys. Res. Lett.* 39 (2), L02805.
128. Laken, B., E. Pallé and H. Miyahara (2012) a decade of the moderate resolution imaging spectroradiometer: is a solar–cloud link detectable? *Journal of Climate* 25 (13), 4430–40.
129. Laken, B. A. and J. Čalogović (2011) Solar irradiance, cosmic rays and cloudiness over daily timescales. *Geophys. Res. Lett.* 38 (24), L24811.
130. Gusev, A. A. (2011) Natural climatic oscillations driven by solar activity. *Geomagnetism and Aeronomy* 51 (1), 131–8.
131. Gusev, A. A. and I. M. Martin (2012) Possible evidence of the resonant influence of solar forcing on the climate system. *Journal of Atmospheric and Solar–Terrestrial Physics* 80, 173–8.
132. Chan, M. A. and J. C. Comiso (2011) Cloud features detected by MODIS but not by CloudSat and CALIOP. *Geophys. Res. Lett.* 38 (24), L24813.
133. Calder, N. (2011) Further Attempt to Falsify the Svensmark Hypothesis. http://calderup.wordpress.com/2011/10/05/further-attempt-to-falsify-the-svensmark-hypothesis.
134. Shaviv, N. J. (2008) Is the Causal Link Between Cosmic Rays and Cloud Cover Really Dead? http://www.sciencebits.com/SloanAndWolfendale.
135. Evan, A. T., A. K. Heidinger and D. J. Vimont (2007) Arguments against a physical long-term trend in global ISCCP cloud amounts. *Geophysical Research Letters* 34, 1–5.
136. Campbell, G. G. (2004) View angle dependence of cloudiness and the trend in ISCCP cloudiness. 13th Conference on Satellite Meteorology and Oceanography, http://ams.confex.com/ams/13SATMET/techprogram/paper_79041.htm, 1–7.
137. Schwadron, N. A., H. E. Spence and R. Came (2011) Does the space environment affect the ecosphere? *Eos* 92 (36), 297–9.
138. Fang, J., J. Zhu, S. Wang, C. Yue and H. Shen (2011) Global warming, human-induced carbon emissions, and their uncertainties. *Science China Earth Sciences* 54 (10), 1458–68.
139. Scafetta, N. and B. J. West (2007) Phenomenological reconstructions of the solar

signature in the Northern Hemisphere surface temperature records since 1600. *Journal of Geophysical Research* 112, D24S03.

140. Scafetta, N. and B. J. West (2008) Is climate sensitive to solar variability? *Physics Today* March, 50–1.
141. Scafetta, N. (2010) Empirical evidence for a celestial origin of the climate oscillations and its implications. *Journal of Atmospheric and Solar–Terrestrial Physics* 72, 951–70.
142. Beer, J., W. Mende and R. Stellmacher (2000) The role of the sun in climate forcing. *Quaternary Science Reviews* 19, 403–15.
143. Shaviv, N. J. (2005) On climate response to changes in the cosmic ray flux and radiative budget. *Journal of Geophysical Research* 110, 1–15.
144. Shaviv, N. J. (2008) Using the oceans as a calorimeter to quantify the solar radiative forcing. *Journal of Geophysical Research* 113, 1–13.
145. El-Sayed Aly, N. (2010) Spectral analysis of solar variability and their possible role on global warming. *International Journal of the Physical Sciences* 5 (7), 1040–9.
146. De Jager, C. and S. Duhau (2011) The variable solar dynamo and the forecast of solar activity; influence on terrestrial surface temperature. In. J. M. Cossia (ed.), *Global warming in the 21st Century*. Nova Science, Hauppauge, NY, 77–106.
147. Soon, W., K. Dutta, D. R. Legates, V. Velasco and W. Zhang (2011) Variation in surface air temperature of China during the 20th century. *Journal of Atmospheric and Solar–Terrestrial Physics* 73, 2331–44.
148. Schwartz, S. E., R. J. Charlson, R. A. Kahn, J. A. Ogren and H. Rohde (2010) Why hasn't earth warmed as much as expected? *Journal of Climate* 23, 2453–64.
149. Lindzen, R. S. (2005) Understanding common climate claims. In R. Raigaini (ed.), *Proceedings of the 34th International Seminar on Nuclear War and Planetary Emergencies*. World Scientific Publishing Co., Singapore, 189–210.
150. Penner, J. E., M. J. Prather, I. S. A. Isaksen, J. S. Fuglestvedt, Z. Klimont and D. S. Stevenson (2010) Short-lived uncertainty? *Nature Geoscience* 3, 587–8
151. Köhler, P., R. Bintanja, H. Fischer, F. Joos, R. Knutti, G. Lohmann and V. Masson-Delmotte (2010) What caused earth's temperature variations during the last 800,000 years? Data-based evidence on radiative forcing and constraints on climate sensitivity. *Quaternary Science Reviews* 29, 129–45.
152. Knorr, G., M. Butzin, A. Micheels and G. Lohmann (2011) A warm Miocene climate at low atmospheric CO_2 levels. *Geophysical Research Letters* 38, 1–5.
153. Schmittner, A., N. M. Urban, J. D. Shakun, N. M. Mahowald, P. U. Clark, P. J. Bartlein, A. C. Mix and A. Rosell-Melé (2011) Climate sensitivity estimated from temperature reconstructions of the Last Glacial Maximum. *Science* 334, 1385–8.

Mining a treasure trove of 50–80-year-old solar data:

the unexpected atmospheric amplifier of solar activity

Werner Weber
INSTITUTE OF PHYSICS, TECHNICAL UNIVERSITY OF DORTMUND

More than 100 years ago Samuel Pierpont Langley, an American astrophysicist and aviation pioneer, proposed carrying out highly precise measurements of solar irradiance. Langley suspected that the sun 'flickered' and solar irradiance varied over time. Even a change of 1 per cent would lead to profound consequences for the climate. According to the Stefan-Boltzmann law, which was known at the time, a 1 per cent increase in solar irradiance would lead to an increase in the average global temperature of approximately 0.7° C. And so a large-scale monitoring programme was launched, undertaken by the Smithsonian Astrophysical Observatory (SAO). After Langley's death in 1906, his assistant, Charles Greeley Abbot, became the project's manager. In the early years of the project (1912–15) the astrophysicist Frederick Eugene Fowle produced groundbreaking papers on the role of atmospheric moisture attenuating solar irradiance, which even today is not well understood [1–5].

In 1923, after several years of preparative measurements and an intense search for appropriate monitoring stations, the main data collection series of the SAO project started. The two most important monitoring stations were Mt. Montezuma in the Atacama desert in northern Chile, at an elevation of 2700 metres and Table Mountain, California at 2200 metres elevation. The main measuring device was the pyrheliometer, invented by Langley. This measured the heat deposited by direct solar radiation, thereby blocking out all diffused sky radiation, the latter being known as the aureole, which is generated by air molecules scattering sunlight (Rayleigh scattering) and by aerosols. The aureole, or diffused radiation (also known as sky radiation), was measured by another instrument, the pyranometer, invented by Abbot. A third series of measurements, of atmospheric moisture content, was determined based on Fowle's findings. For these measurements Claude Pouillet's bolometer was used, and were considered as important as the aureole data. The instruments used in the SAO project over the span of the 30 years of data-gathering were the best available worldwide.

Data for all three measurement series were taken when possible on a daily basis at different times of day and for specified solar angles over the horizon. The path of solar radiation through the atmosphere is different at each angle. The shortest path occurs when the sun is at its zenith (i.e. when it is perpendicular above the observer). From the data sets for the various solar positions it was possible to calculate the solar constant outside of the atmosphere. This is the solar intensity at a position just before the earth where the solar radiation has not yet been attenuated by the atmosphere. All three data series were used for the calculations, which are based on Langley's hypothesis and on the Fowle's findings. The result was the sought-after solar constant, which was not expected really to be a 'constant'. The SAO measurements ran from 1923 to 1954. Abbot ended the data collection because he saw indications that anthropogenic air pollution was beginning to distort the data too much.

Altogether there were approximately 35,000 data groups from which the value of the solar constant was determined. Its value when the sun is inactive, during a solar activity minimum, was on average 1357 $W/m^2 \pm 0.04\ W/m^2$ [6]. During years of strong solar activity the solar constant increased by approximately 0.1 per cent, or 1.4 W/m^2. These values are quite close to present-day satellite data which show a solar constant of 1361 W/m^2 during an inactive solar phase, and also an approximate 0.1 per cent increase during the activity maximum of the 11-year sunspot cycle. Even 50 years ago, the SAO measurements showed that there are only very minor variations in solar irradiance during the solar active years. Therefore, the solar constant is almost a constant, at least for the accuracy that is relevant for the climate. Langley's original suspicion of considerable solar irradiance variation was thus refuted by the SAO data.

An untapped treasure trove of data left behind

The SAO data has been available for many decades. A couple of years ago they were also put on the internet so that everybody could access them. I found them by chance and started investigating them in 2009. At that time I had been aware of Svensmark's papers (see Svensmark in Chapter 5). He had pointed out that cosmic rays are able to produce condensation nuclei (aerosols) in the atmosphere, which in turn could influence solar irradiance at the surface of the earth via cloud formation. It has been known for some decades that an active sun reduces the incoming cosmic rays by up to 20 per cent compared with an inactive sun. Consequently, Svensmark looked for a correlation between solar activity and cloud formation, and found it (see Chapter 6). Such effects could also exist during clear skies when aerosols are very small. Although the aerosols are not visible to the naked eye, they can either scatter sunlight or partly absorb it. Because the number of aerosols produced by cosmic rays should be the greater when the sun is less active, one would naturally assume that less solar irradiance would reach the earth's surface as a consequence. So I decided to check for this effect using the SAO data, not really convinced I would find a significant effect. This stemmed from a principle that I had learned from a prudent experimentalist long ago: 'You better make sure this is not the case.'

So I was very surprised to find that the terrestrial solar irradiance, the first of the three SAO data series, did not decrease by a mere 0.1 per cent during solar activity minima, but rather dropped by 10 – approximately 1 per cent, or 10 W/m2! The values of the other two data series increased correspondingly. On a percentage level, the increase was even larger – up to 10 per cent for sunlight scattering (aureole). Interestingly, in the SAO calculation of the solar constant from the terrestrial data, the opposing trends compensated each other almost completely. Thus, the SAO analysis also leads to a solar constant increase of 0.1 per cent from solar minimum to solar maximum activity, in agreement with the satellite data. As a consequence, all climate models use the 0.1 per cent variation of solar irradiance during the solar activity cycles. However, the terrestrial variation for solar irradiance under clear skies amounts to 1 per cent. This indicates that another atmospheric amplifier mechanism has to exist, one that is controlled by solar activity. This amplifier acts in addition to the mechanism proposed by Svensmark, which acts via cloud cover.

Because the raw data In the SAO project were used only to help calculate the solar constant, they probably did not bother to look into the stark changes in solar irradiation at the earth's surface. For myself, I could not believe the result. A 1 per cent variation in terrestrial irradiance by solar activity would make the sun one of the main drivers of twentieth-century climate change. And to think that such a treasure trove of data had been in the archives for more than 50 years, before being posted on the internet. And in those 50 years nobody had bothered to evaluate the raw data with respect to solar activity dependence. That just couldn't be possible! In climate science, there are thousands of students worldwide, and each year there is a pressing demand for topics for BSc and MSc theses. Trend studies using all kinds of available data are especially suitable as topics. Somebody had to have noticed and known of these results. My first suspicion was that somewhere there had to be a paper containing an analysis and showing that there were flaws in the data and that all my results were no more than an illusion. I looked for evaluations on the quality of the SAO data, but could find hardly anything.

I remained sceptical of the results and thus sought further terrestrial solar irradiance data. I soon found the solar irradiance data for Mauna Loa, which cover the period 1958–2008. They were provided by Ellsworth Dutton. The Mauna Loa observatory is located on the main island of Hawaii at an altitude of 3400 metres. The data comprise approximately 60,000 data points spanning almost four solar cycles. Yet these are only terrestrial irradiance data like those of the first of the SAO data series. The analysis of these data yielded quite similar results to the SAO data, so I decided to publish all my results [7]. Recently, similar results have been reported from Antarctic stations [8]. In truth, the discussion of those important results is now long overdue [9–12].

Clearly, there are processes in the atmosphere that amplify the solar activity variations to a level that has a real impact on the climate. The details of these processes so far are not sufficiently well understood. However, the cosmic rays influenced by the active sun may play a decisive role. The cosmic rays, which are reduced by solar activity, affect the atmospheric layers down to 10 km altitude, which is deep into the troposphere where all relevant climatic processes take place.

As had been shown in principle by Nobel Prize winner Charles Thomson Rees Wilson in 1899, cosmic rays produce ions of oxygen molecules in the atmosphere, which are enclosed by several shells of water molecules and thus form aerosols [13]. Those ions wrapped by water molecules carry either a positive or negative charge and thus can be identified quite easily by mass spectrometry. However, the aerosols can agglomerate quickly to larger aerosols, which together are neutral and thus make them very hard to detect. In addition, they can 'collect' various atmospheric trace gases such as sulphur dioxide, nitric oxides or ozone. These molecules are easily detectable by physical-chemical analyses, yet the charge-carrying molecules remain hidden.

The solar irradiance data which I have analysed indicate that these aerosols stay relatively long in the atmosphere (months or even years). They contribute considerably to the scattering and absorption of sunlight. As mentioned, during a solar activity maximum, the scattering of sunlight is up to 10 per cent less than during a solar minimum. As a consequence a large part of the atmospheric aerosols is probably generated by cosmic rays, which are reduced by up to 20 per cent from solar minima to solar maxima. In addition, the aerosols absorb sunlight in the range of the so-called water lines in the near infrared part of the solar spectrum. Again, this absorption is correspondingly reduced between minima and maxima. Thus the total increase of terrestrial irradiance of 1 per cent during the solar activity maxima can be traced back mainly to the reduced formation of aerosols, which in turn is caused by reduced cosmic ray intensity. The solar activity-caused increase of 0.1 per cent in the solar constant is only a small part of the total increase.

The absorption of the aerosols should give them their own dynamic caused by the effects of thermals produced by the supplied solar heat. As a consequence, the aerosols need not be embedded rigidly in the atmospheric circulation. Instead, they may accumulate in specific zones that are decoupled from the aerosol production zones above 10 km altitude. In such accumulation zones cloud formation may be enhanced because of the larger number of aerosols present there. This effect would be a bridge to Svensmark's observations.

A century since the start of the SAO project, research on properties, residence time and the dynamics of cosmic ray produced aerosols is still at its infancy. It was Fowle who in 1913 proposed that ionic aerosols are important for absorption and scattering of sunlight [3]. Although accurate methods have been available for the analysis of small particles for a couple of decades, they still have not been used in aerosol research. The reason may be that climate research has focused too much on the effects of CO_2, which, like a constraining dogma, has hindered the quest for scientific understanding since the mid 1980s. That dogma may also have been responsible for delaying the analysis of long-available solar irradiance data with respect to solar activity.

References

1. Fowle, F. E. (1912) The spectroscopic determination of aquaeous vapor. *Astrophysical Journal* 35, 149.
2. Fowle, F. E. (1913) The determination of aquaeous vapor above Mt. Wilson. *Astrophysical Journal* 37, 359.
3. Fowle, F. E. (1913) The non-selective transmissiblity of radiation through dry and moist air. *Astrophysical Journal* 38, 392.
4. Fowle, F. E. (1914) Avogadro's constant and atmospheric transparency. *Astrophysical Journal* 40, 435.
5. Fowle, F. E. (1915) The transparency of aquaeous vapor. *Astrophysical Journal* 42, 394.
6. NGDC (2011) Charles Greeley Abbot Solar Constant Database 1902–1954. http://www.ngdc.noaa.gov/stp/solar/solarirrad.html#abbot.
7. Weber, W. (2010) Strong signature of the active sun in 100 years of terrestrial insolation data. *Annalen der Physik* 522 (6), 372–81.
8. Frederick, J. E. and A. L. Hodge (2011) Solar irradiance at the earth's surface: long-term behavior observed at the South Pole. *Atmos. Chem. Phys.* 11, 1177–89.
9. Feulner, G. (2011) The Smithsonian solar constant data revisited: no evidence for cosmic-ray induced aerosol formation in terrestrial insolation data. *Atmos. Chem. Phys.* Discuss. 11, 2297–316.
10. Feulner, G. (2011) Comment on 'Strong signature of the active sun in 100 years of terrestrial insolation data' by Werner Weber. *Annalen der Physik* 523, 946–50.
11. Weber, W. (2011) Reply to the comment of G. Feulner. *Ann. Phys. (Berlin)* 523 (11), 951–6.
12. Hempelmann, A. and W. Weber (in press) Correlation between the sunspot number, the total solar irradiance, and the terrestrial insolation. *Solar Physics.*
13. Wilson, C. T. R. (1899) On the condensation nuclei produced in gases by the action of rontgen rays, uranium rays, ultra-violet light, and other agents. *Philosophical Transactions of the Royal Society of London Series* A 192, 403.

7. A look into the future

A good forecast is half the battle

Forecasts are as old as civilization itself. Estimating future supply and demand, opportunity and risk, has always been important for human society. Will this year's harvest be sufficient to meet demand or do import contracts need to be closed early? How will the economy, prices and population develop over the coming years? Which natural and manmade hazards threaten society? Good forecasts help us anticipate bottlenecks and avoid harm, or limit it through the early implementation of precautionary measures. Prognoses provide us with valuable and urgently needed time to prepare for expected future events. Many things, after all, cannot be done overnight. As soon as an approaching development or danger becomes apparent, personnel and resources need to be organized, decisions have to be made and countermeasures implemented – and the earlier the better. Taking early action protects us against unwelcome surprises.

But speed is not everything. It is also the case that prognoses are less accurate the further into the future the expected event is. The sooner one starts, the more uncertain the prognoses in general. Here things have to be decided by instinct. At what point does the probability of the prognosis exceed a critical value? When is there still enough time to implement countermeasures at a reasonable cost? It can be a real high-wire act. Prognoses first have to be carefully evaluated before comprehensive and costly measures are taken. Fundamentally, one has to distinguish between well-defined, predictable developments and those that are not predictable at all. Lottery draws, for example, are absolutely unpredictable, as is the final score of a sporting event. Predictable within limits is the victory or a loss of a team, if the strengths of the teams are significantly different. The Christmas present you can expect from your mother-in-law is predictable to some extent if she has always given you hand-knitted socks in previous years. A definite prediction is the date of someone's birthday in the coming year or the occurrence of a traffic jam during rush hour just before a long holiday weekend.

Forecasts have to be given with a probability of occurrence. We know this from weather forecasts as the weatherman gives us a 60 per cent chance of showers and does not say with certainty whether it's going to rain or not. But it is precisely this lack of definition that shows the forecast is serious. Very few forecast probabilities are pegged at 100 per cent; for most there is always some degree of uncertainty. Decisions based on prognoses are thus always fraught because the decision maker has only a limited amount of information to hand.

Forecasting methods have developed enormously since the Oracle at Delphi. Nowadays forecasters do not need to make a sacrifice at a fissure in a rock that emits intoxicating fumes. The foundation of the forecasting business has less to do with fortune-telling and more to do with a solid factual basis. A good forecast is based on a well-documented and comprehensive interpretation of past developments. Identified trends are then projected into the future and, if necessary, corrected using further assumptions. Eventually a useful prognosis is made.

The business of forecasting is data-intensive and demanding. There are many pitfalls which can lead to possible errors [1]. In addition to faulty basis data and flawed assumptions, there's a real danger of over- or underrating the power of some of the processes involved. Early results are often overrated because they are fresh in the mind and are often the subject of intense discussion. And because forecasts are made by people, their desires and fears cannot be excluded. Chaotic processes – systems whose dynamics sensitively depend on the start conditions – also pose huge problems as their behaviour is not predictable over the long term [2]. Forecasts may fail entirely when unforeseen events intervene.

The human factor: conflicts of interest everywhere

Critics point out that prognoses are often used to influence individual behaviour or sway public opinion. Prognoses, therefore, should be scrutinized closely, especially when they forecast over the long term, deal with dynamic systems or serve the interest of the forecaster [1]. A good example here is the 2009 swine flu pandemic prognosis made by the World Health Organization. Based on their forecast, countries ordered large quantities of vaccines. Germany, for example, ordered amounts worth triple-digit millions of euros. By the end of 2010 most of the illnesses turned out to be relatively mild; the horror scenarios were not realized. Consequently, most of the vaccines were unused and had to be written off as a loss. In March 2011, then Federal Health Minister Philip Rösler replaced half of the sixteen members of the Permanent Vaccine Commission in a single sweep [3].

Unfortunately, there are also possible conflicts of interest in the current climate discussion. On the one hand, we have the power producers, energy-intensive companies, airlines, car manufacturers and many of the paying public. It is understandable that this group accepts only the CO_2 limitations that are absolutely necessary and refuses any measures that go beyond this due to cost. A few of the more extreme representatives of this group overstep the mark and dismiss the greenhouse effect altogether. Though this would remove the climate problem at a stroke, it would also fly in the face of a vast body of scientific findings that show CO_2 plays some part in climate.

On the other hand, we have some research groups, insurance corporations and finance ministers who are profiting from climate alarmism. Exaggerating the alleged climate risk ensures continued state funding for research institutes, leads to the set-up of more institutions and increases the number of jobs and funding for scientists. The postulated great climate danger opens up new career opportunities, prestige, media limelight and political advisory positions for otherwise anonymous climate scientists. All over the world climate protection agencies are being created, climate representatives are being hired and climate instructors for making presentations at schools are being contracted. There's something in it for insurance companies too. Raising alarm over increasing storm, hail and flood damage is driving up the demand for insurance policies covering such natural hazards while at the same time justifying increasing the cost of insurance premiums. And so much the better if the increase in catastrophes never happens – that way profits will be even bigger!

It becomes especially beneficial when groups with similar interests present a united front. This, for example, is how in May 2011 the General Association of German Insurers (GDV) published the result of a study based on climate models warning of a strong increase in damage from climate-related natural catastrophes by the year 2100 [4]. Among others, the Potsdam Institute for Climate Impact Research (PIK) was involved in the GDV study. The PIK in the past has hardly acted as an independent mediator in the climate discussion and feels at home with the insurance companies in the climate debate, as both stand to profit handsomely from the alleged climate crisis. What should we honestly think of such cooperation and the warnings they issue to the public? One certainly cannot resent the government for seeing a new source of income by issuing CO_2 certificates. Banks and other brokers too are thrilled about the lucrative, high-volume business opportunities of CO_2 certificates trading. Literally trillions are at stake.

The vast array of possible self-interests is indeed a cause for concern. Under these conditions and with the stakes so high, is it really possible to provide independent climate forecasts? The only way out of this impasse is to create a dialogue among the various groups where results can be presented openly. Of great importance is the opportunity to discuss especially controversial points. Questions that still remain open could be researched in jointly formulated projects. Thought control, the proscription of certain ideas and defamation, such as calling opponents 'climate deniers', must have no place in this forum. The basis for such a new beginning in climate science would be, first, to strictly separate science and politics. Only by returning to a factual level can urgently needed transparency be restored. Only then is there hope of narrowing the gap between the climate prognoses of the IPCC and those of the so-called sceptics.

And now the weather

Loosely defined, climate is the average weather over a period of many years. Therefore, it makes sense first to take a close look at weather forecasting. Unlike climate prognoses, people listen to weather forecasts almost every day. Sometimes the weather may not matter to us much as we'll be spending the day indoors. At other times, however, we need and expect sound meteorological advice for our planned outdoor activities. Should we invite friends over for a barbecue or should we watch a DVD and heat a pizza in the oven?

Many of us may not admit it, but the quality of weather forecasts today is really quite good. Only rarely do we have cause to complain. Today twenty-four-hour forecasts have an accuracy of 90 per cent or better. The accuracy for three-day forecasts is in excess of 75 per cent. Five- to seven-day forecasts to some extent are possible. But beyond seven days reliable forecasts are uncertain. This is not due to weaknesses in current technology; problems arise for physical reasons, namely the chaotic processes that weather systems involve. As a result, there is in practice an insurmountable limit when forecasting beyond two weeks. A few especially business-minded meteorologists appear to have succeeded in flouting the laws of nature and offer special services on their websites for an extra fee. These so-called premium weather forecasts may cover periods up to twenty-eight days. Customer complaints made later are, of course, excluded by the fine print in the general terms and conditions.

So if the weather cannot be forecast with any reasonable certainty much beyond a week, how is it possible to project the climate over decades? Here we have to exercise caution as this argument is misleading. Weather is not the same as climate. Climate prognoses have an entirely different character. Climate is the average weather over a period of years. When determining the mean weather, the chaotic elements associated with weather are reduced, which in principle make the forecast of long-term trends possible [5]. The best examples are the seasons. The exact weather for a certain day far into the future cannot be reliably forecast. However, the prognosis that the temperature as a trend will drop from summer to winter is correct.

Weather folklore

For farmers the weather is the most crucial element in their livelihood because the right weather mix is the prerequisite for a good harvest. Rain, sun and warmth at the right times have been among the fundamental elements of successful cultivation for thousands of years. Our ancestors devoted much thought to how best to minimize weather risks long before the invention of satellite-supported weather charts. Based on the weather records kept by earlier generations, our ancestors were able to recognize repeating patterns in weather systems and to decipher their interrelations. Based on weather patterns of certain days or single months, they attempted to predict the subsequent weather. '*When March blows its horn, barns will have hay and corn.*' [6–9]

Weather proverbs often rhymed and today are often quoted, but with a wink of the eye. Yet some proverbs possess a surprising degree of accuracy. Modern scientific studies have discovered a solid core at the heart of about three dozen proverbs that have an accuracy of 75 per cent or better [8]. However, keep in mind that weather proverbs derived from practice are valid only for the local area and cannot be applied in other geographical regions.

As one might expect, many weather proverbs of medieval origin entail superstition. One example is the German Hundertjähriger Kalender (Hundred-Year Calendar) written in the seventeenth century by a monk named Mauritius Knauer. It was based on classic astrological beliefs and planetary constellations. The calendar was supposed to provide weather forecasts for the monastery's agricultural operations, but in reality it was of little value. Yet, the Hundred-Year Calendar is still printed by a number of publishers today.

Climate prognoses of the 1970s: the coming ice age!

Climate scientists made their first really big splash on the global stage in the 1970s, but hardly made the best of impressions. What happened? Temperatures rose steadily from 1910 to 1940 (Figure 4.3). Here the greenhouse gas theory appeared to work well. But starting in 1940 the climate cooled noticeably and simply refused to return to a warming trend. Scientists were baffled. Why didn't nature obey their meticulous models? Nature demonstrated her mischief by delivering some nasty events. The winter of 1968–69 brought with it an ice sheet in the north Atlantic that had not been seen in almost 60 years. During the winter of 1972–73, icebergs were seen as far south as Portugal and in the Arctic Circle the coldest winter temperatures in over 200 years were recorded. Then, in August 1973 a snowstorm ruined a large swathe of the Canadian wheat crop. In autumn of the same year the north German coastline experienced its most severe coastal flooding in half a century [10]. An extremely cold winter left the United States shivering in 1977 [11] (see Chapter 5).

The experts tried to work out what was happening but failed to agree. Some reasoned that the cooling marked a possible end to the warm period awaiting us for the coming century or even millennium. The repeated alternation between glacial and interglacial periods was well known from studies and was the earth's defining climate pattern of the previous two million years. In many cases, the interglacial periods lasted only 10,000 years, and 10,000 years is precisely the time that has elapsed since the last ice age ended. So the scientists were hoping that anthropogenic greenhouse gas emissions would help to avert the approaching new ice age [12–13]. The long periods associated with these natural climatic developments, however, reduced concern on one hand, while on the other, it was well known that massive temperature plunges had occurred during such transitions between the two climate extremes. For example, at the start of our current interglacial 11,000 years ago, a time known as the Younger Dryas, the globe experienced a temporary setback and plunged back into ice age conditions. The polar regions of the Northern Hemisphere cooled a dangerous 10° C within just a few decades [14–15].

Other scientists saw a more short-term threat approaching and warned of a coming cold phase analogous to the Little Ice Age, which persisted from the fifteenth to the nineteenth centuries and caused harvest failures, famine and disease throughout Europe [16]. The global average temperature at the time was only about 1 ° C lower, yet that was enough to spread fear and panic through a society that was poorly developed technically. Unlike the end of the Medieval Warm Period, when a weak sun caused temperatures to drop, mankind in the 1970s was thought to be responsible for the cooling, according to some scientists. Beginning in the mid twentieth century, large volumes of smoke, ash and other particles were being emitted by expanding industrialization, forming a sun-blocking veil across the globe [17–20]. At the same time, jet engine air traffic came under suspicion for augmenting the formation of cooling clouds. The suspected clouding by particles was dubbed global dimming. It was argued that global dimming more than offset the CO_2 warming, and thus dirty air was responsible for the cooling of the 1940s–1970s.

One of the most vehement global cooling alarmists was the American climate scientist Stephen Schneider, a professor at Stanford University who later became an IPCC lead author. In 1971 he calculated that a quadrupling of aerosols would lead to a dramatic global cooling of 3.5° C [19]. If the trend worsened, he saw the danger of a full-fledged ice age on the horizon. Ironically, Schneider later became one of the main alarmists in the global warming campaign. Just after he switched sides, in 1979 Schneider warned in an article in the *Palm Beach Post* that the west Antarctic ice sheet would melt dramatically before the end of the twentieth century and that the sea level could rise several metres. In his projections he pictured with great media effect how cities on the east coast of the United States would be submerged [21]. Thirty years later, New York City, for example, still remains high and dry. Since then the sea level has risen only a manageable 100 mm. Schneider's apocalyptic visions were well wide of the mark.

Another man who warned of cold in the 1970s was the American Nobel Prize winner Linus Pauling. He feared climatic cooling could develop into a global catastrophe and pose the toughest test civilization had ever faced [22]. The American Ad Hoc Panel on the Present Interglacial had similar thoughts and in 1974 projected an annual cooling of 0.15° C until 2015 [23]. Another meteorological scientist, James McQuigg, held out little hope. He estimated the odds of returning to warm days, as in the 1930s, were 'at best 1 in 10,000' [22].

Oddly enough, the consequences that the experts expected to see were the same as those discussed today in connection with global warming. People were told to expect more hurricanes, droughts, floods and famine. Politicians wrung their hands and saw the need to take urgent action, among them US presidents John F. Kennedy, Richard Nixon and Gerald Ford, and Soviet leader Leonid Brezhnev. They even discussed the possibility of closing the Bering Strait between Alaska and Russia in order to lock in the cold Arctic waters. Another geo-engineering suggestion made at the time was covering the ice caps with a black film in order to change its albedo. They also considered putting up mirrors in the earth's orbit to act as additional suns, or to blast underwater mountains in the oceans with atomic devices so that warm water currents could flow more freely. Another possible solution was to crank up CO_2 emissions in order to intensify the greenhouse effect [23].

However, there were some who warned of warming [24]. Others were torn between the two sides [13]. For example, in 1972 Cesare Emiliani, Italian founder of paleo-oceanography, warned of possible apocalyptic dangers on both sides: 'Man's activity may either precipitate this new ice age or lead to substantial or even total melting of the ice caps ...' In those days anyone who could still sleep soundly was considered to be beyond all help.

Today we are much better informed – or at least that's what we like to believe. As we now know, the cold phase of the 1970s was followed by a warming phase (see chapters 4 and 5). Here, climate experts demonstrated their uncanny flexibility, mothballed their warnings of global cooling and wasted no time in rolling out the new global warming scare. Today, just three decades later, new aggravation for the alarmist scientists appears on the horizon once again: the earth simply refuses to warm up; the warming stopped almost 15 years ago. Temperatures have stalled and have oscillated over the last decade about a plateau. Even Hans-Joachim Schellnhuber, Director of the Potsdam Institute for Climate Impact Research, had to admit: 'It is simply a fact that global temperature has stabilized at a high level' [25].

If cooling does occur in the years to come (and there are many signs this will be the case), then the déja vu will be perfect. The cooling phase of the 1960s and 1970s was for the most part not manmade, but had been ushered in by the emerging cold phase of the 60-year PDO oceanic cycle in the 1940s (see Chapter 4). More than 60 years later, the cycle has now run full circle so that today we find ourselves once again at the start of a new cycle. Our surprise over the missing warming of the last decade is indeed somewhat subdued. And why shouldn't it be? Haven't we seen all this before?

Does the IPCC really have everything under control?

Today we just smile when we look back at all the failed predictions of global cooling made during the 1970s. But here we ought not to forget that today's experts have not always been successful with their predictions. Even the German climate scientist Mojib Latif got carried away in the year 2000, proclaiming that Germany should no longer count on cold winters in the future [26]. 'Winters with sub-freezing weather and much snow like those 20 years ago will no longer occur at our latitudes.' Unfortunately, nature ignored the proclamations of climate scientists and went on to produce consecutive cold, snowy winters in 2008–11 in central Europe. Here temperatures were well below the long-term mean for the period 1961–90. In the same year, David Viner of the renowned Climate Research Unit (CRU) of the University of East Anglia declared that British children would soon not know what snow looked like [27]. Shortly thereafter, over the winters that followed, British school children went on to experience a snow-related knowledge gap, not because of a lack of snow, but because of multiple school closures when heavy snow made roads impassable!

In the spring of 2008, Mark Serreze of the American National Snow and Ice Data Center shocked the public with one of his forecasts: the North Pole would be ice-free that year with a probability of 50 per cent [28]. Summer came but nature took no notice of his forecast and thought nothing of exposing the North Pole. In response, Serreze postponed the ice-free North Pole another 27 years to the year 2030 [29]. It's that easy in climate science.

In its 2007 report, the IPCC was convinced that rainfall over East Africa would increase in the coming decades. That much appeared certain. After all, eighteen of their twenty-one climate models had forecast it [30]. Unfortunately, the scientists forgot that the climate has a mind of its own and delivered the opposite result. The Horn of Africa became continuously drier with tragic results: in mid 2011 the worst drought in a century struck Somalia, Kenya and Ethiopia, claiming tens of thousands of victims. Worse, the catastrophe had not come out of the blue and could have been relieved had serious warnings been heeded. Scientists of the Famine Early Warning Systems Network had looked closely at climatic interrelationships a year earlier adding natural climate phenomena like La Niña to their calculations.[31–33]. Using a well thought out methodology, the scientists had correctly forecast the catastrophic drought in the Horn of Africa, but their warnings fell on deaf ears.

Anyone who wishes to gain more insight into the various climate blunders and confusion can find abundant material on the internet.

Here critical observers have assembled lists of controversial points from the climate science world, everything from exaggerations to full-blown scandals [34–35]. There's something for everyone.

The most famous distortion of climate facts was produced by former US Vice President Al Gore. In his Oscar-winning *An Inconvenient Truth*, for which he was awarded the Nobel Peace Prize in 2007, Gore produced an entire plateful of glaring errors and gross exaggerations, which the British politician and journalist Christopher Monckton of Brenchley meticulously refuted in a thirty-five-point list [36]. Nine of these points were even judged as errors by a British High Court in a judgment handed down in October 2007 [37]. The High Court determined that the points presented by Gore were 'in the context of alarmism and exaggeration'. The film's apocalyptic vision, according to the High Court, cannot be viewed as an independent scientific assessment, but rather is as a political statement. For this reason, the High Court ruled that the film could not be viewed in schools without comment, as screening the film without first presenting alternative viewpoints would be a violation of the education law requiring lessons to be politically balanced [38]. The High Court criticized the exaggerated sea level rise of up to 7 metres in 'the near future' and the intentional reversing of cause and effect in explaining the synchronicity between CO_2 and temperature development over the last 650,000 years. Also the shrinking of Lake Chad, the melting of the Kilimanjaro glacier, 'Hurricane Katrina' and drowning polar bears could not be linked to manmade climate change, the High Court ruled [37].

The IPCC too, which was awarded the Nobel Peace prize along with Al Gore in 2007, committed a number of embarrassing errors in its reports. The most prominent of these is the now discredited hockey stick chart, which was prominently featured in TAR. The chart significantly undermined the IPCC's credibility (see Chapter 4). The chart was partly corrected in AR4, but the IPCC once again became mired in controversy at the end of 2009 as three blatant errors in succession came to light. The most widely publicized was probably the projection in Part II of AR4, which claimed that 80 per cent of all Himalayan glaciers would melt completely by the year 2035 [39].

A lapse also occurred in estimating flood risks when the IPCC claimed that 55 per cent of Netherlands was below sea level. The actual percentage is 26 per cent. It certainly would have been economically devastating if costly investment decisions had been made based on these distorted figures. The third blunder was that the IPCC warned that agricultural yields in some African countries 'could fall by up to 50 per cent' by 2020. While this is scientifically plausible for the three Maghreb countries (Algeria, Morocco and Tunisia) it is not true for the other forty-nine African countries. Despite the inaccuracies, these dramatic prognoses found their way into the IPCC Summary for Policymakers.

Regrettably, these errors have a way of persisting, especially when they are purveyed by important representatives of climate science such as Hans-Joachim Schellnhuber. In an interview in a *ZDF* television programme (30 October 2009), he said, 'In the next 30–40 years ... with 2° C of warming ... these [Himalayan] glaciers will for the most part disappear ...' [40]. In the same interview he added, 'the ice ages will no longer naturally occur'. Errors in the 2007 IPCC report also involve estimating the flood risk of the Netherlands and Africa's agricultural yields. Former IPCC Chairman Robert Watson publicly wondered why all three errors had led to an exaggeration of the climate problem and demanded a closer examination of the circumstances leading up to them [41].

To err is human

The German literary giant Johann Wolfgang von Goethe once said, 'Man errs as long as he doth strive.' Thus it is little wonder that the history of science is marked by blunders and exaggerations. Today we know the earth is not flat and that the sun does not orbit the earth. At the end of the nineteenth century leading physicians warned that man would not be able to withstand speeds of 40–50 km an hour. In the early 1970s Paul R. Ehrlich, American butterfly researcher and author of the bestselling book *The Population Bomb*, predicted that famine would kill half the world's population by 1980. He also saw little hope for Great Britain and in September 1971 in a speech delivered at the British Institute for Biology predicted that the country would be threatened by acute poverty and hunger by the year 2000: 'If I were a gambler, I would take even money that England will not exist in the year 2000' [42].

In the late eighteenth and early nineteenth centuries, a bitter scholars' dispute broke out among geoscientists, between the so-called Plutonists and Neptunists, one that closely resembles today's climate dispute between the alarmists and the sceptics. Back then the Neptunists claimed that all stone on earth formed from the early oceans, while the Plutonists postulated that all stone had a volcanic origin. As is the case in this historical example, the truth in the climate dispute lies somewhere between these two extremes.

Another impressive example from the field of natural sciences was the theory of plate tectonics, which for decades was rejected by experts. In 1910 the natural scientist Alfred Wegener concluded that the earth's plates must have drifted over millions of years. Professional circles, however, were aghast and mercilessly ripped his hypothesis to shreds. For 50 years it remained the consensus among scientists that the continents could not move at all. However, surveys of the ocean floor during the 1960s showed that Wegener had been correct. Sadly, he was not around to relish the moment. Today, the theory of plate tectonics is generally accepted and is common knowledge. This example plainly demonstrates that a scientific 'consensus' is no guarantee that a model is correct. Science is not about winning the popular vote. Sometimes, if not generally, an idea from a single person or a scientific minority is all it takes to topple an entire body of knowledge.

Virtual climate in the computer

Compared to the 1970s, our current climate knowledge has multiplied, as has computing power. Climate simulations today are routinely run using computers, which churn out huge quantities of data with many decimal places. However, when you get down to it, all results come out of a closed black box. What actually happens inside the box for many is more or less a mystery. We simply have to trust the experts. But is our quasi-religious belief in mathematical equations justified in climate science? The only recourse we have to find out more about the modelling process is to ask the modellers what processes went into the calculations, and in which form and weighting, and which aspects were not accounted for.

And there are a few additional questions: Do we really know enough today to be able to write an accurate formula for all climate-relevant processes? Does the level of modelling precision suffice to represent the earth's complex climate system processes quantitatively or even just rudimentary trends? Or is the science perhaps not quite at a level where we can realistically represent the immensely complex interplay of atmospheric forces in a realistic manner inside our little black boxes? Where do uncertainties exist and how big are they? Do chaotic elements really have no decisive role?

One thing is certain: climate scientists are striving to cast the climate processes in all their facets into the calculations and to work them into the climate models. However, the various processes are complexly coupled and thus lead to non-linear interactions. Amplifications, beats and interferences are the result. So against this backdrop of complexity, it's more than annoying when a climate scientist like Schellnhuber proclaimed at a press conference before the Copenhagen conference that 'there is an extremely simple, quasi-linear relationship between the global mean temperature and the total amount of CO_2 that will be discharged into the atmosphere in the decades ahead' [25]. Such statements are nothing more than unscientific attempts to take the public for fools.

So far, not so good. What happens when important processes are missing from the climate model equations? As we have seen in Chapter 3, solar activity and temperature development in the past were closely coupled. Yet the current IPCC climate models cannot reproduce their synchronicity. In Chapter 6, we traced the extensive lines of evidence, which clearly illustrate the involvement of one or more solar amplification processes. None of these amplifiers has been considered as a possibility, let alone included in the IPCC climate models. When important control mechanisms are simply left unaccounted for in calculations, the models have no chance of reproducing reality. Even the most powerful and expensive computers are useless against gross errors. The modelling results illustrated in colourful charts with numerous scenarios and alleged high levels of precision are worthless when the fundamental, physical and climatological data are inadequate. This is not about dotting the 'i's and crossing the 't's; rather, it's about getting the basic interrelationships right [43].

That does not only concern the role of the sun in climate, but also the climatic impacts of clouds, water vapour, aerosols, the interaction between oceans and atmosphere and the flow processes of ice sheets [44]. Also inadequately understood are the cycles of trace gases such as CO_2, methane, nitrous oxide and ozone, which cannot be calculated, but have to be estimated in the climate models [45]. All these processes are currently too poorly understood and therefore are not correctly or adequately taken into account in models. Then again, natural oceanic oscillations such as the AMO and the PDO are hardly reproducible in the simulations. When it comes to climate models, it's basically one construction site after another. The often sworn consensus is nowhere in sight.

Falling through the gaps

Because of increased computer capability, the spatial resolution of climate models has steadily improved since 1990. Yet a fully quantitative modelling of the climate by computer would still be elusive, even if, hypothetically, all climate equations were completely to hand. Even with today's supercomputers, modern ocean-atmosphere climate models need a lot of computing time due to their sheer complexity. The spatial resolution of the models is thus still limited to a few hundred square kilometres [45]. All processes that take place within this spatial resolution cannot be expressly formulated. Figuratively speaking, they fall through the gaps of the calculation grid. Because the processes cannot be calculated individually, the model has to be supported by guesstimates.

Very elusive, for example, are the turbulent exchange processes at the boundaries between the earth's surface and the atmosphere, over the ocean as well as over the land [45]. The processes of oceanic deep water and sea ice formation are also too small ever to be depicted in the calculations. Furthermore, the models have serious problems simulating vertical air movements, as well as the interrelationship of airborne particles and cloud formation [46]. For these reasons the various climate models employed in AR4 use very different precipitation distributions and intensities [46]. Reliable statements on the development of precipitation can hardly be generated in this way. Because of the coarse resolution, it is not possible for the climate models to draw any credible conclusions on regional or local levels (e.g. country levels), which would be useful for climate investment decisions [46].

In addition to spatial limitations, there are also temporal limitations. Owing to the unhurried heat transfer between the atmosphere and oceans, the impacts of external climate impulses can be delayed by months, years, decades or even centuries. Due to differences in ocean depths and land/ocean distribution, these time lags differ for the various oceans [47]. As if lurking inside a time machine, historical impulses meander through the climate system and thus end up not being taken into account by the climate models' very short-term consideration.

We could now place our bets on technological development and hope that the high-performance supercomputers of the future will get it all under control. But we are still a long way from that point. Even with an improvement in the models' resolution, the chances of a comprehensive, reliable, long-term forecast are poor because the coupled, non-linear chaotic climate system makes looking into the distant future impossible. Here even the IPCC agrees [48]. Precise mathematical predictions for complex natural processes have always failed in other fields as well [49]. The attempt to support political decision makers with modelled prognoses is laudable, but should not lead to over-confidence where we opt for virtual results over real-world observations.

Problems with checking a model's plausibility

It is clear that models that are used to forecast the climate of the future also have to be able to reproduce the climate of the past (hindcast). And it is precisely here that the IPCC's models run into difficulties. The models can only reproduce certain parts of the real temperature curves. Important natural cycles such as the PDO and the El Niño events cannot be satisfactorily simulated [50–51]. None of the existing climate models predicted the missing warming of the last 10 years. Even the Medieval Warm Period of 1000 years ago cannot be reproduced by any of the current models. That passed unnoticed for years because the IPCC used incorrect historical temperature data in the form of the hockey stick to conduct the reality checks (Figure 1.2; see Chapter 4). Comparisons to false base data inevitably lead to dead ends.

But it is not as if the climate modellers had ignored the past altogether. To the contrary, the target has always been to reach as good an agreement with reality as possible. It is, however, debatable whether synchronicity between a model and reality can confirm the calculation approach. The right answer does not necessarily mean the right approach was used. The models had many freely selectable parameters that could be adjusted in order to get the right curve shape. These controversial and difficult to check adjustment dials are called fudge factors. Some critics accuse the IPCC of having tweaked the models so that they exaggerate the greenhouse warming of CO_2 [52]. A colleague once showed Enrico Fermi, a leading particle physicist, a good agreement between a theory and experimental results. Fermi then asked him about the number of freely selectable parameters, to which his colleague answered there were four. Fermi countered, 'I remember my friend Johnny von Neumann used to say, "With four parameters I can fit an elephant, and with five I can make him wiggle his trunk."'

Interestingly, on p. 774 of TAR, we read: 'In climate research and modelling one has to recognize that we are dealing with a coupled non-linear system and so long-term forecasts of the future climate condition are not possible.' Isn't it odd that this very important statement was not reproduced in the Summary Report for Policymakers, in 2001 or 2007?

Models can be useful

Due to these fundamental limitations, climate models are at best incomplete depictions of a very complex reality. For this reason, and because crucial climate factors such as fluctuating solar activity are insufficiently accounted for in the models, climate models cannot deliver any detailed climate forecasts decades into the future [45]. Yet the simulations do have some value for researching the climate of the future. Climate models help mainly in comparing different scenarios. Here a series of models can be calculated in which only a few parameters are changed. In this way climate trends can be worked out and taken into account in prognoses [44].

Only when one understands the limitations and the potential of the models can the results from the modelling be used appropriately. This is how a 'black box' becomes a 'grey box'. Blind faith in computer results in this context would clearly contradict the fundamental ideas of climate modelling. The comprehensive, no worry climate package will never be reproduced by computer. More important than ever is the critical thinker, who has first to meticulously filter out the real climate signal from the outpouring of digital data and then decide among alternative interpretations.

A new strategy for climate forecasting

It appears highly improbable that the current IPCC climate prognosis is able to reflect reality correctly. As discussed in previous chapters, the IPCC flatly ignored or underestimated the sun and other natural climate cycles. Of course, everyone is allowed to make mistakes and this applies to the IPCC too. Now there is an opportunity to remove these deficits from future assessment reports. We can only hope that this will be done.

So just what would an improved forecasting strategy that brings everything under one roof look like? First, we should send the sinfully expensive modellers on a long holiday, as we need time to think carefully over the absolute fundamentals once again. If the basic data (i.e. the model equations) are doubtful, then all the meticulous calculations become worthless. Garbage in, means garbage out.

Let's go back to the start and pose a question that so far has not been taken seriously enough: What was the climate really like in the past before humans intervened? Is it possible to detect fluctuations that are cyclically repeated? A sober, data-oriented look back here is better than any mathematical high-tech modelling. If a development in the pre-industrial age occurred umpteen times, then how probable is it that the process would suddenly stop during our lifetime? Is it really plausible to think that natural processes have somehow miraculously ceased and the world now takes heed exclusively of manmade climate influences, which, without a doubt, do exist?

All climate scientists should sensibly answer this question with a loud 'No'. The natural fundamental climate patterns of the past also apply today – period. So they must be taken into account in the future, even if we do not fully understand their cause. Once the natural processes are identified and defined, we have to add the anthropogenic climate contributions to these natural dynamics to generate our new climate prognosis. Certain reciprocal actions are possible and have to be researched in detail. What exactly needs to be done?

1. We need a detailed reconstruction and documentation of the temperature history over the various timescales (i.e. the last 10 years, 100 years, 1000 years, 10,000 years, 100,000 years) for as many areas as possible in order to be able to filter out differences caused by climate zones, geographical latitudes, ocean impacts, etc.
2. We should attempt to decide on the climate control factors involved by analysing the historical trends. Here attention must be given to any parallelism, which is always a good starting point but does not necessarily indicate causality. The assignment of cause and effect is not always straightforward. Think for a moment of Al Gore confusing temperature and CO_2 over the last 650,000 years (see Chapter 4).
3. From the historical-geological climate data, natural cycles of every type (sun, internal climate cycles, etc.) must be carefully worked out. At which times and places did these cycles influence the climatic development? Which phase are they currently in? Extending the recognized cyclic provides valuable information on how each natural oscillation will impact on the climate in the future.
4. When generating climate prognoses, it has to be accepted that the timing of certain climate factors simply cannot be predicted. Among these are climate contributions by large volcanic eruptions when ash clouds cause global mean temperatures to drop for a few years. And because of chaotic processes, the exact occurrence of solar cycles and El Niño events does not follow a predetermined annual timetable, but can be given only within a range.
5. Add the anthropogenic effects (CO_2, other climate gases, aerosols, including possible feedbacks) while using a more realistic CO_2 climate sensitivity, as well as various prognoses of CO_2 emission scenarios.

6. Add up the effects of all the contributing natural and anthropogenic climate factors.
7. Employ models to gain a better understanding of the role of individual components and the interrelationship and interaction of the climate control factors.
8. Generate an improved climate prognosis while providing realistic scenario limits. Here the spectrum of possibilities should not be needlessly limited.

Next, we analyse the important elements for an improved prognosis and estimate their potential impact on the future climate. Here we begin with the solar cycles; then look at the internal ocean cycles and finally address CO_2.

Milankovitch earth orbital cycles

As mentioned in Chapter 3, the various earth orbital parameters change cyclically with lengths of about 20,000, 40,000, 100,000, and 400,000 years. This leads to changes in the sun–earth geometry. Thus solar radiation striking the earth varies in intensity, which leads to variations in the amount of energy reaching the earth, independent of the fluctuations in solar activity. The Milankovitch cycles are credited with playing a leading role in the changeovers between interglacial and glacial periods over the last 1.5 million years. A Milankovitch cycle triggers a new glacial period about every 100,000 years, each followed by a short and warm interglacial lasting approximately 10,000 years [53–56]. Our current interglacial has been with us 12,000 years, so it is high time to start considering the possibility of a new glacial period.

To do this we need to include the combined effect of all three Milankovitch cycle types into a so-called insolation curve, which represents a solar radiation curve (Figure 7.1). The curve shows that we are now at an insolation minimum. That means the earth is currently getting less energy than it normally would, for geometric reasons having to do with the earth's orbit and axis tilt.

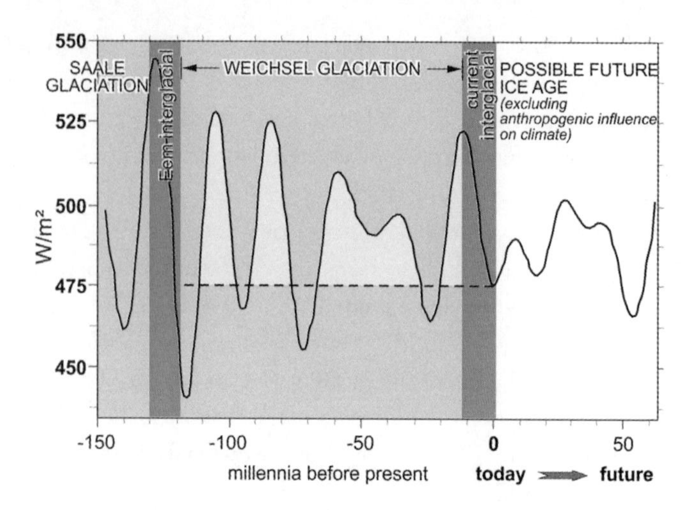

Figure 7.1 Cyclic changes in the earth's orbital parameters (Milankovitch cycles) mean that the distribution of solar energy received on earth changes in both time and geographic area. Due to the uneven land/sea division as well as seasonal effects, these cyclical orbital variations result in changes to the climate. This process occurs independently of the primary changes in solar activity. The graph illustrates the last 150,000 years and a forecast for the next 60,000 years (calculation for mid June insolation at 60° N). Glacial periods began in the past at times of insolation minima, which is where we are today [57].

A look into the past shows that it is precisely at this point in the interglacial (warm period) that a switch to a cold time (ice age) usually occurred. That means today, after 12,000 years of being in an interglacial, the time would in principle be ripe for the next glacial period to start [57]. But things are not as bad as they may appear. The current insolation minimum is considerably above the minimum the earth saw about 100,000 years ago, which ushered in the last ice age [53, 57]. Nevertheless, we should not underestimate the current minimum because 400,000 years ago a similar low-intensity minimum marked the start of a glacial period [57].

So what should we make of the Milankovitch trend? Why aren't we seeing signs of a possible end to the warm times? The languid Milankovitch development is superimposed by shorter-term effects, such as changes in solar irradiation, as well as the manmade greenhouse effect from CO_2. First, the twentieth century was characterized by an unusually high level of solar activity [58–59]. Another reason is because the anthropogenic greenhouse gas emissions have strongly increased since the start of industrialization 150 years ago. Has the slow but powerful Milankovitch cycle not yet had the chance to manifest itself [57]? Will a moderate greenhouse effect by anthropogenic CO_2 interrupt the natural glacial–interglacial cycle?

Over the next decades, and possibly the next centuries, the effects of solar activity and CO_2 will in any case outweigh the effects of the very long-term earth orbital effects, and so it is neither necessary to pursue the Milankovitch cycles further, nor to take them into account in the climate prognosis. Yet, for prognoses over the next thousands of years it is again necessary to consider them.

How much potential for surprise does the sun hold?

As we have seen in Chapter 3, the energy output of the solar power plant fluctuates. The variability is controlled by an entire series of solar cycles, each having a specific period length that ranges somewhere between 11 and 2300 years. Because of the much shorter timescales, solar activity cycles are much more important than the Milankovitch cycles for the climate development of the next decades and centuries. We shall now examine the historical course of each cycle type and extend these cycles into the future. What are the most likely future trends of the various cycles, based on their documented past developments?

At this point we would like to voice a word of caution in order to avoid false expectations. Solar cycle prognoses will neither be exact to the year nor be fully quantitative. The cycles are not strictly periodic [60] because 'chaos' impacts not only the weather and climate, but also solar activity [60–62]. This results in variations in the length and intensity of the solar cycles, but fortunately within certain ranges. For example, the Gleissberg cycle length varies between 60 and 100 years, and not 10 and 500 years. If one uses the historically documented cycle course as a fixed point, then a statistical forecast of the future solar cycle maxima and minima within acceptable error limits is possible. At this point we need not pretend to be more ignorant than we actually are. A rough prognosis is by all means possible and is adequate for our purposes. We shall begin with the shortest of all solar cycles, the Schwabe cycle, which oscillates with an 11-year rhythm.

Weak solar cycle 24

The 11-year Schwabe sunspot cycles have been numbered since 1760. We are currently in solar cycle 24 (Figure 7.2). It is this cycle that has been the cause of sleepless nights for some solar experts. But let's start at the very beginning. During solar cycles 21–23 (1976–2008) the world was in good order. Solar activity during each of these cycles reached high levels with relative sunspot counts of 120–65. Nothing indicated any problems with the solar power plant. As solar cycle 23 underwent its normal winding down in mid 2007, the next cycle was expected to start shortly thereafter [63].

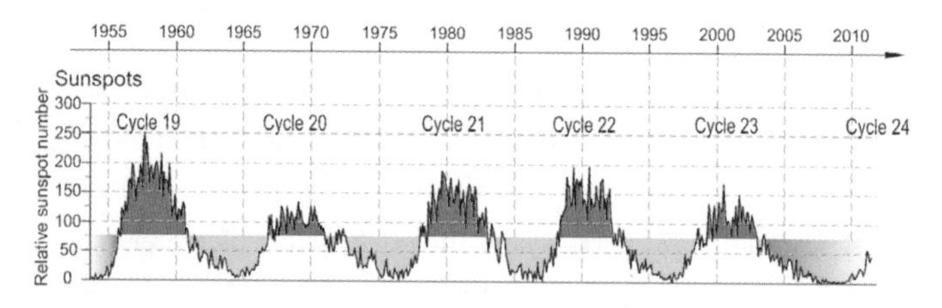

Figure 7.2 Sunspot cycles since 1955. Solar cycle 20 (around 1970) was especially weak. The start of cycle 24, compared to other cycles, was seriously delayed [64].

But the sun simply did not stir; it remained asleep. Sunspots were few. In 2008 there were 266 spot-free days. 2009 was hardly better with 260 days without spots. This was near record territory, as there had been only 3 years with fewer sunspots since 1849 [65]. The solar furnace simply refused to ignite. What was going on?

Because of the solar weak phase, the experts had more than enough trouble determining the exact cycle minimum, which at the same time also marked the official start of cycle 24. After some toing and froing, they finally agreed on December 2008 [66–67]. But the sun ignored the scientists' official solar minimum declaration and continued its slumber. It wasn't until shortly before Christmas 2009 that it ended its long nap and slowly began to return to life (Figure 7.2). Throughout the entire course of solar research history since the middle of the eighteenth century, there had been only four cycles that had started more slowly [68]. Three of these occurred during the so-called Dalton minimum, a cooling phase that lasted from 1790 to 1830, when strongly reduced solar activity is well documented (Figure 7.3). In early 2011, after a delay of about 3 years, the sun finally fired up again, however at only relatively moderate activity levels compared to previous cycles [69].

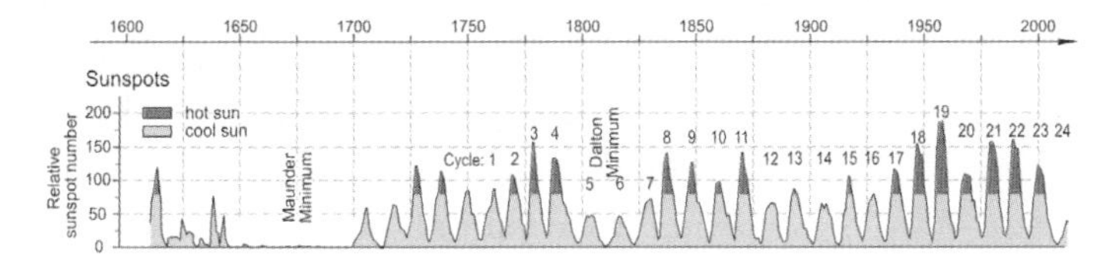

Figure 7.3 Sunspot development over the last 400 years. The number of sunspots has been rising over the long term since the coldest period of the Little Ice Age (Maunder minimum) and reached a peak value in the middle of the twentieth century [70].

The unusual solar activity of the last Schwabe cycle minimum is also reflected by other solar parameters [71–72]. In 2010 the intensity of the solar magnetic field sank to its lowest value in 150 years (Figure 3.5) [73]. Accordingly, cosmic rays reached their highest level in the 50 years in which measurements had been taken [63, 73–75]. Solar winds had also dropped to their lowest level in 50 years [63, 76]. Looking at parts of the extreme UV spectrum, solar radiation here had dropped a full 15 per cent during the recent minimum when compared to the low point of the Schwabe cycle of 1996 [77]. Changes in the UV range were larger than expected [78].

The late start of solar cycle 24 fits well with a weak sun. Late-start, extended cycles like cycles 23–24 were usually associated with fewer sunspots and reduced solar radiation in the past [79–83]. During the low solar activity of the Maunder minimum of 1645–1715, the average Schwabe cycle length was 14 years compared to just 9 years at the start of the Medieval Warm Period, when the sun was very active [63, 79, 84]. It appears that during long cycles the rotating plasma conveyor on the sun's surface weakens and expands further at the poles than it does during active, short solar cycles [85].

Based on these observations, it is possible to make prognoses for the peak value of the current solar cycle 24. NASA is the most important institution for making such forecasts [86]. Interestingly, over the last 3 years NASA has been forced to reduce its sunspot forecast by more than half. In March 2008 NASA officials projected a smoothed sunspot peak of 130–40, and then reduced it again to 100–10 in January 2009. In May 2009 they corrected it once more, this time to 80–90. And in December 2012 NASA reduced it further, now forecasting a value of just 72. It is thought that the cycle peak will occur sometime in autumn 2013, although others believe that the peak may in fact have already taken place in February 2012 without anybody noticing it. For example, Jan Alvestad, a well-known meteorologist, calculated a maximum of 66.9 sunspots for the current cycle [87]. Whatever the case, the current solar cycle will clearly be the weakest Schwabe cycle of the last 100 years [88].

Using the available data, is it possible to make a prediction for solar cycle 25, which will reach its maximum in around the year 2025? [89–90]. From an analysis of the historical sunspot development since the seventeenth century, we know that a single weak cycle rarely occurs by itself but is usually followed by at least two weak cycles (cycles 5–7 during the Dalton minimum and cycles 12–16; Figure 7.3) [59, 91]. Therefore, the current weakening of solar activity could mean a departure from the high solar activity plateau of the last decades.

There are more independent arguments that all point to the same conclusion: the sun is now entering a decades-long quiet period [89, 92–93].

1. The magnetic flux density of sunspots has declined continuously since 1998, and for this reason it is expected that solar cycle 25 will be even weaker than cycle 24 [94–97].
2. The huge plasma currents on and in the sun have slowed down considerably in recent years. NASA thus projected in 2006 that solar cycle 25 could be one of the weakest sunspot cycles in recent centuries [98].
3. The sun's surface is pulsing in a way that indicates its engine is preparing for a long-term quiet period [89–90].
4. Changes in the sun's corona indicate that a slowdown in solar activity is imminent [99–100].

An outlier prognosis hits the bull's eye

The weak and delayed cycle 24 took a number of solar experts by surprise and plainly shows that the field of solar activity prognoses requires much more research [91]. However, no one can claim that they had not been warned. Back in 2003 a small group led by Mark Clilverd, of the British Antarctic Survey, had suspected something was amiss even before there were signs pointing to a slowdown in the Schwabe solar cycles. The team produced a solar activity forecast up to the year 2140 which they refined in 2006 (Figure 7.4) [101–102] in which they predicted that a strongly reduced cycle 24 would mark the start of a solar activity slumber extending until the year 2030, before starting up again and remaining at a more elevated level until 2100, after which another pronounced, extended quiet period would ensue. Clilverd's activity forecast was based on a careful analysis of the entire spectrum of solar cycles available, from the 11-year Schwabe cycle to the 2300-year Hallstatt cycle. By projecting these solar oscillations into the future, he obtained the information needed for correctly predicting the collapse in solar activity towards solar cycle 24 that followed a few years later.

The prognoses made by Clilverd's group came under attack in a heated dispute in 2004, and it was questioned if a reasonable prognosis was even possible [103]. Critics attacked the very idea that a simple extrapolation of the 210-year Suess/de Vries cycle would predict the onset of a pronounced solar minimum. There were no indications to support this and forecasting such a minimum was deemed too adventurous [103]. Just a few years after these pessimistic estimations were made, the solar minimum appears to be in full swing – a victory for the Clilverd group and solar forecasting.

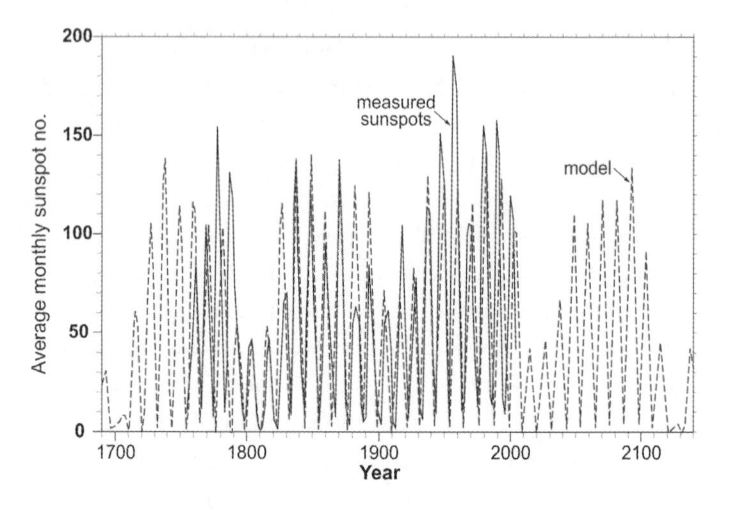

Figure 7.4 Solar activity prognosis made by of the Clilverd group. According to Clilverd, the current decline in solar activity is part of a solar weak phase that will extend to the year 2140 [101].

Like Clilverd, Khabibullo Abdusamatov, Director of the Space Research Laboratory at the Pulkovo Observatory, similarly predicted a weak solar cycle 24 [104]. Using a method similar to Clilverd's, he forecast that the number of sunspots for cycle 24 would not exceed seventy. That agrees quite well with the 2011 forecast. Very few other scientists supported the forecast of a weak cycle 24 [105–107].

Crucial for the next 100 years: the Gleissberg and Suess/de Vries cycles

The solar cycles clearly appear to be more useful than some experts once believed. We shall now take a closer look at the possible future development of the Gleissberg cycle and the Suess/de Vries cycle. The existence of both is well documented through frequency analyses of the long historical solar activity measurement series [98, 108]. As we have discussed, the period of both cycles varies within a certain range. The Gleissberg cycle on average is 87 years, with values of 60–120 years being common. The Suess/de Vries cycle has a mean value of 208 years [60, 109] and fluctuates between 180 and 220 years. How have both cycles performed over recent centuries? When did they each reach their respective minima and maxima?

For the historical analysis we have to recall that the cycles could not always freely unfold because all we can measure is the total curve, combining the contributions of several individual solar cycles. Here the various cycle types strengthen or weaken each other, depending on their course or phase. In an extreme case the peak of a cycle may hardly develop if another cycle type with a different period length is at its minimum. Pure mathematical, automated analysis processes may run into problems here. Thus an additional qualitative visual analysis is necessary. Such superimpositions are most likely why at times deviating reconstructions of the Gleissberg and Suess/de Vries cycles are found in the literature.

Taking all available data into consideration, the last Gleissberg maxima occurred in 1760, 1850, 1940 and 2005 (Figure 7.5) [79, 101, 104, 110–114]. An analysis that goes back further is complicated by the distinct solar Maunder minimum [115], which represents the low point of the 1000-year Eddy cycle (Figure 7.3). The Gleissberg minimum of 1975 was so brief that some observers are assuming an alternative broad maximum for the twentieth century instead, one that spans the years 1940–2005 [71, 115–116]. But most experts agree on the Gleissberg downward trend beginning in 2005 at the latest. The next Gleissberg minimum is therefore expected to occur between 2030 and 2040.

The last minimum of the Suess/de Vries Cycle is assumed to have occurred in 1810 and coincides with the low solar radiation period of the Dalton minimum [117] (Figures 7.3 and 7.6). The subsequent maximum occurred in about 1915, but was not able to materialize fully because the Gleissberg cycle at the time was just coming out of its minimum. The next Suess/de Vries minimum is expected to occur in about 2020 [115]. Currently we find ourselves on a downward Suess/de Vries trend. Here again a tracing of the cycle back to the seventeenth and eighteenth centuries is not possible due to the long Maunder minimum period of inactivity.

The pronounced plateaus of the Gleissberg and the Suess/de Vries cycles resulted in a phase of high solar activity during the second half of the twentieth century. This plateau ended in 2005 as both base cycles began their accelerated downward trend [102, 112, 118]. The projected minima of both cycles fall in 2020–40 (Figures 7.5 and 7.6) and thus will have a compounded effect leading to a low point in solar activity analogous to the Dalton minimum of 1810 [74, 83, 118–127]. This projection is also in harmony with the proposed projection of the Clilverd group (Figure 7.4) [101]. The decline may have begun with Schwabe cycle 24, which we currently find ourselves in.

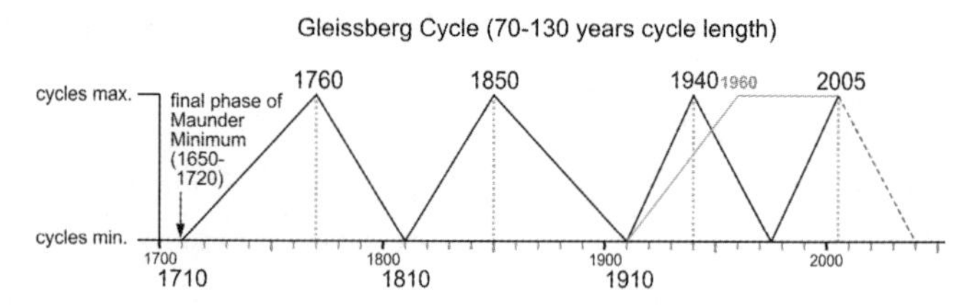

Figure 7.5 Schematic historical maxima and minima of the 87-year Gleissberg solar activity cycle. Currently we are on the downward flank of the Gleissberg cycle.

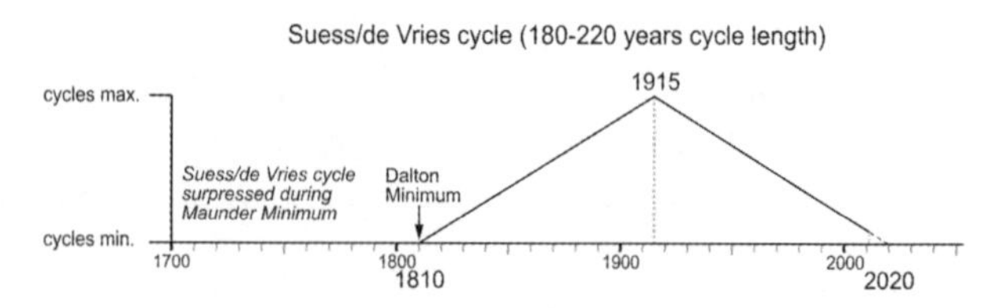

Figure 7.6 Schematic development of the 210-year Suess/de Vries solar activity cycle. The last minimum coincides with the Dalton minimum in around 1810. The Suess/de Vries cycle is currently on a downward trend.

The 1000-year cycle is now at the end of a steep climb

As we have seen in Chapter 3, the solar 1000-year cycle is well documented during our current postglacial period and has had a marked impact on the global climate. The temperature development of the Roman Warm Period [128], the Little Ice Age and today's Modern Warm Period are all components of the 1000-year cyclic. The last maxima of the solar 1000-year (Eddy) cycle took place in AD 0, 1000 and 2000 (Figure 7.7). The next minimum is expected to occur around the year 2500.

At the moment we find ourselves on the plateau area of the Eddy maximum. This maximum is superimposed with the more frequent Gleissberg and Suess/de Vries cycles. Thus it is difficult to determine whether the 1000-year Eddy cycle is already on its downward flank or whether it will remain at its plateau a little longer. One of these superimposed effects is the recent solar activity drop of the Schwabe solar cycle 24. Therefore, the exact point of change of the 1000-year cycle is not easy to determine.

There are many signs that suggest that the long-term peak of solar activity was left behind during the second half of the twentieth century and we do not anticipate any significant increase in solar activity over the levels we have seen during the last few decades. During the twentieth century, the sun's activity reached an unusually

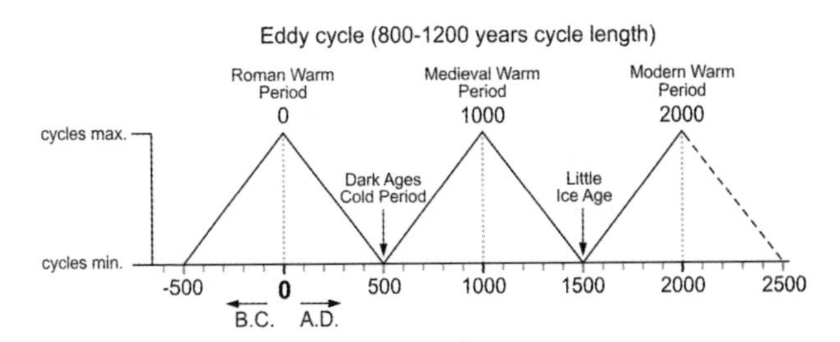

Figure 7.7 Historical maxima and minima of the 1000-year Eddy solar activity cycle (schematic sketch). We currently find ourselves at the plateau area of the maximum, and thus for the coming centuries we do not anticipate a further increase in solar activity, contrasting with the steep climb of the nineteenth and twentieth centuries. At some point during the course of the next 100 years, the long-term decline will begin and will lead from the current Eddy maximum to the Eddy minimum, which is expected to occur around the year 2500.

high maximum when compared to the entire 10,000 years of the postglacial period [58, 129]. From an historical analysis of solar activity, we know that the sun has never maintained such a high level of activity for a long time and that phases of high solar irradiation rarely lasted more than 50 years. For 85–90 per cent of the time, the sun usually runs at considerably lower activity levels [60, 71].

Consequently, we do not anticipate any further warming contribution from the 1000-year Eddy cycle in the future. Rather, we should expect a negative, cooling contribution from the Eddy cycle by the year 2100 at the latest, once it embarks on its downward path towards a minimum by the year 2500. This is the opposite of the development of the last two centuries when the 1000-year cycle powerfully drove the solar total irradiation upwards. Much more decisive for the climate equation for the next century are Gleissberg and Suess/de Vries developments. Both cycles are now on a rapid decline and are getting set for a pronounced intermediate minimum. This minimum is likely to be of mid-category intensity as was the case during the Dalton minimum in 1810 [63, 71]. More extreme and longer solar periods of inactivity of the sort we saw during the Maunder minimum will probably have to wait until the year 2500, as they are controlled by the 1000-year Eddy cycle.

Internal oceanic climate variability

In addition to solar activity, an historical analysis of the temperature development shows that climate system internal cycles also have a pronounced influence on climate [130–131] (see Chapter 4). Particularly important are the PDO, the AMO [132–133], the NAO and the El Niño/Southern Oscillation (ENSO). Like solar activity cycles, these climate system internal oscillations are not strictly periodic and so precise annual predictions are not possible. But their periods do fluctuate within certain well-known parameters and so statements on general trends with a certain amount of acceptable tolerance are possible [134]. The causes of the internal oceanic oscillations are still unknown. However, the phenomenon is well documented and therefore the empirically established natural fluctuations must be an integral part of climate prognoses.

Pacific Decadal Oscillation

The PDO is characterized by changes in the surface temperature of the northern Pacific Ocean. Positive and negative phases alternate every 20–30 years, which yields a full cycle length of 40–60 years. The PDO and the global mean temperature have demonstrated an extraordinary synchronicity with an amplitude of a few tenths of a degree over the last 500 years [135–138] (Figure 7.7 and see Chapter 4).

The decline from the last high peak of the PDO cycle started in about 2000, when the global temperature stopped rising. Since then the PDO has declined and is entering its cold negative phase. Now we have a large area of cold water in the eastern Pacific (Figure 4.6). A similar cold water configuration was present during the previous PDO cold phase (1945–77), a time when a considerable global temperature drop occurred over the globe [139–140]. In 1979 a few scientists recognized the 60-year PDO cycle and correctly forecast that the 1945–77 cold phase would be followed by a warming until the year 2000, followed by another cooling [141]. But at the time no one took their forecasts seriously. The current PDO cold phase is expected to persist until 2030–40 (Figure 7.8) [139, 142–143] and so coincides with the projected solar minimum phase. A warm PDO phase is expected to start again in the year 2040 and will then make a positive contribution to global temperature development.

The Atlantic Multidecadal Oscillation

The AMO describes the alternation of warm and cold areas in the north Atlantic and is characterized by a cycle. The AMO development can be reconstructed for the past few hundred years [145–147]. The cycle length varies between 50 and 80 years and hence is similar to the PDO. Although the two ocean cycles are not synchronous, they appear to be linked [148–150]. The AMO experienced a rapid

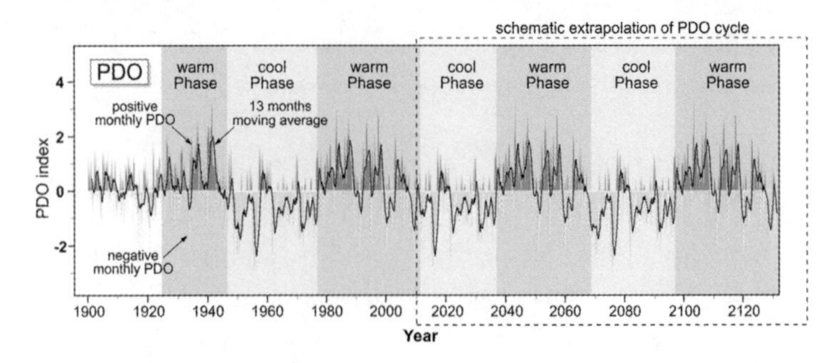

Figure 7.8 Development of the PDO since 1900 along with a schematic illustration of a suggested progression until the year 2130. For the purpose of illustration, earlier segments of the PDO were copied and added to the historical curve. The objective of the prognosis here is solely to determine the approximate positions of the positive and negative PDO phases and is not meant as a precise forecast, which is not possible due to natural variability [144].

increase from the late 1970s until about the year 2000, in remarkable synchronicity with the global warming trend [151–152] (Figure 7.9). The 'summer of the century' in Europe in 2003 as well as the decline in Arctic sea ice over recent years were due in part to the positive (warm) AMO [153–154]. Since 2000 the positive AMO has been at a plateau, which, according to typical AMO periodicity, could last until 2030. A significant warming or cooling contribution from the AMO is therefore not expected in the next 20 years. Its anticipated decline beginning in 2030 could then drag global temperatures down with it.

A Norwegian research team led by Odd Helge Otterå recently discovered that the AMO has to be significantly controlled by external drivers [145]. They were able to show that the phases of the AMO over the last 600 years were controlled mainly by fluctuations in solar activity and large volcanic eruptions. The AMO represents the largest main factor of influence in north Atlantic sea surface temperatures, which in turn also impact on the global temperature curve. A solar influence on the AMO was also found in other studies [147].

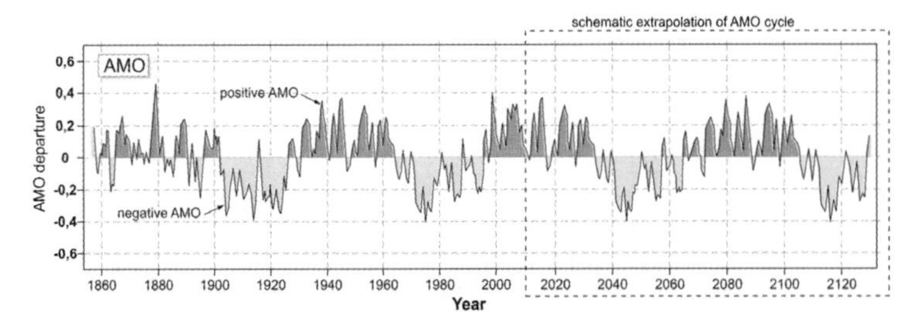

Figure 7.9 The development of the AMO since 1860 and the schematic future progression of the natural cycles until the year 2130. For the purpose of illustration, earlier segments of the AMO development were copied and added to the historical curve. The aim of the prognosis is solely to determine the approximate positions of the positive and negative AMO phases and not to make a precise annual forecast, which is not possible due to natural variability [155].

North Atlantic Oscillation

The NAO is an atmospheric air exchange between the Icelandic low and the Azores high in the Atlantic which is controlled by air pressure. Climate in north-western Europe and the Arctic region is strongly affected by the NAO and the NAO history can be reconstructed over many thousands of years [156]. The duration of the NAO cycle varies widely with time and operates on several timescales. Besides the short-term fluctuations in the range of 2–5 years, the NAO also fluctuates with superimposed oscillations of 12–15 years (decadal oscillation) and also of about 70 years [157]. The NAO increased sharply between 1970 and 1990 [158] and has most likely added to the climate warming of the late twentieth century (Figure 7.10). Since 1990 the NAO has been on a rapid decline, and this could be one of the reasons for stagnating global temperatures since 2000. Mojib Latif and his team projected in 2008 that because of the decline of an NAO-related cycle, global temperatures would not rise until 2015 [159]. Stefan Rahmstorf challenged Latif's prediction and, together with a colleague, bet 5000 euros that temperatures would continue to climb [160]. The way things look now, Rahmstorf should consider himself fortunate that Latif doesn't view the science lab as a casino, and didn't accept the wager. Incidentally, the NAO is suspected of being influenced,

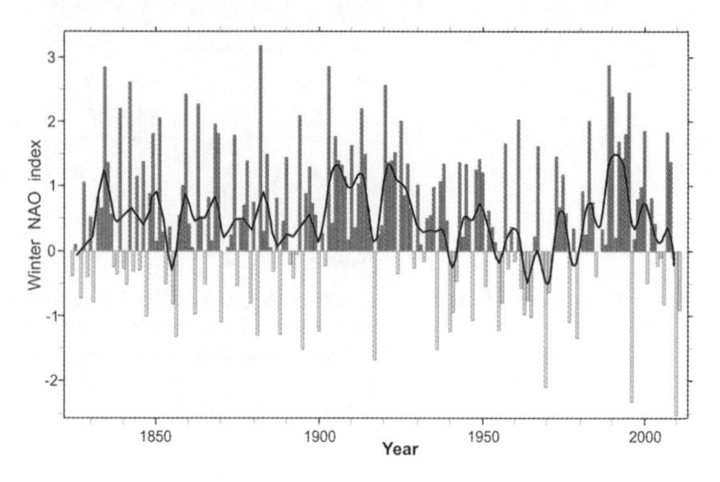

Figure 7.10 Development of the NAO since 1820. The strong warming phase of 1977–2000 occurred during the rapidly rising flank of the NAO [173].

or even largely controlled, by solar activity changes via a complex set of mechanisms [161–172].

El Niño and the Southern Oscillation

A one-year strong warming of the sea surface water takes place in the equatorial eastern Pacific every 2–7 years. This is the so-called El Niño. The corresponding cooling event is called La Niña, and occurs between El Niños [174] (see Chapter 4). El Niño and La Niña events influence the climate far beyond the Pacific region and have a profound impact on global climate values. During an El Niño the global mean temperature rises several tenths of a degree Celsius. But when the El Niño ends, the temperatures quickly return to normal levels.

The last two major El Niños occurred in 1997–98 and 2009–10 and produced pronounced peaks in the warming curve when viewed over the last 150 years (Figure 4.1). Because of their irregularity, El Niño and La Niña events cannot be accurately forecast over the long term. However, they do occur within a typical repetition pattern of every 2–7 years and so have to be taken into account. These internal

Pacific cycles therefore distort the long-term temperature prognoses for single years by a few tenths of a degree upward or downward.

El Niño and La Niña events are closely coupled to the atmospheric Southern Oscillation, which is an alternating air mass exchange between the south Asian low pressure zone and the south-east Pacific high pressure region and is controlled by air pressure [175]. Because of the strong impact that the ENSO cycles have on global climate values, the Southern Oscillation Index (SOI) is a good impulse generator for influencing the short-term global mean temperatures after deducting volcanic events from the temperature curve. Note that the temperature generally follows the SOI curve after a lag of 5–7 months (Figure 7.11) [176]. The degree of synchronicity between the SOI and temperature offers promising possibilities for forecasting global temperatures for half a year in advance [177]. These short-term prognoses have been well confirmed in the past [178]. Another ENSO prognosis method is based on the analysis of sea surface temperatures in the north-western Pacific and attempts a one-year forecast [179]. There are similar prognosis possibilities available for the winter temperatures of the Northern Hemisphere [180].

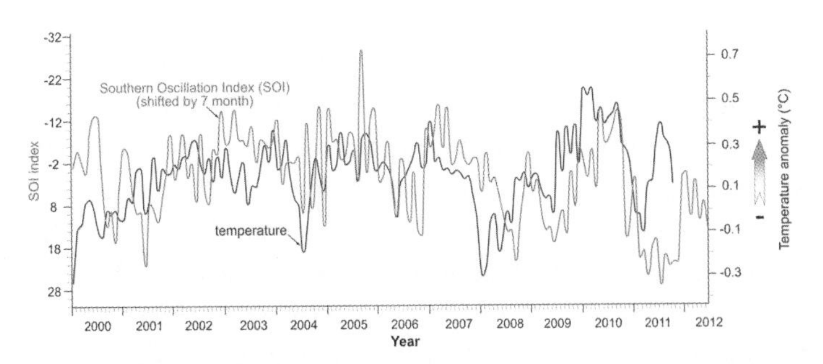

Figure 7.11 Half-year forecasts of annual global temperature (red) through the SOI (green). This is possible because it has been empirically established that temperature follows the SOI curve after a lag of 5–7 months [181].

Last but not least: carbon dioxide

The atmospheric CO_2 concentration since the pre-industrial era, beginning in 1750, has risen steadily from 280 ppm to the current level of around 390 ppm (Figure 3.8). One reason is the increased emissions of CO_2 from burning fossil fuels (coal, oil and natural gas). Currently, the CO_2 concentration in the atmosphere is rising almost 2 ppm a year and atmospheric concentrations will continue to rise owing to the continued consumption of fossil fuels. The rate mainly depends on global economic growth and the industrialization associated with it, especially in emerging and developing countries. Another important factor is the rate at which fossil fuels are replaced by low CO_2 or non-CO_2 energy [182–183].

Overall, the climatic impact of the projected CO_2 rise depends on the climate sensitivity of CO_2. Just how powerful is CO_2 in causing warming? The IPCC models assume temperature increases of 2.0–4.5° C for each doubling of CO_2 [184]. But because this significantly underestimates the role of natural climate control factors, more realistic values need to be applied (see Chapter 6). If one eliminates the water vapour amplification effects and attributes the observed 0.8° C of warming since the end of the Little Ice Age to the pre-industrial climate drivers, then a CO_2 climate sensitivity of 1.0–1.5° C appears probable for each doubling of CO_2 [185]. Progressive corrections of the exaggerated CO_2 climate sensitivity have already begun and IPCC values are gradually crumbling. An American–Spanish research team recently published a study in Science in which they reduced the possible range of the climate sensitivity to 1.7–2.6° C per CO_2 doubling [186]. Further downward corrections may well be shown in future studies.

For the purpose of calculation, we use the IPCC A1B emissions scenario, which many experts consider to be realistic. In this scenario CO_2 emissions increase during the first half of the twenty-first century because of the industrialization of today's developing and emerging countries, but later stabilize because of a more balanced energy mix and expected technological progress in the second half of the century. They may even drop somewhat by then (Figure 7.12) [182, 184]. In the A1B scenario the atmospheric CO_2 concentration reaches a peak of about 700 ppm by the year 2100 [187], less than doubling today's 390 ppm, which would result in a warming of about 0.8–1.3° C using the given CO_2 climate sensitivity. To estimate the total climate change, this CO_2-dependent warming contribution has to be added to the temperature contributions of the other climate control factors, such as solar cycles and internal oceanic climate cycles.

Under the bottom line

It is clear that the temperature movements in the pre-industrial period, as well as over the past 150 years, have always been the result of a combination of the different climate controlling processes [128, 143, 188]. Claiming that CO_2 is the main factor for the future of the climate makes no sense.

We shall now carry out a schematic trend estimate for the future climate. The basis is the projected course of the solar activity cycles and the climate system internal factors such as the PDO, AMO and NAO. To these we will add a realistic CO_2 emission warming contribution. Short-term cooling effects through large volcanic eruptions as well as the El Niño/La Niña-related temperature ups and downs will not be considered in our longer-term estimation. Note that our model is a schematic representation, and future climate models will have to be used to adjust and refine this prognosis.

The world in 2035

As the solar cycles show, we can expect cooling from the sun over the next few decades. The Gleissberg and Suess/de Vries cycles will reach their low points between 2020 and 2040, and that means solar activity may reach a low level comparable to the Dalton minimum, which occurred in around 1810. At that time the temperature was nearly 1° C lower than it is today [189–191] and at least half of that was due to the weak sun. The PDO, which impacts the global temperature trend significantly, will also be at its low point in 2035 (Figure 7.8), and thus it too will add a cooling effect [140]. Another negative temperature contribution can be expected from the AMO, which will begin to drop in around 2020 (Figure 7.9). Internal climate cycles are generally responsible for about 0.2–0.3° C of the temperature dynamic. If one calibrates the various natural climate cycles to the documented geological data series of the past, then a total cooling contribution from the natural climate control factors of 0.4–0.6° C is expected to occur by the year 2035 compared to today.

This cooling will be absorbed to some extent by the warming effect of the anthropogenic greenhouse effect. According to the IPCC A1B emission scenario, the CO_2 concentration of the atmosphere will reach 450 ppm by the year 2035 (Figure 7.12) [192]. Applying a realistic CO_2 climate sensitivity of 1.0–1.5° C for each doubling of CO_2 means an increase to 450 ppm will yield a temperature rise of 0.2–0.3° C. In summary under the bottom line, and taking all the important natural and anthropogenic climate factors into account, we can anticipate a modest global cooling of 0.2–0.3° C by 2035 compared to today (Figure 7.12). Others also project cooling [193–200]. Claims that the next solar minimum barely has a measurable effect on world temperatures [201–202] are questionable because they are based on an unrealistically small climate effect of solar activity changes, which is incompatible with real data reconstructed for the past 10,000 years using geological methods (see Chapter 3).

The world in 2100

In the second half of the twenty-first century the sun will once again become more active and the earth will warm, but not as it did in the 1980s and 1990s [101, 126]. This phase will end in about 2100, a time when the sun will once again reach a point of low activity last witnessed during the Dalton minimum in around 1810 [101–102]. This solar slumber will produce a cooling of about

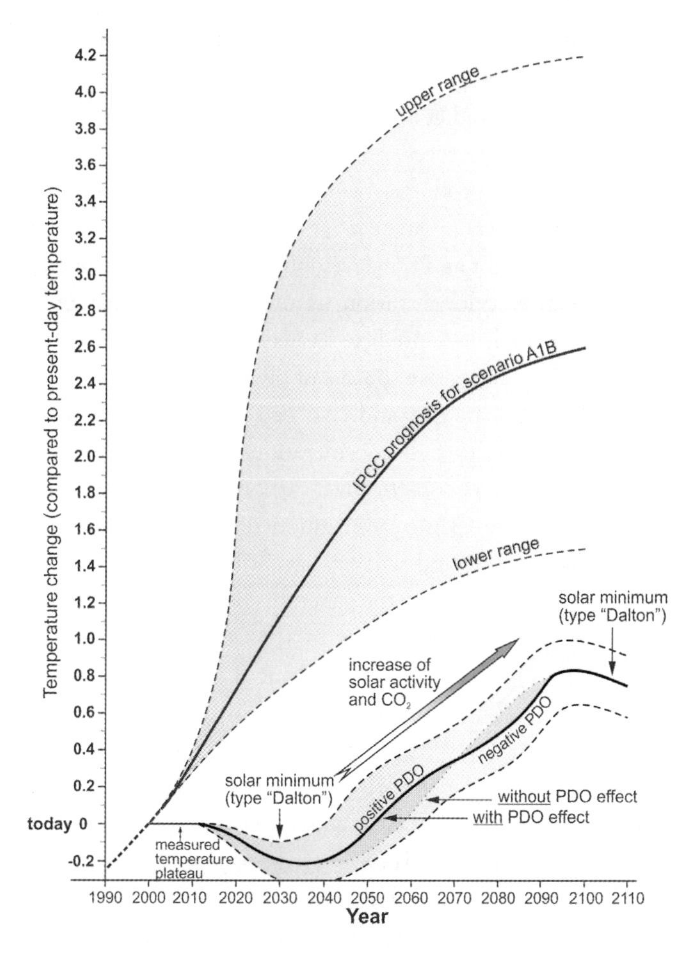

Figure 7.12 Our schematic temperature prognosis until the year 2100 uses a combination of anthropogenic and natural climate factors. As a comparison, the current IPCC prognosis is shown. The values apply for the A1B IPCC CO_2 emissions scenario of the AR4 report.[12]

[12] The datum reference for the illustrated temperature chart is represented by the temperature plateau of the past 10 years. If one takes the average temperature of the interval 1980–99 as the reference point, temperatures are increasing by 0.85–1.25° C by 2100 according to our prognosis. This contrasts with a warming of 1.7–4.4° C according to the AR4 models.

0.3–0.4° C compared to today. The PDO will also reach a low to intermediate level (Figure 7.8) [139, 193,196–197].

It is difficult to predict atmospheric CO_2 concentration for the year 2100. First, oil reserves will probably be exhausted by then [203] and natural gas production will be approaching its end phase. Will fossil fuels be playing any part at all? Research and technology do not remain stagnant and the development of new kinds of energy technology, such as nuclear fusion, is uncertain. For example, 90 years ago who would have predicted today's computer age? Thus it is unclear whether alternative sources of oil and gas reserves such as shale oil, oil sand, shale gas, and coal bed methane (coal gas) will be tapped at all, due to cost considerations, or will have become obsolete thanks to more cost-effective, carbon-free energy sources.

The atmospheric CO_2 concentration of the year 2100 can be estimated only roughly, and will fall somewhere between 500 and 1000 ppm [192]. If we assume the generally expected A1B emissions scenario, then CO_2 concentration in the atmosphere will climb to 700 ppm by 2100 (Figure 7.12). Using a realistic CO_2 climate sensitivity of 1.0–1.5° C for each doubling of CO_2, a CO_2-induced warming of about 0.8–1.3° C will result. Taking sun, PDO and CO_2 together will yield a temperature increase of 0.6–1.0° C compared to today, depending on the CO_2 climate sensitivity (Figure 7.12). This is in stark contrast to IPCC's prediction of a more dramatic temperature increase of 2.55° C [184].

The IPCC was not unaware of the discussion of the sun entering a slumber, and thus has scrambled to remove all doubt over its projection using a 'rescue publication'. Stefan Rahmstorf, together with his colleagues in Potsdam, fed the new solar forecast into a solar-unfriendly computer model and now believe they have been able to show that a weak sun will only marginally put the brakes on the IPCC forecast of a strong warming trend until the year 2100 [202], The authors, who are very close to the IPCC and keep true to its tradition, failed to take any solar amplification into account (see Chapter 6). They merely calibrated their temperature forecast to variants of Mann's hockey stick [204–207] and assumed an exaggerated CO_2 climate sensitivity. Because of these reality-divorced assumptions, the model's results are worthless. Using the same dubious approach, the Potsdam group also tried to play down the sun's climate relevance over the last millennium [208].

Some scientists are predicting an even smaller temperature increase than ours of only 0.2–0.5° C by the year 2100 compared to today [139, 196–197]. This may be possible because CO_2 climate sensitivity might be even lower than what we have assumed in our calculations, and/or because the sun is responsible for more than 50 per cent of the long-term climate activity, and/or the IPCC A1B emissions scenario is too high owing to a possible steady decline in the use of fossil fuels later this century.

The well-known 2-degree target limit in any case will not be reached, and done so without a hectic, socially explosive transformation of the entire industrial base. The 2-degree target describes the international climate policy of limiting global warming to less than 2 degrees Celsius compared to the level before the start of industrialization [209]. Here the 0.8° C warming observed so far has to be added to the expected warming from today until the year 2100.

One has to recall that at least half, if not two-thirds, of the warming so far is attributed to the sun and is modulated by internal oceanic climate cycles. The reference point of the 2-degree target happens to be the Little Ice Age, a natural cold phase. The subsequent warming is a natural process that routinely took place about every 1000 years during the postglacial period. Conflating the anthropogenic and natural causes for the 2-degree target limit is certainly not leading us anywhere.

The world in 500 years' time

Because of the 1000-year Eddy cycle, solar activity will have reached its low point in 500 years. Analogous to the last irradiative minimum of this cycle, the Maunder minimum of the Little Ice Age, a cooling contribution of about 1° C has to be calculated with respect to today's temperature [189].

The CO_2 concentration in the atmosphere that we can expect at this point is unclear. It has to be assumed that oil, gas and coal reserves will be exhausted by then, and so no further CO_2 emission will come from these sources. In addition, it is unclear if fossil fuels will even have to be exhausted because inexpensive carbon emission-free sources of energy may be found long before the year 2500.

Let us assume that conventional fossil fuels for the most part will be fully exhausted in the year 2500. That would give an atmospheric CO_2 concentration of up to 2000 ppm [210]. However, it has to be assumed that methods for technically processing CO_2 will have been developed by then and thus the 2000 ppm level will never be reached.

If filtering out does not occur, however, CO_2 in the atmosphere will double twice with respect to today's concentrations. Using a CO_2 climate sensitivity of 1.0–1.5° C per doubling of CO_2 compared to today, we arrive at a positive temperature contribution of 2–3° C for this maximum 2000 ppm of CO_2. If one tallies the climate contributions of the weakening sun and the anthropogenic greenhouse effect, then in the year 2500 we will reach a temperature of 1.0–2.0° C above today's level. And if we limit emissions so that atmospheric concentrations never exceed 450 ppm, then the natural solar cooling of 1° C will be fully effective and global temperatures will fall to Little Ice Age levels.

As no significant emissions of CO_2 from fossil fuels are expected by the year 2500, we can assume that the climate by then will have begun to stabilize. Once the main emission phase of CO_2 into the atmosphere ends, the CO_2 concentration will slowly decline because the land and oceans will steadily absorb much of the CO_2.

IPCC projections reality check

Since 1990 the IPCC has published four reports with temperature prognoses. We now compare the IPCC projected temperature developments with the real measured curves (Figure 7.13). The temperature prognosis in the first report was the most aggressive of all the IPCC forecasts. But the real measured temperatures turned out to be almost always below the prognoses. Starting in 2000 the gap began opening up as the warming of the real world came to a halt and the projected curves continued their upward climb (Figure 7.13).

Because of the divergence problem, the prognosis in the 1995 report was significantly scaled back, a little too far, as the middle forecast of the 1995 second report was always somewhat below the real measured temperatures. So in the TAR in 2001 a stronger warming was again projected.

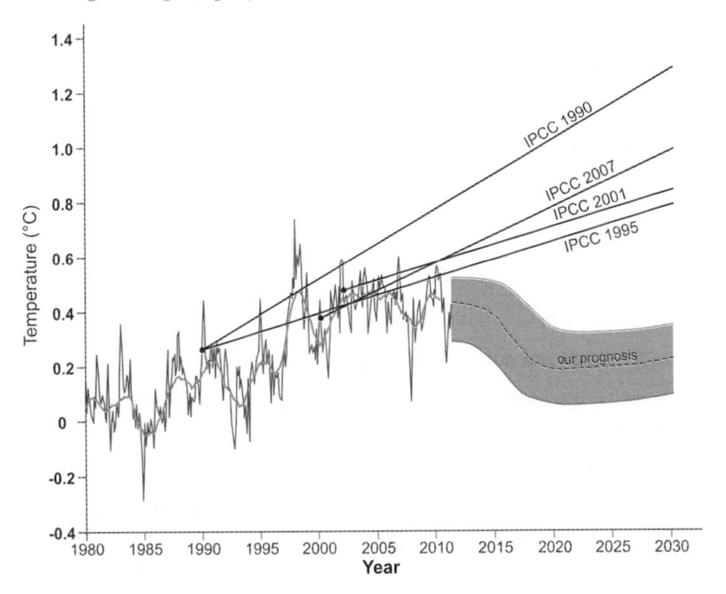

Figure 7.13 Real temperature development since 1980 compared to the mean prognoses of the various IPCC reports. Also shown is our temperature prognosis for the next 20 years [211].

In the first years the prognosis wasn't that bad (Figure 7.13). The measured temperatures fluctuated within the IPCC's projected corridor. But beginning in the year 2004 the real measured values were once again well below the projected trends most of the time. The warming prediction of the 2007 AR4 deviates only slightly from TAR and thus suffers from the same problem. Because of the missing warming in the first decade of the twenty-first century, the difference between the prognosis and the real measured values quickly opens up (Figure 7.13).

It is clear that none of the model projections are going to work. After just a few years the prognoses will have to be replaced and adjusted to reflect the real world. Real prognosis capability has eluded all models thus far. None of them predicted the missing warming since 2000. That alone would not be bad should the missing warming reappear over the coming years and the temperature trends upwards. But that doesn't appear to be the case, as solar activity and the climate internal ocean cycles are on a downward trend. The model projection curves will therefore most likely diverge even further over time.

That raises the question how all this could have happened at the IPCC. What went wrong? The strong temperature increase from 1978 to 1998 apparently stunned the scientists so much that they simply extended the trend into the future in a kneejerk reaction. They failed to notice that the positive flank of the 60-year PDO cycle was at play and that solar activity had been rising since its low during the 1970s (Figures 4.7 and 4.8). They simply failed to recognize that the rapid warming was a special case, not the norm. And when the PDO reached its positive peak in the year 2000, the temperature increase ended. The recent dramatic collapse of solar activity now heralds a cooling phase over the coming few decades, one that will initially not be offset by CO_2.

The time for simplistic IPCC climate prognoses is over. In its next report the IPCC should abandon its linearly rising forecast once and for all. But unfortunately, this is not going to happen. A draft of the fifth report (AR5) was leaked in December 2012 and revealed that temperatures in the latest IPCC climate models will still be mainly driven by CO_2. According to the draft, the global mean surface air temperature will be warming by 0.4–1.0° C during 2016–35 relative to the reference period of 1986–2005. For the end of the twenty-first century, AR5 expects a warming of 1° C in the best case to 4.8° C in the worst case [212]. This is a little less than in the fourth IPCC report but again is very one-dimensional.

These forecasts will ultimately have to be replaced by refined curves characterized by the up and down trends of various natural cycles superimposed over each other. Carbon dioxide will remain an integral component of the climate equation, but its dominant role will have to be abandoned. Instead, it will have to become a partner in a mix of climate control factors which have been playing a role on our climate for millions of years. Man is influential and has profoundly changed the world over the course of time. But at the same time we should not overestimate our abilities and believe we are able to overcome natural forces and processes.

References

1. Wikipedia Prognose. http://de.wikipedia.org/wiki/Prognose.
2. Wikipedia Chaos. http://de.wikipedia.org/wiki/Chaos.
3. Berndt, C. (2011) Fragwürdige Runderneuerung. *Süddeutsche Zeitung*, 2 March. http://www.sueddeutsche.de/wissen/staendige-impfkommission-fragwuerdige-runderneuerung-1.1066713.
4. GDV (2011) Auswirkungen des Klimawandels auf die Schadensituation in der deutschen Versicherungswirtschaft. http://www.gdv.de/wp-content/uploads/2012/01/Klimakonferenz_2011_PIK_Studie_Hochwasser.pdf.
5. Skeptical Science Chaos Theory and Global Warming: Can Climate Be Predicted? http://www.skepticalscience.com/chaos-theory-global-warming-can-climate-be-predicted.htm.
6. Bauernregeln.net http://www.bauernregeln.net.
7. Malberg, H. (1993) *Bauernregeln: Aus meteorologischer Sicht*. Springer, Berlin.
8. Kumpfmüller, J. and D. Steinbacher (2006) *Die besten Wetter- und Bauernregeln*. Wilhelm Heyne Verlag, Munich.
9. Wikipedia Weather lore. http://en.wikipedia.org/wiki/Weather_lore.
10. *Der Spiegel* (1974) Katastrophe auf Raten. http://www.spiegel.de/spiegel/print/d-41667249.html 33.
11. Kukla, G. J., J. K. Angell, J. Korshover, H. Dronia, M. Hoshiai, J. Namias, M. Rodewald, R. Yamamoto and T. Iwashima (1977) New data on climatic trends. *Nature* 270, 573–80.
12. Kukla, G., R. K. Matthews and J. M. Mitchell Jr. (1972) The end of the present interglacial. *Quaternary Research* 2, 261–9.
13. Duphorn, K. (1976) Kommt eine neue Eiszeit? *Geologische Rundschau* 65, 845–64.

14. Severinghaus, J. P., T. Sowers, E. J. Brook, R. B. Alley and M. L. Bender (1998) Timing of abrupt climate change at the end of the Younger Dryas interval from thermally fractionated gases in polar ice. *Nature* 391, 141–6.
15. Alley, R. B. (2000) The Younger Dryas cold interval as viewed from central Greenland. *Quaternary Science Reviews* 19, 213–26.
16. Fagan, B. (2000) *The Little Ice Age*. Basic Books, New York.
17. McCormick, R. A. and J. H. Ludwig (1967) Climate modification by atmospheric aerosols. *Science* 156 (3780), 1358–9.
18. Bryson, R. A. and W. M. Wendland (1970) Climatic effects of atmospheric pollution. In S. F. Singer (ed.), *Global Effects of Environmental Pollution*. Springer, New York, 130–8.
19. Rasool, S. I. and S. H. Schneider (1971) Atmospheric carbon dioxide and aerosols: effects of large increases on global climate. *Science* 173, 138–41.
20. Yamamoto, G. and M. Tanaka (1972) Increase of global albedo due to air pollution. *Journal of the Atmospheric Sciences* 29 (8), 1405–12.
21. *Palm Beach Post* (1979) Picture grim if polar ice melts. http://stevengoddard.wordpress.com/2011/05/23/1979-ncar-forecast-sea-level-may-rise-15-25-feet-before-the-year-2000. 8 January, A3.
22. Kulke, U. (2009) Als uns vor 30 Jahren eine neue Eiszeit drohte. http://www.welt.de/wissenschaft/umwelt/article5489379/Als-uns-vor-30-Jahren-eine-neue-Eiszeit-drohte.html.
23. Behringer, W. (2007) *Kulturgeschichte des Klimas*. C. H. Beck Verlag, Munich.
24. Peterson, T. C., W. M. Connolley and J. Fleck (2008) The myth of the 1970s global cooling scientific consensus. *Bulletin of the American Meteorological Society* September, 1325–37.
25. *Phoenix* (2009) 2 November. http://tinyurl.com/ce8kv5e.
26. *Der Spiegel* (2000) Nie wieder Schnee? 1 April. http://www.spiegel.de/wissenschaft/mensch/0,1518,71456,00.html
27. *The Independent* (2000) Snowfalls are now just a thing of the past. 20 March. http://www.independent.co.uk/environment/snowfalls-are-now-just-a-thing-of-the-past-724017.html 20.
28. *The Independent* (2008) Exclusive: scientists warn that there may be no ice at North Pole this summer. 27 June. http://www.independent.co.uk/environment/climate-change/exclusive-scientists-warn-that-there-may-be-no-ice-at-north-pole-this-summer-855406.html.
29. *The Age* (2008) Arctic may be ice free by 2030. 20 September. http://www.theage.com.au/national/arctic-may-be-ice-free-by-2030-20080919-4k8u.html.
30. IPCC (2007) *Climate Change 2007: The Physical Science Basis*. Kapitel 11.2.3.2, S. 869. Cambridge University Press, Cambridge, and New York.
31. Funk, C. (2011) We thought trouble was coming. *Nature* 476 (7).
32. US Geological Survey (2010) A climate trend analysis of Kenya – August. Fact Sheet 2010–3074, http://tinyurl.com/ck6kffg, 1-4.
33. Lyon, B. and D. G. DeWitt (2012) A recent and abrupt decline in the East African long rains. *Geophys. Res. Lett.* 39 (2), L02702.
34. notrickszone.com http://notrickszone.com/climate-scandals.
35. thegwpf.org http://www.thegwpf.org/false-alarms.html.

36. Monckton of Brenchley, C. (2007) 35 Inconvenient truths – the errors in Al Gore's movie SPPI – *Science & Policy Institute*. http://scienceandpublicpolicy.org/monckton/goreerrors.html, 1–21.
37. *The Times* (2007) Al Gore's inconvenient judgment. 11 October. http://business.timesonline.co.uk/tol/business/law/article2633838.ece.
38. Focus Online (2007) Al Gore – Vom Wahlverlierer zum Nobelpreisgewinner. 12 October. http://www.focus.de/politik/ausland/al-gore_aid_135654.html.
39. BBC (2009) Himalayan glaciers melting deadline 'a mistake'. 5 December. http://news.bbc.co.uk/2/hi/south_asia/8387737.stm.
40. *ZDF* (2009) Die lange Nacht des Klimas, 30 October. http://www.youtube.com/watch?v=aBDqCju6rSM.
41. The Times (2010) UN must investigate warming 'bias', says former climate chief. 15 February. http://myreader.co.uk/msg/1210715.aspx.
42. Dixon, B. (1971) In praise of prophets. *New Scientist and Science Journal* 16 September, 606.
43. WUWT (2011) The big self parodying 'climate change blame' list. http://wattsupwiththat.com/2011/04/03/the-big-self-parodying-climate-change-blame-list.
44. Archer, D. and S. Rahmstorf (2010) *The Climate Crisis*, 1st edn. Cambridge University Press, Cambridge.
45. Umweltbundesamt Klimaschutz – Antworten des UBA auf populäre skeptische Argumente. http://www.umweltbundesamt.de/klimaschutz/klimaaenderungen/faq/antworten_des_uba.htm.
46. Schiermeier, Q. (2010) The real holes in climate science. *Nature* 463, 284–7.
47. Santer, B. and S. Solomon (2010) Stephen H. Schneider (1945–2010). *Eos* 91 (41), 372.
48. IPCC (2001) *Climate Change 2001: The Scientific Basis.* Section 14.2.2.2. Cambridge University Press. http://www.ipcc.ch/ipccreports/tar/wg1/505.htm.
49. Pilkey, O. H. and L. Pilkey-Jarvis (2007) *Useless Arithmetic – Why Environmental Scientists Can't Predict the Future*. Columbia University Press, New York.
50. Scafetta, N. (2010) Climate change and its causes – a discussion about some key issues. *Science & Public Policy Institute*. Original Paper.
51. Lindzen, R. S. (2009) The climate science isn't settled. *The Wall Street Journal*, 30 November. http://tinyurl.com/yh9t9j8.
52. Akasofu, S.-I. (2009) Two Natural Components of the Recent Climate Change. http://people.iarc.uaf.edu/~sakasofu/pdf/two_natural_components_recent_climate_change.pdf.
53. Berger, A. and M. F. Loutre (2002) An exceptionally long interglacial ahead? *Science* 297, 1287–8.
54. Ganopolski, A. and R. Calov (2011) The role of orbital forcing, carbon dioxide and regolith in 100 kyr glacial cycles. *Clim. Past Discuss.* 7, 2391–411.
55. Tzedakis, P. C., J. E. T. Channell, D. A. Hodell, H. F. Kleiven and L. C. Skinner (2012) Determining the natural length of the current interglacial. *Nature Geosci* advance online publication.
56. Tzedakis, P. C., E. W. Wolff, L. C. Skinner, V. Brovkin, D. A. Hodell, J. F. McManus and D. Raynaud (2012) Can we predict the duration of an interglacial? *Climate of the Past* 8, 1473–85.

57. Müller, U. C. and J. Pross (2007) Lesson from the past: present insolation minimum holds potential for glacial inception. *Quaternary Science Reviews* 26, 3025–9.
58. Solanki, S. K., I. G. Usoskin, B. Kromer, M. Schüssler and J. Beer (2004) Unusual activity of the sun during recent decades compared to the previous 11,000 years. *Nature* 431, 1084–7.
59. Russell, C. T., J. G. Luhmann and L. K. Jian (2010) How unprecedented a solar minimum? *Reviews of Geophysics* 48, 1–16.
60. Usoskin, I. G. (2008) A history of solar activity over millennia. *Living Rev. Solar Phys.* 5 (3), 1–87.
61. Berndtsson, R., C. Uvo, M. Matsumoto, K. Jinno, A. Kawamuran, S. Xu and J. Olsson (2001) Solar–climatic relationship and implications for hydrology. *Nordic Hydrology* 32 (2), 65–84.
62. Usoskin, I. G., S. K. Solanki and G. A. Kovaltsov (2007) Grand minima and maxima of solar activity: new observational constraints. *Astronomy & Astrophysics* 471, 301–9.
63. Miyahara, H., K. Kitazawa, K. Nagaya, Y. Yokoyama, H. Matsuzaki, K. Masuda, T. Nakamura and Y. Muraki (2010) Is the sun heading for another Maunder minimum? Precursors of the grand solar minima. *Journal of Cosmology* 8, 1970–82.
64. NGDC (2011) Sun Spots (monthly averages). ftp://ftp.ngdc.noaa.gov/STP/solar_data/sunspot_numbers/international/monthly/monthly.plt.
65. *Focus* (2010) Forscherstreit um die Sonne. http://www.focus.de/wissen/wissenschaft/klima/tid-17053/klimatologie-forscherstreit-um-die-sonne_aid_469274.html 2.
66. NOAA (2009) Solar Cycle Progression. http://www.swpc.noaa.gov/solarcycle.
67. astronomie.info (2010) Sonnenflecken und Polarlichter. http://news.astronomie.info/sky201002/sunactivity.html.
68. Turner, R. (2011) Solar cycle slow to get going: what does it mean for space weather? *Space Weather* 9, 1–2.
69. NASA (2011) Solar activity heats up. http://science.nasa.gov/science-news/science-at-nasa/2011/14apr_thewatchedpot.
70. NGDC (2011) Sonnenflecken, Jahresmittelwerte. ftp://ftp.ngdc.noaa.gov/stp/solar_data/sunspot_numbers/international/yearly/yearly.plt.
71. Li, K. J., W. Feng, H. F. Liang, L. S. Zhan and P. X. Gao (2011) A brief review on the presentation of cycle 24, the first integrated solar cycle in the new millennium. Ann. *Geophysicae* 29, 341–8.
72. Janardhan, P., S. K. Bisoi, S. Ananthakrishnan, M. Tokumaru and K. Fujiki (2011) The prelude to the deep minimum between solar cycles 23 and 24: interplanetary scintillation signatures in the inner heliosphere. *Geophysical Research Letters* 38, 1–5.
73. McDonald, F. B., W. R. Webber and D. V. Reames (2010) Unusual time histories of galactic and anomalous cosmic rays at 1 AU over the deep solar minimum of cycle 23/24. *Geophysical Research Letters* 37, 1–5.
74. Stozhkov, Y. and V. Okhlopkov (2010) New Maunder minimum in solar activity and cosmic ray fluxes in the nearest future. 22nd European Cosmic Ray Symposium, Turku, Finland, 3–6 August, 1–11.
75. Jian, L. K., C. T. Russell and J. G. Luhmann (2010) comparing solar minimum 23/24 with historical solar wind records at 1 AU. *Solar Physics*.

76. Haaland, S., K. Svenes, B. Lybekk and A. Pedersen (2012) A survey of the polar cap density based on Cluster EFW probe measurements: Ssolar wind and solar irradiation dependence. *J. Geophys. Res.* 117 (A1), A01216.
77. Solomon, S. C., T. N. Woods, L. V. Didkovsky, J. T. Emmert and L. Qian (2010) Anomalously low solar extreme-ultraviolet irradiance and thermospheric density during solar minimum. *Geophysical Research Letters* 37.
78. DeLand, M. T. and R. P. Cebula (2012) Solar UV variations during the decline of cycle 23. *Journal of Atmospheric and Solar-Terrestrial Physics* 77, 225–34.
79. Friis-Christensen, E. and K. Lassen (1991) Length of the solar cycle: an indicator of solar activity closely associated with climate *Science* 254 (5032), 698–700.
80. Thejll, P. and K. Lassen (1999) Solar forcing of the Northern Hemisphere land air temperature: new data. *Danish Meteorological Institute Scientific Report* 99-9, 1–18.
81. Skeptical *Science* (2010) What does solar cycle length tell us about the sun's role in global warming? http://www.skepticalscience.com/solar-cycle-length.htm.
82. Hathaway, D. H. and R. M. Wilson (2004) What the sunspot record tells us about space climate. *Solar Physics* 224, 5–19.
83. Richards, M. T., M. L. Rogers and D. S. P. Richards (2009) Long-term variability in the length of the solar cycle. *Publications of the Astronomical Society of the Pacific* 121 (881), 797–809.
84. Miyahara, H., Y. Yokoyama and Y. T. Yamaguchi (2009) Influence of the Schwabe/Hale solar cycles on climate change during the Maunder minimum. In A. G. Kosovichev, A. H. Andrei and J.-P. Rozelot (eds.), *Solar and Stellar Variability: Impact on Earth and Planets.* Proceedings IAU Symposium No. 264.
85. Dikpati, M., P. A. Gilman, G. de Toma and R. K. Ulrich (2010) Impact of changes in the sun's conveyor-belt on recent solar cycles. *Geophysical Research Letters* 37, 1–6.
86. appinsys.com (2010) Solar Cycle Prediction. http://www.appinsys.com/NASASolar.htm.
87. Alvestad, J. (2012) Monthly solar cycle data, www.solen.info/solar.
88. NASA Solar Cycle Prediction. http://solarscience.msfc.nasa.gov/predict.shtml.
89. Becker, M. (2011) Forscher sagen lange Sonnenpause voraus. *Der Spiegel.* http://www.spiegel.de/wissenschaft/weltall/0,1518,768731,00.html.
90. Hill, F., R. Howe, R. Komm, J. Christensen-Dalsgaard, T. P. Larson, J. Schou and M. J. Thompson (2011) Large-scale zonal flows during the solar minimum – where is cycle 25? [abstract]. American Astronomical Society, SPD Meeting 2011, Las Cruces, NM, 12–16 June, http://tinyurl.com/6bgvyjd.
91. Lindholm Nielsen, M. and H. Kjeldsen (2011) Is cycle 24 the beginning of a Dalton-like minimum? *Solar Physics* 270 (1), 1–8.
92. Owens, M. J., M. Lockwood, L. Barnard and C. J. Davis (2011) Solar cycle 24: implications for energetic particles and long-term space climate change. *Geophysical Research Letters* 38, 1–5.
93. Lockwood, M., M. J. Owens, L. Barnard, C. J. Davis and F. Steinhilber (2011) The persistence of solar activity indicators and the descent of the sun into Maunder minimum conditions. *Geophys. Res. Lett.* 38 (22), L22105.
94. Penn, M. J. and W. Livingston (2010) Long-term evolution of sunspot magnetic fields. IAU Symposium No. 273, http://arxiv.org/abs/1009.0784, 1–8.
95. Dambeck, H. (2010) Experten prophezeien lange Sonnenschwäche. *Der Spiegel,* http://www.spiegel.de/wissenschaft/weltall/0,1518,717893,00.html.

96. Livingston, W. C., M. Penn and L. Svalgard (2011) A decade of diminishing sunspot vigor [abstract]. American Astronomical Society, SPD Meeting 2011, Las Cruces, NM, 1–16 June, http://tinyurl.com/68lr76m.
97. Page, L. (2011) Good news for Mars astronauts – less good for carbon traders, perhaps. *The Register*, 14 June. http://www.theregister.co.uk/2011/06/14/ice_age.
98. NASA (2006) Long Range Solar Forecast http://science1.nasa.gov/science-news/science-at-nasa/2006/10may_longrange.
99. Southwest Research Institute Planetary Science Directorate (2011) What's down with the sun? Major drop in solar activity predicted. Press release. http://www.boulder.swri.edu/~deforest/SPD-sunspot-release/SPD_solar_cycle_release.txt.
100. Altrock, R. C. (2011) Whither goes cycle 24? A view from the Fe XIV Corona [abstract]. American Astronomical Society, SPD Meeting 2011, Las Cruces, NM, 12–16 June, http://tinyurl.com/659z7yj.
101. Clilverd, M. A., E. Clarke, T. Ulich, H. Rishbeth and M. J. Jarvis (2006) Predicting solar cycle 24 and beyond. *Space Weather* 4, 1–7.
102. Clilverd, M. A., E. Clarke, H. Rishbeth, T. D. G. Clark and T. Ulich (2003) Solar activity levels in 2100. *Astronomy & Geophysics* 44, 5.20-25.22.
103. Tobias, S., N. Weiss and J. Beer (2004) Long-term prediction of solar activity – a discussion. *Astronomy & Geophysics* 45 (2), 2.6.
104. Abdusamatov, K. I. (2007) Optimal prediction of the peak of the next 11-year activity cycle and of the peaks of several succeeding cycles on the basis of long-term variations in the solar radius or solar constant. *Kinematics and Physics of Celestial Bodies* 23 (3), 97–100.
105. Mackey, R. (2007) Rhodes Fairbridge and the idea that the solar system regulates the earth's climate. *Journal of Coastal Research Special Issue* 50, 955–68.
106. Svalgaard, L., E. W. Cliver and Y. Kamide (2005) Sunspot cycle 24: smallest cycle in 100 years? *Geophysical Research Letters* 32, 1–4.
107. Badalyan, O. G., V. N. Obridko and J. Sýkora (2001) Brightness of the coronal green line and prediction for activity cycles 23 and 24. *Solar Physics* 199, 421–35.
108. Gray, L. J., J. Beer, M. Geller, J. D. Haigh, M. Lockwood, K. Matthes, U. Cubasch, D. Fleitmann, G. Harrison, L. Hood, J. Luterbacher, G. A. Meehl, D. Shindell, B. van Geel and W. White (2010) Solar influences on climate. *Reviews of Geophysics* 48, 1–53.
109. Damon, P. E. and C. P. Sonett (1991) Solar and terrestrial components of the atmospheric 14C variation spectrum In C. P. Sonett, M. S. Giampapa and M. S. Mathews (eds.), *The Sun in Time*. University of Arizona Press, Tuscon, 360–88.
110. Landscheidt, T. (1981) Swinging sun, 79-year cycle, and climatic change. *Journal of Interdisciplinary Cycle Research* 12 (1), 3–19.
111. McCracken, K. G., G. A. M. Dreschhoff, D. F. Smart and M. A. Shea (2001) Solar cosmic ray events for the period 1561–1994. 2. The Gleissberg periodicity. *Journal of Geophysical Research* 106 (A10), 21599–609.
112. Kahler, S. W. (2008) Prospects for future enhanced solar energetic particle events and the effects of weaker heliospheric magnetic fields. *Journal of Geophysical Research* 113, 1–16.
113. Demetrescu, C. and V. Dobrica (2008) Signature of Hale and Gleissberg solar cycles in the geomagnetic activity. *Journal of Geophysical Research* 113.
114. Feynman, J. and A. Ruzmaikin (2011) The sun's strange behavior: Maunder minimum or Gleissberg cycle? *Solar Physics* 272 (2), 351–63.

115. Mouradian, Z. (2002) Extended Gleissberg cycle. Proceedings of the Regional Meeting on Solar Physics 'Solar Researches in the South-Eastern European Countries: Present and Perspectives', 24–28 April 2001, Bucharest. Observations Solaires, 56–60.
116. Garcia, A. and Z. Mouradian (1998) The Gleissberg cycle of minima. *Solar Physics* 180, 495–8.
117. Archibald, D. (2011) The Current US Drought is not a Surprise. http://wattsupwiththat.com/2011/04/17/the-current-us-drought-is-not-a-surprise.
118. De Jager, C. and S. Duhau (2011) The variable solar dynamo and the forecast of solar activity: influence on terrestrial surface temperature. In J. M. Cossia (ed.), Global Warming in the 21st Century. Nova Science Publishers, Hauppauge, NY, 77–106.
119. Abdassamatov, H. I. (2010) The Sun Dictates the Blimate. ppt-presentation, http:/www.heartland.org/events/2010Chicago/program.html.
120. Murphy, G. and L. Hecht (2009) Deepest Solar Minimum in Nearly a Century: Goodbye Global Warming. http://www.larouchepac.com/node/9916.
121. Mörner, N.-A. (2011) Arctic environments by the middle of this century. Energy & Environment 22 (3), 207–18.
122. Abdusamatov, K. I. (2005) Long-term variations of the integral radiation flux and possible temperature changes in the solar core. Kinematics and Physics of Celestial Bodies 21 (6), 328–32.
123. Abdusamatov, H. I. (2012) Bicentennial decrease of the total solar irradiance leads to unbalanced thermal budget of the earth and the Little Ice Age. Applied Physics Research 4 (1), DOI: 10.5539/apr.v5534n5531p5178.
124. Bonev, B. P., K. M. Penev and S. Sello (2004) Long-term solar variability and the solar cycle in the 21st century. The Astrophysical Journal 605, L81–L84.
125. Lockwood, M. (2010) Solar change and climate: an update in the light of the current exceptional solar minimum. Proc. R. Soc. A 466, 303–29.
126. Barnard, L., M. Lockwood, M. A. Hapgood, M. J. Owens, C. J. Davis and F. Steinhilber (2011) Predicting space climate change. *Geophysical Research Letters* 38, 1–6.
127. Ahluwalia, H. S. and J. Jackiewicz (2012) Sunspot cycle 23 descent to an unusual minimum and forecasts for cycle 24 activity. *Advances in Space Research* 50 (6), 662–8.
128. Stewart, M. M., I. Larocque-Tobler and M. Grosjean (2011) Quantitative inter-annual and decadal June–July–August temperature variability ca. 570 BC to AD 120 (Iron Age–Roman period) reconstructed from the varved sediments of Lake Silvaplana, Switzerland. *Journal of Quaternary Science* 26, 491–501.
129. Max-Planck-Gesellschaft (2004) The Sun is More Active Now than over the Last 8000 Years. http://www.mpg.de/495993/pressrelease20041028.
130. Swanson, K. L., G. Sugihara and A. A. Tsonis (2009) Long-term natural variability and 20th century climate change. *PNAS* 106 (38), 16120–30.
131. Swanson, K. L. and A. A. Tsonis (2009) Has the climate recently shifted? *Geophysical Research Letters* 36, 1–4.
132. Lee, S.-K., W. Park, E. van Sebille, M. O. Baringer, C. Wang, D. B. Enfield, S. G. Yeager and B. P. Kirtman (2011) What caused the significant increase in Atlantic Ocean heat content since the mid 20th century? *Geophysical Research Letters* 38, 1–6.

133. Muller, R. A., J. Curry, D. Groom, R. Jacobsen, S. Perlmutter, R. Rohde, A. Rosenfeld, C. Wickham and J. Wurtele (submitted) Decadal Variations in the Global Atmospheric Land Temperatures. http://berkeleyearth.org/Resources/Berkeley_Earth_Decadal_Variations.
134. OSS Atlantic Multidecadal Oscillation (AMO). http://ossfoundation.us/projects/environment/global-warming/atlantic-multidecadal-oscillation-amo.
135. Easterbrook, D. (n.d.) Evidence of the cause of global warming and cooling: Recurring global, decadal, climate cycles recorded by glacial fluctuations, ice cores, ocean temperatures, historic measurements and solar variations. http://myweb.wwu.edu/dbunny/research/global/easterbrook_climate-cycle-evidence.pdf.
136. Easterbrook, D. J. The Looming Threat of Global Cooling. http://myweb.wwu.edu/dbunny/research/global/looming-threat-of-global-cooling.pdf.
137. Easterbrook, D. J. (2011) Geologic evidence of recurring climate cycles and their implications for the cause of global climate changes – the past is the key to the future. In: D. J. Easterbrook (ed.), *Evidence-Based Climate Science.* Elsevier, Oxford, 3–51.
138. D'Aleo, J. and D. J. Easterbrook (2011) Relationship of multidecadal global temperatures to multidecadal oceanic oscillations. In D. J. Easterbrook (ed.), *Evidence-Based Climate Science. Elsevier*, Oxford, 161–84.
139. Akasofu, S.-I. (2009) Natural Components of Climate Change During the Last Few Hundred Years. http://people.iarc.uaf.edu/~sakasofu/natural_components_climate_change.php.
140. Mochizuki, T., M. Ishii, M. Kimoto, Y. Chikamoto, M. Watanabe, T. Nozawa, T. T. Sakamoto, H. Shiogama, T. Awaji, N. Sugiura, T. Toyoda, S. Yasunaka, H. Tatebe and M. Moric (2010) Pacific decadal oscillation hindcasts relevant to near-term climate prediction. *PNAS* 107 (5), 1833–7.
141. Alexander, G. (1979) Prediction: warming trend until year 2000, then very cold. *St Petersburg Times* (USA). 1 January. http://stevengoddard.wordpress.com/2011/05/26/1979-before-the-hockey-team-destroyed-climate-science.
142. Easterbrook, D. (2006) The cause of global warming and predictions for the coming century. Geological Society of America Philadelphia Annual Meeting (22–25 October), *Abstracts with Programs* 38 (7), 235.
143. Zhen-Shan, L. and S. Xian (2007) Multi-scale analysis of global temperature changes and trend of a drop in temperature in the next 20 years. *Meteorol Atmos Phys* 95, 115–21.
144. Climate Charts and Graphs (2011) Pacific Decadal Oscillation (PDO). http://processtrends.com/images/RClimate_pdo_trend_latest.png.
145. Otterå, O. H., M. Bentsen, H. Drange and L. Suo (2010) External forcing as a metronome for Atlantic multidecadal variability. *Nature Geoscience* 3, 688–94.
146. Vásquez-Bedoya, L. F., A. L. Cohen, D. W. Oppo and P. Blanchon (2012) Corals record persistent multidecadal SST variability in the Atlantic Warm Pool since 1775 AD. *Paleoceanography* 27 (3), PA3231.
147. Kuhnert, H. and S. Mulitza (2011) Multidecadal variability and late medieval cooling of near-coastal sea surface temperatures in the eastern tropical North Atlantic. *Paleoceanography* 26 (4), PA4224.
148. Knudsen, M. F., M.-S. Seidenkrantz, B. H. Jacobsen and A. Kuijpers (2011) Tracking the Atlantic multidecadal oscillation through the last 8,000 years. *Nature Communications* 2 (178), 1–8.

149. Hetzinger, S., J. Halfar, J. V. Mecking, N. S. Keenlyside, A. Kronz, R. S. Steneck, W. H. Adey and P. A. Lebednik (2012) Marine proxy evidence linking decadal North Pacific and Atlantic climate. *Climate Dynamics* 39 (6), 1447–55.
150. Wyatt, M. G., S. Kravtsov and A. A. Tsonis (2012) Atlantic multidecadal oscillation and Northern Hemisphere's climate variability. *Climate Dynamics* 38 (5–6), 929–49.
151. Chylek, P., C. K. Folland, G. Lesins, M. K. Dubey and M. Wang (2009) Arctic air temperature change amplification and the Atlantic Multidecadal Oscillation. *Geophysical Research Letters* 36, 1–5.
152. Chylek, P., C. K. Folland, G. Lesins and M. K. Dubey (2010) Twentieth century bipolar seesaw of the Arctic and Antarctic surface air temperatures. *Geophysical Research Letters* 37, 1–4.
153. Della-Marta, P. M., J. Luterbacher, H. v. Weissenfluh, E. Xoplaki, M. Brunet and H. Wanner (2007) Summer heat waves over western Europe 1880–2003, their relationship to large-scale forcings and predictability. *Clim. Dyn.* 29, 251–75.
154. Titz, S. (2008) Im Wechselbad des Klimas. Spektrum der Wissenschaft, http://www.spektrum.de/artikel/960476 8, 54–60.
155. Wikipedia (2009) Monthly values for the AMO Index, 1856–2009. http://en.wikipedia.org/wiki/File:Amo_timeseries_1856-present.svg.
156. Olsen, J., N. J. Anderson and M. F. Knudsen (2012) Variability of the North Atlantic oscillation over the past 5,200 years. *Nature Geoscience* 5, 808–12.
157. Wikipedia Nordatlantische Oszillation. http://de.wikipedia.org/wiki/Nordatlantische_ Oszillation.
158. Latif, M., C. Böning, J. Willebrand, A. Biastoch, J. Dengg, N. Keenlyside and U. Schweckendiek (2006) Is the thermohaline circulation changing? *Journal of Climate* 19, 4631–7.
159. Keenlyside, N. S., M. Latif, J. Jungclaus, L. Kornblueh and E. Roeckner (2008) Advancing decadal-scale climate prediction in the North Atlantic sector. *Nature* 453, 84–8.
160. Dambeck, H. (2008) Forscher wetten 5000 Euro gegen pausierende Erderwärmung. *Der Spiegel*. http://www.spiegel.de/wissenschaft/natur/0,1518,553296,00.html.
161. Zanchettin, D., A. Rubino, P. Traverso and M. Tomasino (2008) Impact of variations in solar activity on hydrological decadal patterns in northern Italy. *Journal of Geophysical Research* 113, D12102.
162. Lockwood, M., R. G. Harrison, T. Woollings and S. K. Solanki (2010) Are cold winters in Europe associated with low solar activity? *Environ. Res. Lett.* 5, 1–7.
163. Meyers, S. R. and M. Pagani (2006) Quasi-periodic climate teleconnections between northern and southern Europe during the 17th–20th centuries. *Global and Planetary Change* 54, 291–301.
164. Berger, W. H. (2008) Solar modulation of the North Atlantic oscillation: assisted by the tides? *Quaternary International* 188, 24–30.
165. Thejll, P. A. (2001) Decadal power in land air temperatures: Is it statistically significant? *Journal of Geophysical Research* 106, 31,693–31,702.
166. Versteegh, G. J. M. (2005) Solar forcing of climate. 2: Evidence from the past. *Space Science Reviews* 120, 243–86.
167. Li, Y., H. Lu, M. J. Jarvis, M. A. Clilverd and B. Bates (2011) Nonlinear and nonstationary influences of geomagnetic activity on the winter North Atlantic Oscillation. *Journal of Geophysical Research* 116, 1–15.

168. Woollings, T., M. Lockwood, G. Masato, C. Bell and L. Gray (2010) Enhanced signature of solar variability in Eurasian winter climate. *Geophys. Res. Lett.* 37 (20), L20805.
169. Swingedouw, D., L. Terray, C. Cassou, A. Voldoire, D. Salas-Mélia and J. Servonna (2011) Natural forcing of climate during the last millennium: fingerprint of solar variability. *Climate Dynamics* 36 (7-8), 1349–64.
170. Alvarez-Ramirez, J., J. C. Echeverria and E. Rodriguez (2011) Is the North Atlantic Oscillation modulated by solar and lunar cycles? Some evidences from Hurst autocorrelation analysis. *Advances in Space Research* 47 (4), 748–56.
171. Helama, S. and J. Holopainen (2012) Spring temperature variability relative to the North Atlantic Oscillation and sunspots – a correlation analysis with a Monte Carlo implementation. *Palaeogeography, Palaeoclimatology, Palaeoecology* 326–8, 128–34.
172. van Loon, H., J. Brown and R. F. Milliff (2012) Trends in sunspots and North Atlantic sea level pressure. *J. Geophys. Res.* 117 (D7), D07106.
173. Osborn, T. (2011) North Atlantic Oscillation index data. http://www.cru.uea.ac.uk/~timo/datapages/naoi.htm.
174. Landscheidt, T. (1999) Solar Activity Controls El Niño and La Niña. http://www.john-daly.com/sun-enso/sun-enso.htm.
175. Latif, M. (2008) *Bringen wir das Klima aus dem Takt? Hintergründe und Prognosen.* Fischer Taschenbuch Verlag, Frankfurt a.M.
176. McLean, J. D., C. R. de Freitas and R. M. Carter (2009) Influence of the southern oscillation on tropospheric temperature. *Journal of Geophysical Research* 114, 1–8.
177. Leyland, B. (2010) El Niño/La Niña Effect (SOI) Predicts Global Cooling by the End of 2010. http://joannenova.com.au/2010/08/is-the-cold-weather-coming.
178. McLean, J. D. (2011) Our ENSO – temperature paper of 2009 and the aftermath. http://mclean.ch/climate/ENSO_paper.htm.
179. Wang, S.-Y., M. L'Heureux and H.-H. Chia (2012) ENSO prediction one year in advance using western North Pacific sea surface temperatures. *Geophys. Res. Lett.* 39 (5), L05702.
180. Cohen, J. and J. Jones (2011) A new index for more accurate winter predictions. *Geophysical Research Letters* 38, 1–6.
181. Leyland, B. (2011) Global Warming. http://web.me.com/bryanleyland/Site_3/Climate_Change.html.
182. IPCC (2000) IPCC Special Report Emission Scenarios, Summary for Policymakers. IPCC Special Report, http://www.ipcc.ch/pdf/special-reports/spm/sres-en.pdf.
183. National Academy of Sciences (2011) *America's Climate Choices.* The National Academies Press, Washington, DC.
184. IPCC (2007) Climate Change 2007: *The Physical Science Basis. Contribution of Working Group I to the Fourth Assessment Report of the Intergovernmental Panel on Climate Change.* Cambridge University Press, Cambridge and New York.
185. Ziskin, S. and N. J. Shaviv (in press) Quantifying the role of solar radiative forcing over the 20th century. *Advances in Space Research.*
186. Schmittner, A., N. M. Urban, J. D. Shakun, N. M. Mahowald, P. U. Clark, P. J. Bartlein, A. C. Mix and A. Rosell-Melé (2011) Climate sensitivity estimated from temperature reconstructions of the last glacial maximum. *Science* 334, 1385–8.
187. Kasang, D. Veränderung des mittleren Klimas. http://bildungsserver.hamburg.de/zukuenftige-klimaaenderungen/2081672/mittleres-klima-artikel.html.

188. Krivova, N. A., L. E. A. Vieira and S. K. Solanki (2010) Reconstruction of solar spectral irradiance since the Maunder minimum. *Journal of Geophysical Research* 115, 1–11.
189. Ljungqvist, F. C. (2010) A new reconstruction of temperature variability in the extra-tropical northern hemisphere during the last two millennia. *Geografiska Annaler*: Series A 92 (3), 339–51.
190. Loehle, C. (2007) A 2000-year global temperature reconstruction based on non-treering proxies. *Energy & Environment* 18 (7–8), 1049–58.
191. Loehle, C. and J. H. McCulloch (2008) Correction to 'A 2000-year global temperature reconstruction based on non-tree ring proxies'. *Energy & Environment* 19 (1), 93–101.
192. Skepticalscience.com (2011) IEA CO_2 Emissions Update 2010 – Bad News. http://www.skepticalscience.com/iea-co2-emissions-update-2010.html.
193. Easterbrook, D. J. (2008) Global Cooling is Here – Evidence for Predicting Global Cooling for the Next Three Decades. http://tinyurl.com/6r8tbz.
194. Archibald, D. C. (2007) Climate outlook to 2030. *Energy & Environment* 18 (5), 615–19.
195. Archibald, D. (2010) The Past and Future of Climate. http://www.davidarchibald.info, 142.
196. Appinsys.com (2009) Global Temperature Prediction from Recurrent Cycles. http://www.appinsys.com/GlobalWarming/PredictionFromCycles.htm.
197. Orssengo, G. (2010) Predictions of Global Mean Temperatures and IPCC Projections. http://wattsupwiththat.com/2010/04/25/predictions-of-global-mean-temperatures-ipcc-projections.
198. Casey, J. L. (2011) *Cold Sun*. Trafford Publishing, Bloomington, IN.
199. Landscheidt, T. (2003) New Little Ice Age instead of global warming? *Energy & Environment* 14 (2–3), 327–50.
200. Wang, S., X. Wen and J. Huang (2010) Global cooling in the immediate future? *Chinese Science Bulletin* 55 (33), 3847–52.
201. Jones, G. S., M. Lockwood and P. A. Stott (2012) What influence will future solar activity changes over the 21st century have on projected global near-surface temperature changes? *J. Geophys. Res.* 117 (D5), D05103.
202. Feulner, G. and S. Rahmstorf (2010) On the effect of a new grand minimum of solar activity on the future climate on Earth. *Geophysical Research Letters* 37, 1–5.
203. BMWi (2006) Verfügbarkeit und Versorgung mit Energierohstoffen. http://tinyurl.com/dxowmp.
204. Mann, M. E., Z. Zhang, M. K. Hughes, R. S. Bradley, S. K. Miller, S. Rutherford and F. Ni (2008) Proxy-based reconstructions of hemispheric and global surface temperature variations over the past two millennia. *PNAS* 105 (36), 13252–7.
205. Shindell, D. T., G. A. Schmidt, M. E. Mann, D. Rind and A. Waple (2001) Solar forcing of regional climate change during the Maunder minimum. *Science* 294, 2149–52.
206. Shindell, D. T., G. A. Schmidt, R. L. Miller and M. E. Mann (2003) Volcanic and solar forcing of climate change during the preindustrial era. *Journal of Climate* 16, 4094–107.
207. Waple, A. M., M. E. Mann and R. S. Bradley (2002) Long-term patterns of solar irradiance forcing in model experiments and proxy based surface temperature reconstructions. *Climate Dynamics* 18, 563–78.

208. Feulner, G. (2011) Are the most recent estimates for Maunder minimum solar irradiance in agreement with temperature reconstructions? *Geophysical Research Letters* 38, 1–4.
209. Wikipedia 2-Grad-Ziel. http://de.wikipedia.org/wiki/2-Grad-Ziel.
210. Kump, L. R. (2002) Reducing uncertainty about carbon dioxide as a climate driver. *Nature* 419, 188–90.
211. HadCRUT3 Datenquelle historische Temperaturen: HadCRUT3 (Monatsmittel werte und gleitendes 24-Monatsmittel).
212. IPCC (2012) Climate Change 2013: The physical science basis. summary for policymakers. *IPCC WGI Fifth Assessment Report*, First Order Draft. http://www.stopgreensuicide.com/SummaryForPolicymakers_WG1AR5-SPM_FOD_Final.pdf.

8. How climate scientists are attempting to transform society

Energy policy used to be decided by three pillars: economics, supply and environmental protection. But today European and German policies are predominantly decided by only one: climate protection. Every energy policy measure is now subject to the guiding narrative of climate protection. Supply reliability and economics have taken a back seat. Moreover, it all boils down to a lot more than just abstract energy policy measures; it is also about controlling the behaviour of the global citizenry. In 2008 IPCC Chairman Rajendra Pachauri asked people to eat less meat. There are probably good reasons for doing so, but protecting the climate had never been among them. Greenpeace advises citizens to eat vegetable quiche instead of roast pork and that we also refrain from travelling by air [1]. Professor Leggewie of Essen, a member of Hans-Joachim Schellnhuber's Advisory Council on Global Change (WBGU), recommends that Europeans refrain from eating asparagus out of season: 'The enjoyment of the pleasures of individual freedom must not lead to the blockade of third parties,' he said sombrely of those who he believes emit too much CO_2. 'Of course nobody has to buy asparagus from Chile during the winter when this involves excessive emissions of greenhouse gases' [2].

There is a growing tendency not to just leave it to more calls to act, but to interfere with people's lives through restrictive laws and regulations. That leading representatives of current climate policy are demanding we go even further, all in the name of a just cause, is illustrated by a high-level position paper titled *World in Transition – A Social Contract for Sustainability* (literal translation from the German: *Societal Contract for a Great Transformation*). The position paper, which Schellnhuber himself called a 'master plan for a transformation of society' [3], was released by the WBGU in April 2011 [4]. The nine-member WBGU is an influential advisory board to Chancellor Merkel on environment and climate protection. Among its functions the expert report examines some essentials for rapidly steering global society onto a climate-friendly development path, one that would quickly relinquish coal and nuclear energy. The most important feature of the society-transforming plan is a powerful green state that would be able to force sustainable lifestyles and consumption cutbacks on world citizenry.

These are the WBGU's core statements:

1. The current economic model ('fossil industrial metabolism') is normatively no longer tenable. 'The transformation to climate compatibility is ... morally as compelling as the abolition of slavery and the condemnation of child labour.' Decarbonizing the global economy has to proceed quickly. Going without nuclear energy and coal must be implemented simultaneously and immediately.
2. 'The WBGU views the sustainable global conversion of the economy and society as "The Great Transformation". The named central transformation fields of production, consumption patterns and lifestyles must be modified so that global greenhouse gas emissions over the course of the next decades drop to an absolute minimum and that climate compatible societies can begin to take over.'
3. All nations must put their self-interest aside and submit themselves to a new form of collective responsibility for the climate (Global Contract for Society for a Climate-Compatible and Sustainable Global Economic Order) 'The global citizenry agree to innovation expectations that are normatively bound to the postulates of sustainability, and in return give up the spontaneous wishes for maintaining the status quo. The guarantor of this virtual contract is the guiding state.'

4. 'The energy transformation to sustainability can only succeed when ... non-sustainable lifestyles, especially in the industrial and emerging countries, are stigmatized and frowned upon by society.'
5. In Germany, climate protection must become the state's overriding objective to which the decision-making of the legislature, executive and judicial branches are aligned. 'To institutionally anchor future interests, the WBGU proposes supplementing parliamentary legislative processes with a deliberative "Future Chamber". To prevent interest- and party-based interference, the make-up of such a chamber could be determined by random selection, e.g. a lottery process.'

The WBGU's assessment reveals that the rapid decarbonization of the global economy while going without coal or nuclear energy is a utopian target. It demands the highest levels of idealism, altruism from its citizens and society, and it expects a willingness to sacrifice that blows away any realistic dimensions of normal life. Therefore, it is no surprise that this cannot be realized through democratic means. Why should people voluntarily give up their claims to material prosperity and security?

Realistically, the WBGU shows that the decarbonization of society can be achieved only through the firm hand of an all-powerful state on both a national and international scale. This would be implemented principally by new institutions. Germany's WBGU calls for a Global Security Council for Sustainability. The suggested Future Council for Germany, for example, would supplement legislative authority, and would not be democratically appointed.

Although the WBGU calls for 'civil participation', it does so only if it serves to implement the objective of 'climate protection'. Or, as poignantly described by Professor Emeritus Carl Christian von Weizsäcker, who criticizes the WBGU, 'More democracy yes, but only if it serves our purposes' [5]. The climate protection objective of a powerful green state could in itself no longer be questioned. The demanded 'stigmatization of non-sustainable lifestyles' would rapidly be achieved. Those who do not agree with the sustainability movement would be marginalized and excluded from the new green state order, and that would include anyone sympathetic to today's industrial society.

The powerful green state would present a societal contract. The WBGU claims there is a general desire for climate protection and decarbonization. It bases this above all on the superior moral insight it believes it has gained through its expert knowledge. This puts the WBGU on a par with the state-philosophical tradition of Jean-Jacques Rousseau. As known from the history of western states, authoritarian and utopian Jacobinism had its roots in the concept of *volonté générale* – the general will of the people. The vision of the powerful green state is in the same tradition.

With respect to scale, the WBGU places the decarbonization of the global economy in the same league as the Neolithic and industrial revolutions. But it is not correct when it claims that the deliberate, planned, radical restructuring of the economic and social systems – as is the case with decarbonization – is without historical precedent. There are examples in recent history, such as, to some extent, the industrialization of the Soviet Union during the 1920s and 1930s, or the Great Leap Forward, as well as the Cultural Revolution, in Mao's China [6].

No matter whether planned or not, revolutionary transformations of large-scale economic systems always involve austerity and hardship for the generations who experience them at first hand. A radical transformation always means that productive economic structures are toppled and new ones have to be constructed. Each 'great transformation' for that reason is necessarily a period of investment and abstinence from consumption. This is clearly illustrated by history (e.g. through the 'social question' in the Industrial Revolution with its temporary immiseration of the working classes).

The price of the WBGU's utopian Jacobinism, which would achieve climate protection quickly, radically and through great sacrifice, is astronomically high. Democracy, freedom of choice and the right to material prosperity cannot be sacrificed. The same is true for the variant that Schellnhuber advocates: 'Perhaps we have to innovatively further develop the democratic institutions. Personally, I could well imagine assigning 10 per cent of all parliamentary seats to ombudsmen who would exclusively represent the interests of future generations' [7].

Fortunately, it is becoming increasingly clear that the driving force behind the 'Great Transformation' is stalling because global warming hasn't happened over the last 15 years and the number of respected scientific opinions (outside of the WBGU and the Potsdam Institute) agreeing that we are headed for a protracted phase of cooling or stagnation is growing. In the end, the entire basis for the legitimacy of the 'Great Transformation' could disappear completely. Until then, we will be forced to scale back a fundamental component of state decision-making ... in a word, democracy! Many of the decisions that have been made so far are irrevocable. The high costs serve to destroy our prosperity because this money is lost permanently and therefore is not available for other crucial areas such as education, infrastructure and development aid.

Throwing democratic achievements overboard in favour of protecting the climate has become fashionable, not only in Germany, but in other countries too. James Hansen, Director of the NASA Goddard Institute for Space Studies in New York, regularly and openly expresses his doubts that democracy can halt global warming. Not long ago he praised China's autocratic regime, calling it a beacon of hope. There, sustainable living could be decreed by law [2].

Hansen's sociopolitical stance became clear in 2008 when he endorsed Keith Farnish's book *Time's Up!*. In this eco-fascist work, calls are made to eradicate cities, blow up dams and carry out acts of sabotage in order to stop the greenhouse gas machinery: 'The only way to prevent global ecological collapse and thus ensure the survival of humanity is to rid the world of industrial civilization' [8]. Hansen's comment can be read on the website of *Green Books*: He writes, 'Keith Farnish has it right: time has practically run out, and the "system" is the problem. Governments are under the thumb of fossil fuel special interests – they will not look after our and the planet's well-being until we force them to do so, and that is going to require enormous effort' [9]. Hansen later backpedalled, claiming that he was not supporting anarchy. But in the end he only made sure that his government affiliation would no longer be used along with his name when advertising the book – a well-nigh meaningless gesture when compared to the dehumanizing demands mentioned.

This scarcely dented Hansen's reputation in the climate scene, as he's a 'rock star' among American climate scientists. His popularity and high reputation within the circles of environmental organizations is understandable: he always has the most radical prognoses and demands. This applies to his prognosis for the sea level rise (5 metres by the year 2100; see Chapter 5) or his demand that the atmospheric CO_2 concentration be reduced to 350 ppm (today it is 390 ppm) because such CO_2 concentrations have never existed over the last 400,000 years [10]. He concludes from this that the 2° C increase, the upper target of climate policy, had never been exceeded during this period. Here, it has long since been shown that CO_2 during the transitions between glacials and interglacials of the last million years followed natural temperature change.

As others have shown and as we have illustrated in this book, the 2-degree target in this century is not in danger. However, this EU-decreed target has been manoeuvring the ruling climate theory into a very awkward position. Undoubtedly, climate policy is now snared in its own trap – a seemingly inescapable dilemma. If the IPCC climate theory is correct, and at the same time the 2-degree target really has to be met, then the scale of the necessary sociopolitical intervention in society would be so great that it would be impossible to get the citizenry's backing. Consequently, the sociopolitical proposals for intervening in social and economic systems are becoming ever shriller and more extreme, as is clearly demonstrated by the WBGU.

If the IPCC were right, the 2-degree target would be impossible to attain because annual emissions are going to rise to 42 billion tons by 2030 no matter what we do, despite the strictest of reduction measures in Europe (see Chapter 9). This is mainly due to China's rapid CO_2 emissions increase. As Figure 8.1 shows, an annual emissions reduction of 9 per cent a year would have to take place beginning in the year 2020. But China has already declared that it will not implement any absolute emissions reduction until 2030. Therefore, it is simply impossible to meet the 2-degree target.

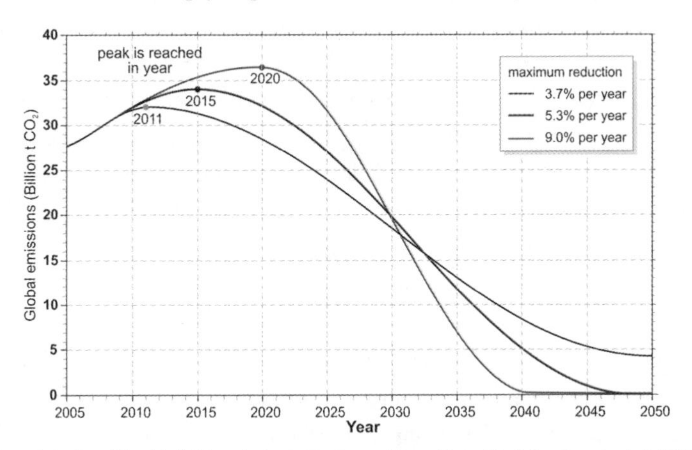

Figure 8.1 Possible global CO_2 emission paths the world would need to follow in order to fulfil the 2-degree target with a high probability, according to the WBGU [11].

It is thus illusory to believe that anthropogenic emissions could be reduced to zero between 2030 and 2040. How did Europe even come up with the 2-degree target in the first place? The economist William Nordhaus once determined that the global mean temperature over the last several hundred thousand years had never been 2° C warmer than it is today. And because of this, he postulated in 1975 [12] that this should also be the case in the future. Derived from data from the earth's history, he also suspected that a doubling of CO_2

would lead to a 2° C warming. Carlo Jaeger of the Potsdam Institute describes how the 2-degree target was arrived at: 'Nordhaus believed at the time that a warming of 2° C meant the same as doubling the pre-industrial CO_2 concentration (which we know today is based on false assumptions) and thus took this doubling as a preliminary benchmark. This is what a large majority of the modellers did for the IPCC as well, and they produced new estimations of the impacts of such a doubling over the following years' [13].

In 1995 Schellnhuber made the idea his own and through the WBGU demanded that the 2-degree target be adopted by European and German politicians. At the time the WBGU advised then Federal Environment Minister Angela Merkel, which led to far-reaching consequences. In 1996 the EU set the 2-degree target in the European Council of Ministers as the official target of European climate policy [13]. But agreement on binding steps to reduce emissions of greenhouse gases failed to materialize at the UN Climate Conference in Durban in December 2011, and UN members also failed to agree on concrete action in Doha in December 2012. Some EU states would love to formulate new (reduced) targets, but they are constrained because the influence of climate scientists, the IPCC and environmental organizations is too powerful in international climate politics, and any changes would be snuffed out in short order [14]. As a result, the EU is trapped by the climate scientists' generally accepted demand for an 80–95 per cent reduction in CO_2 emissions by the year 2050.

However, this is simply not going to work, nor is it necessary, as we have shown. According to Silke Beck of the Helmholtz Centre for Environmental Research, Leipzig, 'At stake is the very reputation of the part of climate science that provides statements of its findings to the public and claims to have the authority to define with respect to policy and the public' [15]. Climate scientists are about to find out that 'climate policy in western democracies is about politics' [14].

Hopefully, in the future a far less important role will be played by the climate Cassandras like Schellnhuber and Hansen, who are convinced that over the long term we ought to take the atmosphere back to a cooler state, like the one that prevailed in the Late Neolithic when humans were sedentary [7].

Politics will surely be aligned along the more realistic lines of citizens. In the end even Greenpeace will have to accept this. No organization has 'climate change' written across its flag more than Greenpeace. For 20 years this organization has been staging headline-grabbing, spectacular stunts against the use of fossil fuels. Greenpeace writes about its own director for renewable energies, Sven Teske: 'Sven Teske needs global dimensions and powerful opponents to be at his best. For more than half of his life he has clashed with industry bosses, politicians, and even people of his own kind. And not once has he ever lost sight of his target: bringing renewable energy to a breakthrough – and to do so worldwide' [16]. This occasionally included being on board the Greenpeace ship Argus as a campaigner at the coal-importing seaport of Rotterdam in order to block shipyard cranes and prevent the unloading of foreign coal freighters [17].

One really has to wonder that Teske was a lead author of the IPCC 2011 Special Report on the state of renewable energies [18]. Together with the European Association for Renewable Energies and the Deutsches Zentrum für Luft und Raumfahrt (German Centre for Aerospace and Aviation) Teske published a study titled *The Energy (R)evolutions Scenario* [19]. It is precisely this *Energy (R)evolutions* scenario that the 2011 Special Report quotes in many places. Teske's scenario also provides the opening sentence of IPCC press releases, namely, 'Almost 80 per cent of the worldwide energy supply could originate from renewable energy sources by the middle of the century – if political measures supported it' [20]. In reality, this figure is reflected in only one of 160 studies. According to the other studies, only an average share of 27 per cent will be achievable by the year 2050. Yet only Teske's 80 per cent value made it to the headlines in the media. As is the case for IPCC members, lead authors are selected by the governments, and Teske was chosen by the German government – specifically by the Federal Ministry of Environment and the Federal Research Ministry, according to *Der Spiegel* [21–22]. Climate policy thus gets formed as follows: Teske mutates to lead author of the IPCC's scientific report and then quotes himself. The result? What he had been demanding in various brochures since 2006 winds up becoming the IPCC's essential demand for world policy.

This is by no means an isolated incident as Donna Laframboise discovered [17, 23]. Many key IPCC players are closely connected to Greenpeace and WWF. For example, the lead author for a special IPCC report about carbon capture and sequestering (CCS) was Gabriela von Görne, who was also a reviewer of the 2007 report. As a climate expert on the Greenpeace website in 2004, she announced, 'As long as the course of the direction [the expansion of renewable energies] is not set, we have to keep our hands off CO_2 storage. Otherwise this will become a dangerous dead end: one carries on as normal, burns the coal, and thinks everything is OK. But it isn't ...' [24]. It is also telling that one of the lead authors of the 2007 IPCC report, Bill Hare, has been director of climate policy at Greenpeace since 2000 and advises the organization today. He was one of the authors that summarized the Synthesis Report, including the recommendations for policy makers [25]. Isn't it odd that only his project work at the Potsdam Institute for Climate Impact Research was mentioned in the author overview, and not his association with Greenpeace [26]. Hare has also been selected as a lead author for the next IPCC report. Malte Meinshausen is shown by many papers to have been responsible for Greenpeace until 2003 [25]. In 2005 he received his PhD and was immediately selected as author for three (!) chapters. He has been at the Potsdam Institute since 2006 and until 2011 led the team that dealt with the 2-degree target [27].

It is remarkable how a small group of scientists, through the UN, has managed to create the impression that *what they have discovered to be correct is the scientific consensus.* Worse, policy makers have been using these poorly based findings as guidance on policy.

That may work for a while, but reality is now destroying the theories that have long been the sacred cows of these scientists, who allowed themselves to become intoxicated by their political influence and whose views have long been closed by dogma.

References

1. Greenpeace (2007) Erste Hilfe für das Klima. http://www.greenpeace.de/themen/klima/klimawandel_aufhalten/artikel/erste_hilfe_fuer_das_klima.
2. *Frankfurter Allgemeine Zeitung* (2011) Die herzliche Ökodiktatur. 16 May. http://www.faz.net/artikel/C30770/gruene-revolution-die-herzliche-oekodiktatur-30337021.html.
3. *Der Spiegel* (2011) Leading climatologist on Fukushima: 'We are looting the past and future to feed the present.' 23 March. http://www.spiegel.de/international/germany/0,1518,752474,00.html.
4. WBGU (2011) Welt im Wandel: Gesellschaftsvertrag für eine Große Transformation. http://www.wbgu.de/veroeffentlichungen/hauptgutachten/hauptgutachten-2011-transformation.
5. von Weizsäcker, C. C. (2011) Die Große Transformation: ein Luftballon. 30 September. http://fazjob.net/ratgeber_und_service/beruf_und_chance/umwelttechnik/?em_cnt=120030&em_cnt_page=1.
6. von Weizsäcker, C. C. (2011) Die Große Transformation: ein Luftballon. *FAZ*, 29 September. http://www.faz.net/frankfurter-allgemeine-zeitung/wirtschaft/carl-christian-von-weizsaecker-die-grosse-transformation-ein-luftballon-11374355.html.
7. *Der Spiegel* (2010) Tritt in den Hintern. 16 August. http://www.spiegel.de/spiegel/print/d-73290108.html.
8. Farnish, K. (2009) *Time's Up! An Uncivilized Solution to a Global Crisis*. Green Books, Totnes.
9. Green Books (2009) http://www.greenbooks.co.uk/Book/206/Times-Up.html.
10. Hansen, J., M. Sato, P. Kharecha, D. Beerling, R. Berner, V. Masson-Delmotte, M. Pagani, M. Raymo, D. L. Royer and J. C. Zachos (2008) Target atmospheric CO_2: Where should humanity aim? *The Open Atmospheric Science Journal* 2, 217–31.
11. WBGU (2009) Solving the Climate Dilemma: The Budget Approach, http://www.wbgu.de/fileadmin/templates/dateien/veroeffentlichungen/sondergutachten/sn2009/wbgu_sn2009_en.pdf.
12. Nordhaus, W. D. (1975) Can we control carbon dioxide? IIASA Working paper WP-75-63, http://www.iiasa.ac.at/Admin/PUB/Documents/WP-75-063.pdf, 1–47.
13. Jaeger, C. C. and J. Jaeger (2010) Warum zwei Grad? http://www.european-climate-forum.net/fileadmin/ecf-documents/publications/articles-and-papers/jaeger_jaeger__warum-zwei-grad.pdf, Aus Politik und Zeitgeschichte: Klimawandel 32–3, 7–15.
14. Geden, O. (2010) Abkehr vom 2 Grad Ziel. Arbeitspapier der Forschungsgruppe 1, Stiftung Wissenschaft und Politik (SWP), 2010/02, http://www.swp-berlin.org/fileadmin/contents/products/arbeitspapiere/Arbeitspapier_2_Grad_Ziel_formatiert_final_KS.pdf.

15. Beck, S. (2009) *Das Klimaexperiment und der IPCC. Schnittstellen zwischen Wissenschaft und Politik in den internationalen Beziehungen.* Metropolis, Marburg.
16. Greenpeace (2008) Mit Energie geladen. http://www.greenpeace-magazin.de/index.php?id=4998, greenpeace magazin 1/08.
17. No Frakking Consensus (2010) Greenpeace and the Nobel-Winning Climate Report. http://nofrakkingconsensus.com/2010/01/28/greenpeace-and-the-nobel-winning-climate-report.
18. IPCC (2011) Special Report on Renewable Energy Sources and Climate Change Mitigation. http://srren.ipcc-wg3.de/report.
19. Teske, S., T. Pregger, S. Simon, T. Naegler, W. Graus and C. Lins (2010) Energy [r] evolution – a sustainable world energy outlook. *http://www.energyblueprint.info.*
20. IPCC (2011) Press release: Potential of Renewable Energy Outlined in Report by the Intergovernmental Panel on Climate Change. http://srren.ipcc-wg3.de/press/content/potential-of-renewable-energy-outlined-report-by-the-intergovernmental-panel-on-climate-change.
21. *Der Spiegel* (2011) Bericht zu Öko-Energien – Blogger werfen Weltklimarat Interessenkonflikt vor. 20 June. http://www.spiegel.de/wissenschaft/natur/0,1518,769471,00.html.
22. Teske, S. (2011) Der neue IPCC-Bericht unter Beschuss – weil ein Greenpeace-Angestellter Mitautor war? http://blog.greenpeace.de/blog/2011/06/22/der-neue-ipcc-bericht-unter-beschuss-weil-ein-greenpeace-angestellter-mitautor-war.
23. Laframboise, D. (2011) *The Deliquent Teenager Who Was Mistaken for the World's Top Climate Expert.* Ivy Avenue Press, Toronto.
24. Greenpeace (2004) CO_2-Speicherung ist das Letzte. http://www.greenpeace.de/themen/klima/nachrichten/artikel/co2_speicherung_ist_das_letzte.
25. No Frakking Consensus (2011) Peer into the Heart of the IPCC, Find Greenpeace. http://nofrakkingconsensus.com/2011/03/14/peer-into-the-heart-of-the-ipcc-find-greenpeace.
26. IPCC (2007) Climate Change 2007: Synthesis Report. IV.1 Core Writing Team members. http://www.ipcc.ch/publications_and_data/ar4/syr/en/annexessiv-1-core.html.
27. PIK (2011) Malte Meinshausen. http://www.pik-potsdam.de/members/mmalte.

9. A new energy agenda emerges

The most important factors impacting the development of the worldwide energy supply are a growing population and the simultaneously rising gross domestic product of emerging countries. Moreover, concern over a shortage of oil and gas plays an important role, as it risks leading to rising world market prices. From the perspective of the industrialized countries, dependency on energy imports is also a driving factor. Seventy per cent of oil reserves and 40 per cent of today's natural gas reserves are located within a strategic ellipse that extends from the Persian Gulf to Russia. Many geopolitically unstable countries are found there. The approaches for securing an energy supply vary from country to country. Currently, China is buying oil and gas reserves all over the world, from Iran to Sudan, Bolivia to Syria, and establishing access to other important raw materials and natural resources to secure its own supply.

Unlike China, the main driver of the energy agenda in Europe is climate change. In Europe, and especially Germany, the assumption that climate gases emitted by man will lead to a near uncontrollable global warming is the most important factor determining energy policy. The quest to avoid CO_2 emissions dominates while all other considerations are cast aside. How else does one explain why the EU has mandated a biofuel target of 10 per cent by 2020 without even considering the impacts that agricultural land shortages have on food crops, and why the domestic supply of fossil fuels such as lignite is being scaled back by CO_2 certificate trading? That and much else would be justified if this were truly successful in offsetting the effects of global warming. All this assumes that the fundamental claims that manmade CO_2 emissions catastrophically influence the climate are justified and that these EU measures are successful in the first place.

Today 87 per cent of global energy consumption depends on fossil fuel-based energy [1]. If the target proclaimed at the UN conference in Copenhagen to limit global warming to a maximum of 2° C is to be met no matter what the cost, then the industrial countries will have to reduce their emissions by at least 80 per cent by the year 2050. Yet, emerging countries such as China, currently by far the world's largest emitter, are not considering cutting back until 2030. With per capita emissions of 7.2 tons, China has already overtaken most countries (such as France at 5.7 tons per person) and will overtake the EU in 2014 and the United States by the end of the decade [2]. Climate policy even determines the distribution of economic growth, and with it growth in national prosperity over the coming decades. That is precisely the objective, says Ottmar Edenhofer of the Potsdam Institute for Climate Impact Research: 'Through climate policy, we are *de facto* redistributing global wealth. That the countries who own reserves of coal and oil are not keen on this is obvious. One has to disabuse oneself of the illusion that international climate policy is environmental policy' [3].

With regard to a reduction in CO_2 emissions, Europe has for the most part been ploughing a lonely furrow, accompanied only by Australia and New Zealand. These nations together represent 14.3 per cent of global emissions, while 85.7 per cent comes from countries that have no intention of limiting their emissions, among them China, India, the United States, Russia, Japan, Canada, Saudi Arabia, Korea, Brazil, Mexico, South Africa and Indonesia (Figure. 9.1). Their share rose in 2011 to 86.3 per cent.

One thing is certain: reducing CO_2 emissions is expensive. In his report commissioned by the UK government, Lord Stern proposed in 2006 that 1 per cent of the world's GDP should be used annually to stabilize atmospheric CO_2 concentrations at 550 ppm [5]. Two years later he increased his demand: 'In order to remain under 500 ppm ... it would cost approximately 2 per cent of the GDP' [6]. That's

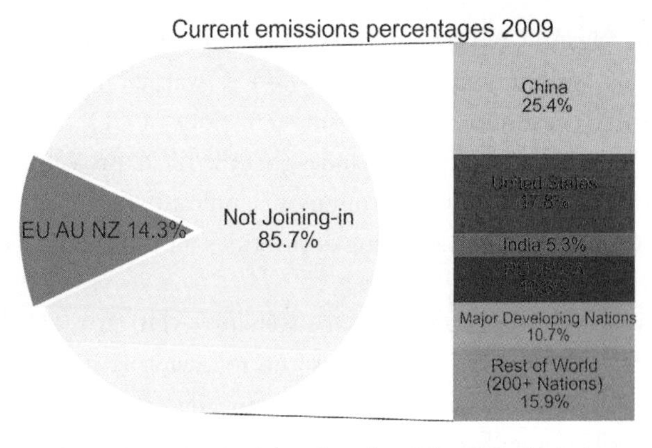

Figure 9.1 Share of emissions by Europe, Australia and New Zealand, who have announced a reduction in emissions while the rest of the world prefers not to commit [4].

an unimaginably massive figure! The German government alone would be saddled with a 50 billion euros a year bill [7]. For the EU as a whole annual outlays of 270 billion euros would be needed to achieve the corresponding CO_2 reduction by 2050 [8]. In addition, according to the Copenhagen conference, 100 billion dollars a year are to be transferred by 2020 to developing countries via a 'Green Climate Fund' [9]. The United Nations goes so far as to calculate that the developing countries need 500 billion dollars annually [10].

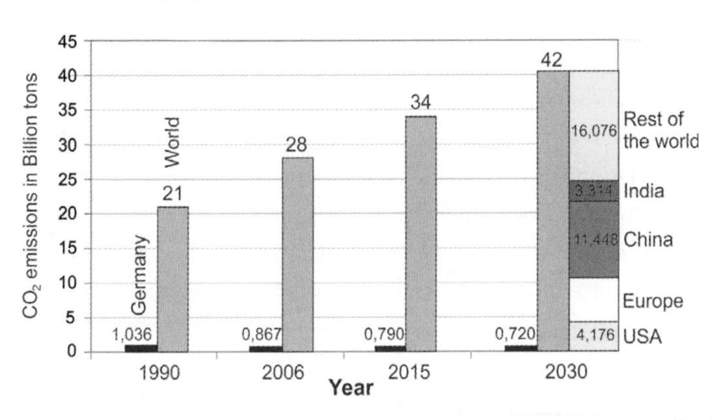

Figure 9.2 Projected increase in the annual CO_2 emissions rate [11].

This gigantic redistribution machine can be attributed to a single hypothesis: it is going to be 1.8°C–4° C warmer in the twenty-first century because of the increasing emissions of CO_2, according to the IPCC [12]. But if one concludes that the IPCC projections are unfounded and that only a warming of significantly less than 2° C is to be expected, then the setting of priorities for the energy agenda changes dramatically.

In the IPCC report, there is one scenario (A1B) that we believe is realistic and sensible. It includes the following [13]: 1) strong global economic growth of 3 per cent annually on average; 2) the world population will reach its maximum of approximately 9 billion by 2050 and then will drop back to 7 billion by 2100; 3) energy efficiency will climb each year at a rate of 1.3 per cent; and 4) a mix of energy technologies and resources will be employed.

In this scenario, annual CO_2 emissions will climb from approximately 25 billion to about 60 billion metric tons in the year 2050, and then will drop back to 40 billion metric tons by 2100. Note that other greenhouse gases are not considered here, nor are emission changes due to deforestation and altered land use. This development corresponds to the basic scenario of the International Energy Agency until the year 2030 [14], as well as projections from Shell and BP [1]. This will all lead to an atmospheric CO_2 concentration of about 700 ppm by the end of the century. However, in contradiction to what the IPCC assumes, this will not lead to a warming of about 2.8° C[13] [12], but only 0.85–1.25° C, according to our calculation (see Chapter 7). The energy agenda of the twenty-first century thus will no longer be driven solely by CO_2, which indeed becomes just a second-level parameter under the comprehensive ecological, economic and social demands on energy supply.

---

[13] With respect to the mean temperature of 1980–99.

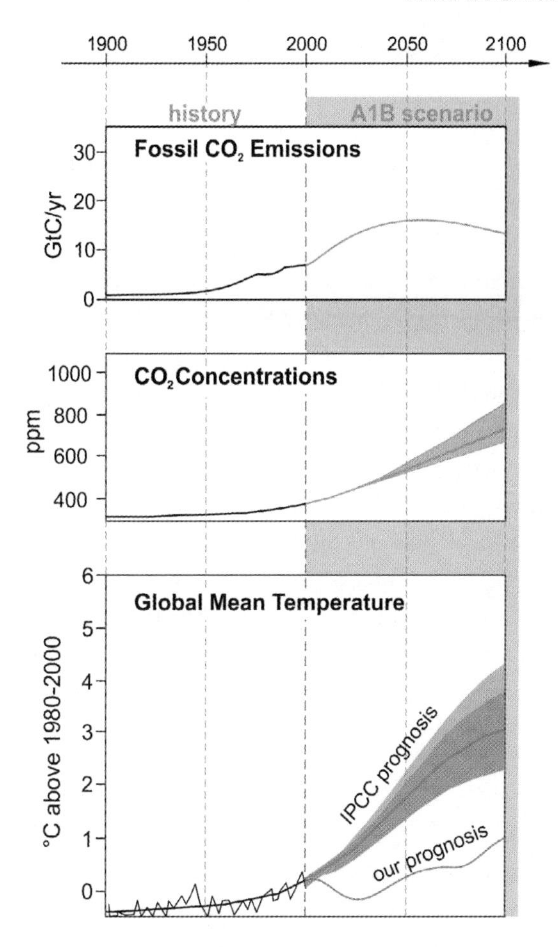

Figure 9.3 IPCC prognoses for emissions, CO_2 concentration in the atmosphere and global temperature. Our prognosis is added for comparison. Source: modified from AR4 [15].

How the world's hunger for energy can be satisfied

If one looks at the 2009 Shell energy scenario [16] for the year 2050, the 'Scramble', the first thing one notices is that there is growth in the end-use energy consumption in the emerging countries, namely China and South America, and, beginning in 2010, the consumption of energy in the OECD countries remains constant. This agrees with the latest estimates of the International Energy Agency, who show that 90 per cent of the growth will occur in non-OECD countries [17]. China will account for a third of additional kilowatt-hours and for half of the fuel consumption that will be added. Since 2011 China has become the greatest coal-importing nation in the world.

Interesting are the shifts in the energy mix itself. Coal and renewable energies in the Shell study cover the lion's share of growth. But what is remarkable for an oil company is the statement that a peak in oil supply is expected to occur between 2025 and 2030. The concentrations of climate gas CO_2 will shoot past 450 ppm, which the IPCC considers tolerable, and reach 550 ppm. But this in any case should not be considered troubling. In our opinion, from today's perspective, the value will probably reach 700 ppm without exceeding the 2° C temperature rise.

It is clear today how quickly prognoses for single fuels can shift because of technological breakthroughs, changed political conditions or discrete events. Just two years after the Shell study, Shell showed in its 'Signals and Signposts' report a shift in the energy mix to the detriment of coal and the benefit of natural gas [18]. What accounted for the switch? At the start of the decade all the talk was about a coal renaissance because the chances of capturing and sequestering CO_2 using CCS technology were becoming evident. At the end of the decade there were new signals from the EU indicating that coal would be very much burdened by the costs of CO_2 certificates (for climate protection reasons) to the point that no one would want to pay a penny for coal in Europe. Environmental groups and Green political parties were highly successful in preventing the development of modern coal power plants and the storage of captured CO_2 underground. Here low CO_2 coal technology could have led to CO_2 reduction, especially in coal-burning countries such as China and India, if only Europe had been able to show that electricity could be generated without economic disadvantage.

As soon as coal reached a dead end (and the horror of the Chernobyl nuclear power plant accident had long since faded) a renaissance of nuclear energy was ushered in. Nuclear energy came to

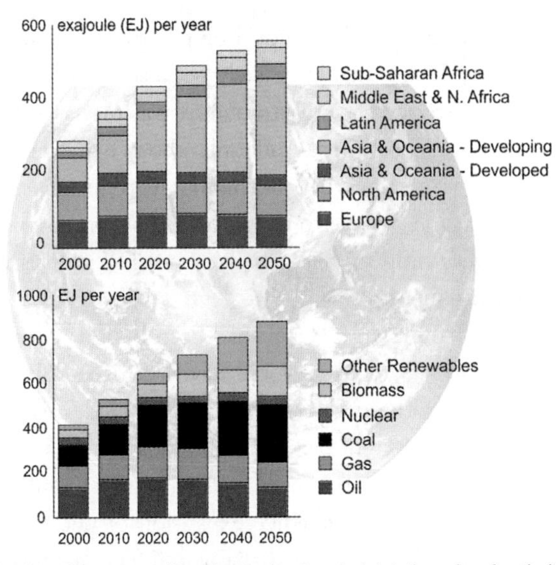

Figure 9.4 The demand for energy will continue to rise strongly. Asia is the region of particular growth. Renewable energies and coal will supply the demand until 2050 [16].

be viewed as a low-cost and CO_2-free source of energy. All European countries, with the exception of Austria and Denmark, began to plan new reactors. Even Sweden and Switzerland wanted to replace their existing nuclear power plants with new ones. In 2010 Germany reversed an earlier decision to abandon nuclear power, which had been enacted by the Socialist–Green coalition government in 2000. The 2000 law called for a globally unique limit on nuclear plant operating time of 32 years. The 2010 U-turn allowed the operating life of nuclear reactors to be extended by 8–14 years. At the same time the Netherlands extended the operating time of its only nuclear power plant to 60 years and planned to build two new nuclear power plants.

The nuclear renaissance persisted until a horrendous earthquake and tsunami struck Japan in March 2011, forcing several nuclear reactor blocks at the Fukushima nuclear power plant to be shut down. A few days later the German government announced it would abandon nuclear energy – this time permanently. Will other countries follow Germany's lead or drop plans to build more nuclear power capacity? This is highly unlikely because of the uniqueness of the Japan's catastrophe. The reactor failure was preceded by political failure. Building a protective wall only about 8 metres high in a region where a tsunami wave more than 10 metres high strikes every 30 years [19] had very little to do with the specific risks of nuclear power and everything to do with irresponsible politics and poor company management.

If the Fukushima power plant had been built to German safety standards, its nuclear reactors would have been the only things left functioning in an otherwise completely devastated area. We would have seen entirely different pictures being posted around the world. China wasn't the least bit bothered by the images streaming in from Japan and continued on a massive expansion of nuclear energy on the very day Fukushima petrified the world. The People's Republic wishes to satisfy its vast appetite for energy by multiplying its nuclear power generation capacity. Currently there are about 11 gigawatts on stream, but that will increase eightfold by 2020. By the year 2015 China will begin the construction of nuclear power plants with 40 gigawatts capacity. Germany, needless to say, doesn't want to entertain any thought of this.

Until recently, a remarkable symbiosis between nuclear and renewable energy had been the plan. As renewable energy, such as wind and solar energy, takes precedence over other forms of energy, the resulting fluctuations in their output would be compensated by so-called balancing power. When winds are strong, the electricity generated exceeds demand; therefore, other sources of power can be reduced. When the wind abates, the other power plants need to be fired up, and quickly. No technology can do this as well and on such a great scale as nuclear energy. Nuclear energy can be throttled down to 40 per cent of its capacity in a matter of minutes. Indeed, up to 10,000 megawatts of flexible power for balancing out the renewable energy output could be provided in Germany before the nuclear exit was announced in March 2011 (Figure 9.5).

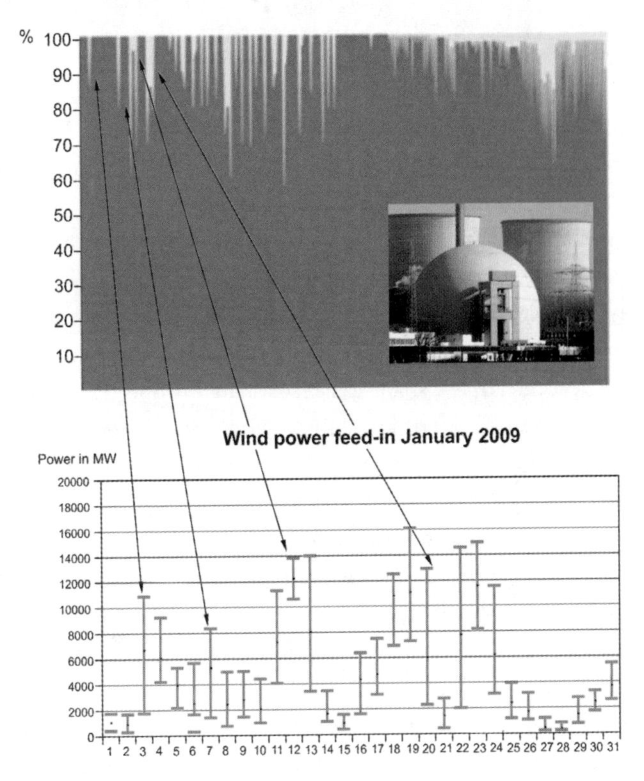

Figure 9.5 Nuclear power plants can be regulated within limits with respect to power output. During periods of strong wind power generation (lower diagram, all of Germany) nuclear power plants were simply throttled back; here as an example is the nuclear power plant in Esensham in January 2009 (upper characteristic line) [20].

If political acceptance for coal and nuclear energy reaches a dead-end in some industrial countries, there will be hope for other energy sources to step in and back up the renewable energies. The International Energy Agency is convinced 'that without a doubt natural gas will play a crucial role in the energy supply of the world' [21]. Forty-four per cent more natural gas will be in demand by 2035 than in 2008. At about 6 per cent a year, the demand in China is growing the most rapidly [14]. About 35 per cent of the global increase in natural gas production is due to unconventional reserves such as shale gas. And it is still unknown to many that, for a number of years, there has been a new technology in use for extracting natural gas, one that could revolutionize energy supplies. Shale gas has enabled the United States to become independent of the world gas market. Shale gas reserves have the potential to be climate friendly, even if it is difficult to extract.

One decisive advantage is its availability. A wide, prospective shale gas belt extends across Europe from the United Kingdom through the Netherlands and Germany to Poland. In northern and western Germany, especially in North Rhine Westphalia, one finds the second largest shale gas reserve in Europe. ExxonMobil estimates there are 2.1 trillion cubic metres of recoverable gas in North Rhine Westphalia alone. That's enough to meet Germany's gas demand for electric power and heating supply for 100 years. Pumping this gas would not be without considerable intrusion into the ground structure. The sub-surface rock formation would have to be shattered using shock waves to release the gas. This is known as fracking. And who, if not the famous Ruhr industrial region, would better know the risks of an underground intrusion for removing natural resources and bringing them to the surface? Unfortunately, a campaign quickly coalesced to prevent energy-rich North Rhine Westphalia from entering a new age of prosperity [22–24]. Critics also gained support outside of Germany. In June 2011, France decided to become the first country in the world to ban fracking of the bedrock altogether [25]. Could protecting the local nuclear industry there have anything to do with that?

More renewable energies, more low CO_2 natural gas and reduced CO_2 reduction targets by recalculating the faulty IPCC prognoses would give us hope that dramatically increasing energy consumption will not lead to an unacceptable temperature increase after all. Whatever energy mix we eventually reach, with more or with less nuclear energy, with coal power, with or without CCS, we will exceed the EU target stabilization level of 450 ppm CO_2 [26] (including other greenhouse gases) in the atmosphere. But because the 2° C increase is no longer at risk of being breached, in our view, the global community gains lots of room for manoeuvre for achieving a reduction in total CO_2 emissions and other climate gases in an optimum and reasonable way over the next 40 years. Nature has provided us with the time to achieve a sustainable energy supply, and this not only involves the generation of electricity.

The demand for oil for transport will also rise continuously. 'Total net growth comes from non-OECD countries, almost half from China alone where growth will come predominantly from its transportation sector' [14]. With this in the background, political knee-jerk decisions such as the introduction of biofuels, which are mainly due to the general hysteria surrounding CO_2, have to be regarded much more critically. The EU target is to supply 10 per cent of its fuel through bioethanol based on wheat, rye or sugar cane, including rapeseed oil as a replacement for diesel fuel. As is often the case when it comes to climate protection, Germany hastily enacted a law requiring that the share of biofuels be 6.25 per cent by 2010 and 8 per cent by 2015. The CO_2 saving from these measures, however, is highly doubtful. Because of the IPCC-induced horror scenarios, 20 per cent of Germany's agricultural areas are planted with energy crops today. The CO_2 avoidance effect of this measure is estimated to be trivial, yet very expensive. The avoidance cost for biodiesel is estimated to be almost 200 euros per metric ton of CO_2. For biodiesel based on wheat, the cost is pegged at about 500 euros per metric ton of CO_2, which is about as expensive as the avoidance costs of photovoltaic panels in Germany. Biomass from wood for generating electricity is, on the other hand, at 30–50 euros per metric tonne for avoidance cost, comparably inexpensive [27]. Furthermore, the Scientific Council of the European Environment Agency says that the energy savings calculation of biomass is flawed because you have to deduct the CO_2 that is produced by another use of land [28].

When one considers that the increase in fuel demand in China over the coming 25 years alone will be five times more than Germany's annual fuel consumption, and that the additional demand for fuel over half a year in China will more than offset the biofuel volume in Germany, one inevitably concludes that the hysterical climate debate in Germany urgently needs to return to rationality. There are in principle no objections to biofuel or biogas made of organic residual materials. However, Reinhard Hüttl, chairman of the Bio Economy Council of the German government, warns in no uncertain terms that the use of agricultural land in Germany is for the most part exhausted. Adverse effects on the soil and on land-lease prices are already profound [29]. Elsewhere, energy crops have already led to widespread deforestation of tropical rain forests and to increases in food commodity prices and thus social instability in poor countries.

But the sun is on our side and is offering us the time we need to organize everything sustainably in a reasonable way. There are very sensible solutions for drastically reducing automobile oil consumption: efficient diesel engine and Otto engine-powered automobiles, electro-mobility, natural gas and biogas as fuels. But time is needed for converting the vehicle fleet – time that climatologists have dogmatically refused to grant up to now. For natural gas, we need time to tap into the huge reserves; for electric cars we are only at the dawn of battery development and, as things stand, a hybrid solution (plug-in hybrids) will have to offset the disadvantages of the time-consuming charging of batteries. One million electric vehicles will be on the road in Germany by 2020. The number is planned to increase to six million by 2030 – still only just over 11 per cent of today's automobile count.

Renewable energy for generating power

Renewable energy technologies have written a new chapter in the history of energy over the last 20 years. In 1990 renewable energies were confined to the niches of research politics, for example, the German government attracted attention in 1989 with a 100 megawatt (!) programme for wind energy. In 1991 the Energy Feed-In Act followed, thus guaranteeing wind energy plant investors feed-in rates. After the 1992 Rio conference and the 1997 Kyoto Protocol, the subsidizing of renewable energy became a fixture of German politics. Many countries in Europe, and later the world, adopted the German model. In the early 1990s it was small manufacturing companies that drove the development of 150 kilowatt wind turbines into 1.5 MW turbines by the end of the decade. In 2002 the first 2 MW class of wind turbines came out, and just a few years later the 3 MW units followed. Today units as large as 6 MW of capacity have been tested and will be used for offshore wind farms.

By using larger rotors – currently up to 130 metres in diameter – and a total height of 180 metres, the cost of generating electricity has been reduced to less than a third and is currently down to 6–9 ct/kWh. At the beginning of 2011, 200,000 MW had been installed worldwide at a gigantic investment volume of 250 billion euros. Despite mass production and higher efficiency, costs over the last few years have not been reduced much as the requirements for wind plants have become considerably more stringent. Because wind turbines have become a major component in power generation in many countries, they, like all conventional power plants, have to fulfil technical requirements concerning grid compatibility, frequency stability and reactive power compensation. But the triumphal procession remains uninterrupted: wind energy will make a major contribution to fulfilling the ambitious goals of the EU, United States and China.

Photovoltaic systems have experienced a considerable drop in prices over the last 20 years. At the end of the twentieth century in northern Europe, electricity production costs for photovoltaic systems were 50 €ct/kwh, and so it was more of a niche system for use in isolated locations. Today prices have fallen by more than half to about 20 €ct/kwh. Although the cost basis is still far from competitive, photovoltaic systems have considerable potential as an intelligent solution for supplying regions in the sun-rich south that are remote from the grid. Today nearly two billion people have no access to a power grid. Here photovoltaic systems can be competitive compared to conventional energy, such as diesel-powered generators.

The worldwide expansion of renewable energies will depend on state subsidies for quite some time, as wind, biomass and solar energy are still not competitive compared to conventional energy sources. With the exception of a few geothermal power plants in Iceland, Italy and Turkey, only hydroelectric power is able to compete without subsidies.

Electricity prices are heavily dependent on gas prices because gas-fired power plants are the most expensive (other than oil-fired power plants) at times of peak demand and thus determine the price of power. Therefore, the expansion of renewable energy will depend on the price of natural gas in the future. The discovery of the huge reserves of shale gas worldwide will influence the growth of renewable energy in two ways: on the one hand, the fluctuating energy generation from solar and wind sources needs gas-fired power plants to balance out at times when the wind subsides and the sun doesn't shine. On the other hand, renewable energy will have a difficult time gaining a foothold in the market without subsidies if natural gas prices remain low. Naturally, the price of CO_2 is important here. In the European Union the price of CO_2 is less than 10 euros per metric ton. A further increase essentially depends on the worldwide CO_2 reduction targets.

Photovoltaic systems have the highest level of acceptance and the highest costs

Because of the generous subsidies for photovoltaic systems, today more than 30,000 MW of solar capacity are mounted on German roofs and as stand-alone systems. The amount of sunshine in Germany permits full use of the modules for about 700–1000 hours a year. Usually, feed-in is into the local power grid, which is not suitable for taking in a large amount of power on site. This requires considerable investment in street cable networks and corresponding transformer stations in order to be able to transport any surpluses.

So far, photovoltaic systems do not deliver any reactive power, which is necessary for long-distance transmission. Because the energy is paid to photovoltaic producers by all power customers for a period of 20 years, the amounts for the subsidies are piling up and are now currently more than 100 billion euros [30].

This is creating social disparity. As a rule, high income earners and homeowners invest more in solar panels. However, tenants and low wage earners, who cannot afford such systems or do not have roof space to install modules, end up paying for the massive subsidies via their electricity bills. This social transfer of wealth from the bottom to the top amounts to 6 billion euros a year in Germany alone. In addition, there is a cross-border transfer between states. This is how citizens in the state of North Rhine Westphalia pay more than 1 billion euros each year to solar power investors in Bavaria.

Photovoltaic systems today are subsidized at a rate that is five times more than the average wholesale market price. Today almost 50 per cent of the world's capacity is installed in grey and rainy Germany, and thus this has great consequences on power prices [31]. Every German household pays 180 euros a year in subsidies for renewable energy, half of it for photovoltaic [30]; 6 billion euros went to owners of roofs fitted with solar panels. If this were used to save CO_2 cheaply around the world, 600 million metric tons of CO_2 at a price of 10 euros per metric ton could be avoided. That would be two-thirds of Germany's entire annual total output. Much would have been achieved if these investments in solar energy had been made in countries where the sun shines three times longer, as in North Africa. There, for the same money, three times more CO_2 could have been avoided. That photovoltaic systems create secure jobs for Germany over the long term is also not the case. Today 70 per cent of the solar modules are imported, mostly from China and other Asian countries [32]. So under the green label, we have managed to force the average German wage earner (most not owning any solar modules) to finance the transfer of jobs to East Asia through higher electricity rates.

Our critical examination of the CO_2 avoidance targets does not mean we are arguing against energy efficiency and renewable energies. There are good reasons for expanding wind, hydropower, biomass and solar energy. As wind, hydro and solar power do not have to be imported, they can reduce our exposure to volatile fossil energy price fluctuations on global markets.

In the meantime the IEA is building on the greater use of renewable energies [14]. Power generation from these sources will triple between 2008 and 2035: 'The share of renewable energies in global power production will increase during this period from 19 per cent to almost a third, and thus will catch up with coal. This increase will mainly come from wind and hydroelectric power ... Power generation from photovoltaic systems is growing rapidly, even though its share of global energy production will reach only 2 per cent by 2035' [14].

Thus renewable energies will play an important role in European energy production [11]. This is why the contribution of renewable energies to Europe's energy supply (electricity, heat, transportation) will be at least 20 per cent by 2020. For the electric power sector, this will mean a share of 35 per cent. Also by 2020 greenhouse gas emissions in Europe will be 20 per cent lower than in 1990 and energy consumption will be reduced by 20 per cent through energy efficiency measures. The EU member states have strengthened this objective and backed them up with concrete measures in the national action plans they have submitted to the EU Commission. Only Belgium, Italy and Luxemburg have stated they will not be able to reach the targets on their own. Most states have reported they even intend to exceed their national targets (Figure 9.6).

It is still questionable whether all member states will reach the targets in 2020. A considerable shortfall for the savings amounts given in the national action plans can be expected. The shortfall results mainly because of the construction speed we have seen thus far due to the protracted approval processes. The intense discussions about the size of subsidies for renewable energy in Spain, the Netherlands,

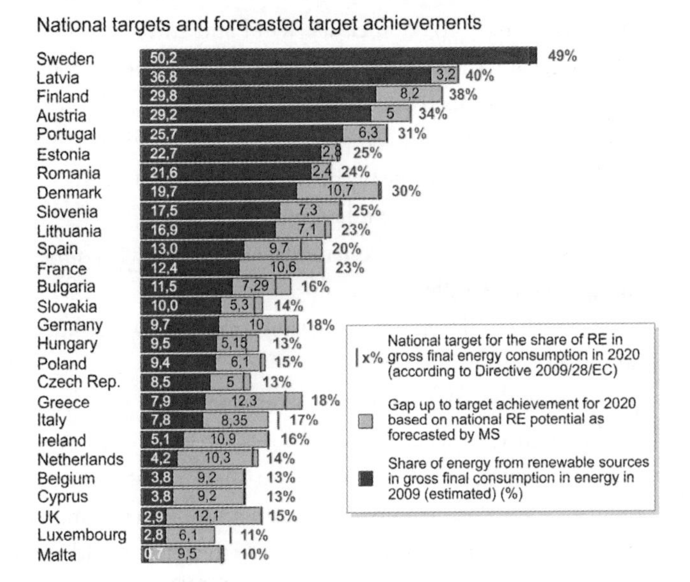

Figure 9.6 National targets for fulfilling the European 20 per cent target for the expansion of renewable energies compared to the year 2009 share [11, 33–34].

the Czech Republic and Greece only increase scepticism. The debt crisis does the rest. Even if the targets in each country are missed by 10–25 per cent, the construction rate for renewable energy systems in each country will have to be several times faster than the rates we've seen so far: Germany, France, Italy and Spain will need to double the speed, the Netherlands and Great Britain will need to triple it, and Poland will have to make its construction speed five times faster than it is at present.

Even more ambitious is the German government's proposed Energy Concept 2050 [35]. It postulates a greenhouse gas reduction of 80–95 per cent compared to 1990. This target can only be reached through enormous increases in efficiency, especially in buildings and transportation. In the electric power sector, the German government today has firmly decided what share each single energy generation type will have. Eighty per cent of the electric power generation will come from renewable energies by the year 2050. Natural gas and coal will only be used to compensate wind and sun shortfalls. That is a bold target for the world in 40 years. Even more remote from reality are the proposals from some political and societal forces such as the Greens, who deem the 80 per cent target is too half-hearted. It has to be 100 per cent to satisfy the demands of the Greens.

This leads us to pose the following questions:

1. How will grid power fluctuations associated with renewable energies be controlled?
2. How can the additional costs arising from renewable energy be minimized?
3. Will there be the political acceptance for the necessary infrastructure projects?

Without first having concrete answers to these questions, it would be foolhardy simply to give up our existing, well-functioning energy supply system. For this reason it is necessary to work on and research other alternatives such as nuclear fusion and CCS technology, as a full supply from renewable energies in the next decades is a long way from reality. Furthermore, the untapped potential of energy efficiency should be increasingly exploited.

More wind means more volatility

The lion's share of renewable energy construction will involve wind energy. Even if the full-capacity hours and the annual power production of each turbine of offshore parks increases, the problem of fluctuating feed-in power will remain and will only get worse. Figure 9.7 illustrates the current challenges for the current power plant park and stability of supply. During January 2010, over the course of one month, Germany's supply of wind power fluctuated between close to zero and 20,000 MW.

In addition to the volatile supply of wind, there is yet another effect that needs to be acknowledged. There is an asynchronous relationship between supply and demand that occurs repeatedly. This is illustrated in Figure 9.7, which depicts the first week of October 2008. The effect here serves as an example, and has in fact since been reproduced on numerous occasions. A strong supply of wind energy along with a weak demand for power occurs in the middle of the chart. Because of the oversupply of wind energy that occurred at this point, the price on the energy market sank to below zero.[14] This power was then 'given away' to other European countries, even though it had already been paid for by German power consumers.

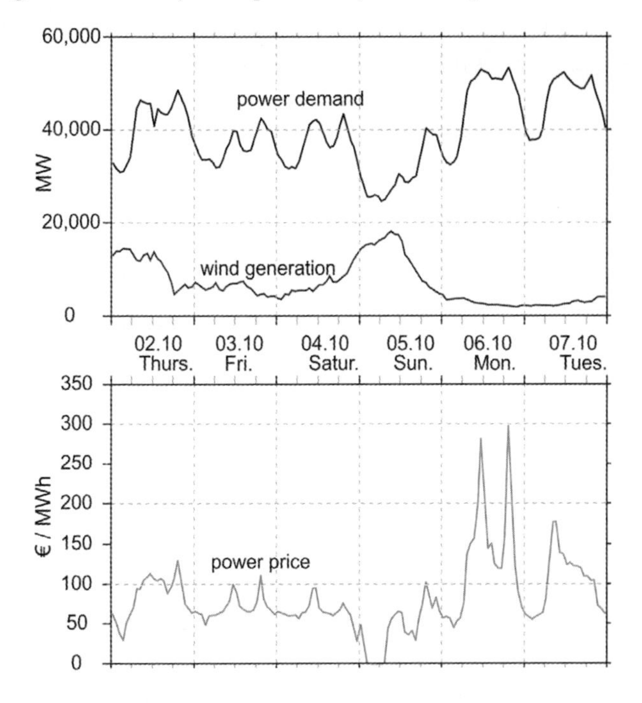

Figure 9.7 The black line shows the daytime and night-time power consumption (network load), the green line depicts the changing power generation from wind. If reduced demand occurs during the weekend and a high amount of wind energy is produced, then the price (lower graph, brown line) drops to near zero, or even becomes negative (negative power prices occurred during the course of the year 2008). Once the wind died down, the price skyrocketed because conventional power plants could not be fired up quickly enough (2–7 October 2008).

---

14 In October 2008, negative prices were not allowed on the power exchange market, which is why in this example the price could not fall below zero.

Storage systems are essential – capabilities must be realistically estimated

Because of volatility and the inability to plan wind energy, technical possibilities have to be created so that power surpluses occurring on certain days can be transmitted and stored, and then tapped later during calm days, which occur often enough in northern Europe. Here an improvement in the European power grid is essential so that power can be transmitted from power generation points to areas that need it. In other words, storage capacities must be expanded. Periods of a few days with low or no wind occur frequently. But what happens when there is a period of ten days with little wind, which occurs in Germany about twice a year? If we assume a somewhat reduced annual consumption of about 450 terawatt-hours (twh), then the average daily consumption is about 1.25 twh, which translates to 12.5 twh for a ten-day period. Currently in Germany the pump-storage power plant capacity is 7000 MW, with which about 0.04 twh (40,000 megawatt-hours) of power can be generated before it completely runs empty. That means in order to buffer a ten-day period of low wind, 312 (12.5/0.04) times more than today's available pump-storage capacity would first have to be installed.

For Germany and the region of the Alps, this is completely unrealistic due to the problem of public acceptance. Therefore, many studies focus on Norway, ignoring the question of whether Norway is ready to convert its landscape into a giant pump storage system on a scale that is capable of supplying Germany with power. Most storage possibilities currently involve dams with natural water feed-in that can deliver power when shortages arise, but are not available as storage basins in times of surpluses. In addition, if a need should arise due to a wind lull in Germany, Norway would first have to meet its own demand. Although this is just a simple example, it illustrates that potential storage in Norway is not at all easy and the difficulties cannot be underestimated.

Today there are other storage possibilities available, such as batteries for automobiles. Possible storage in electric cars, however, should not be overrated. The estimated one million cars that are expected in 2020 could consume 3 twh of electrical energy, which corresponds to 0.5 per cent of all electric power consumption. The storage capacity then is 10 gigawatt-hours. Ten years later the number could be six times higher. By then it will no longer be possible to reduce electrical power consumption further year on year. Rather, we will need more electric power in order to become less reliant on oil imports.

Another way to make surplus wind and solar energy available during windless and sunless days is to compress air into underground caverns. In times of surplus power, air would be compressed then later fed through turbines to produce electricity. But here we find ourselves only at the early stages of development. Worldwide storage technologies have not been given enough political attention. This applies to chemical storage systems that generate methane from CO_2 with the help of renewably produced hydrogen [36], or hydrogen electrolysis itself. Today all these technologies are inefficient and so would add to the current price of power. Finally, the ability to control demand for power is overestimated because the consumer will be unwilling to give up convenience for minimal savings, for example by delaying running the washing machine or dishwasher to times of surplus wind. For commercial businesses, this would be even more important.

Grid expansion and public acceptance are decisive factors

The current power grid in Germany is built around its existing power plants. There are no high voltage transmission lines in Germany that are longer than 60 km from junction to junction. There is an extraordinarily high level of grid stability: if one line goes down, others can take over the loads. In the future a greater share of power generation is going to come from wind energy in the northern regions of the country, onshore and offshore. Once the phase-out of nuclear energy is completed, more than 10,000 MW will have to be transmitted from northern to southern Germany. Three new direct current transmission lines will have to be built, which will take 10 years. However, bottlenecks will occur even earlier as 2800 km of new high voltage power lines will be necessary and will require an investment of about 20 billion euros. This does not take the necessary investments in low and medium voltage lines into account.

A decisive factor for comprehensively converting the energy supply system is time. The approval and construction time of a high voltage line is between 7 and 12 years, at best. This is foreseeable, because of the current unfinished construction of high voltage lines between eastern and western Germany, a steady power supply around Berlin, Hamburg and south Germany can no longer be assured. The network operator in the East German states (50 Hertz) already has to intervene in the generation of power 200 days a year in order to prevent the grid collapsing. Here the grid operators see 'the room to manoeuvre and the available measures for maintaining system stability as being mostly exhausted, especially in the winter half-year' [37–38]. Two high voltage lines have been in planning since German reunification 20 years ago: in the north, between Schwerin and Krümmel, and in the south through the Thuringia Forest – the so-called Rennsteig. But because of local protest and the snail's pace of the approval process, only the northern line had been completed by December 2012. For the Rennsteig line, it did not help that the planning was changed and the lines had to run underground as the proposed swathe through the forest was not acceptable to local citizens protest groups.

In addition to the investment volumes and protracted approval periods, yet another factor plays a central role: political and societal acceptance for all infrastructure projects. Included here are the production systems (wind, biomass and biogas, geothermal, hydroelectric power, etc.) as well as the networks and storage systems. Near Atdorf in the southern Black Forest, the largest German pump-storage system is now in the planning stages. The system is scheduled to begin operation in 2018. The water in the upper reservoir is to be fed through a 700-metre long, 7-metre diameter vertical hydraulic penstock which will deliver the water to an underground 1400 MW-rated capacity power plant. However, environmental groups, citizens and the local Green Party are mobilizing to oppose the project. There are similar protests against wind parks, biogas plants and geothermal power plants throughout the country.

Only when all elements of the energy supply infrastructure are accepted will it be possible to achieve the desired overhaul of the energy supply system. Even then an energy supply that is 80 per cent (let alone 100 per cent) from renewable energy by the middle of the century is unrealistic when one takes all these aspects into account. If we wish to cover half of Germany's needs with renewable energy, then the country faces an enormous, but achievable challenge if a sensible cost allocation is adhered to and not too much money is spent on the expansion of photovoltaic modules. The other 50 per cent could then come from low CO_2 emission and flexible power plants, which would supplement the renewable energies and provide the necessary reserves. Here a mix of low-emission coal and highly flexible gas-fired power plants would make sense. This would lead to a power generation that is low in CO_2, competitive and less dependent on imports. In a word: sustainable. Only when we can prove that an energy supply system conversion is achievable without losses in prosperity or jobs will countries such as China and India take the German energy route.

How do the other steps to a sustainable energy supply appear?

There is no doubt that the world's expanding population will lead to a rapidly increasing demand for energy. So the challenge is much greater. We do not only want the coming additional supply to be based on low CO_2 energy over the long term, but we wish also eventually to convert the existing supply. The warming of 0.8–1.3° C in this century with respect to today is a significant yet manageable change in climate. We have recognized that the technological development of renewable energies is increasing year after year, but the additional costs are still considerable. We also know that the grid conversion and development of storage capacity could take decades. If we add CCS from coal-fired power plants, which will also entail a great amount of time, as will the search for new shale gas reserves, then the conclusion is plain to see: to achieve everything without putting our prosperity in jeopardy, we need time.

But so far European energy policy has faced a dilemma: some powerful elements demand that the conversion be completed for the most part by 2020, and at the latest 2030, and this is based solely on ecological reasons. The UN global conference leaves no doubt that CO_2 emissions in the industrial countries must be cut by 20 per cent by 2020 and 80 per cent by 2050, and that too much is at stake to wait any longer. So wouldn't it be a welcome stroke of luck if it turned out that CO_2's share of the warming were only half as much, or quite possibly even less, and that the 2-degree target will certainly be reached, even if we allow ourselves more time to convert the energy supply system? Would it not be a wonderful coincidence if declining solar activity over the next decades dampened the warming, as a considerable number of renowned scientists now say is likely to be the case?

Would it not make more sense if conservation of resources once again gained in importance instead of hectic, knee-jerk show projects like burning food crops as biofuel, installing solar panels where the sun rarely shines or banning incandescent lights before a truly viable alternative is available (e.g. LED lamps)? Then transferring hundreds of billions for ineffectively reducing CO_2 emissions worldwide would be spared. Financial resources would be freed up to implement adaptation measures in climatically challenged regions (e.g. Bangladesh, the Sahel), which are already necessary due to the natural and anthropogenic warming.

This does not mean that the strategy of decarbonizing our energy supply is to be stopped *ad acta*. First, a warming of 0.8–1.3° C is a considerable change. Second, we cannot be certain to what extent the sun will have an impact in the second half of this century. But we do know that we have more time to achieve decarbonization by implementing new, renewable technologies, through higher energy efficiency, better material consumption and by improving the generation of conventional fossil energy in a rational, cost-effective way. Enormous efforts in research will be necessary to develop the technology. New technologies should be promoted through start-up financing, but have to survive in the market over the longer term without subsidies. To achieve emissions targets through market mechanisms, a global emissions trading with upper limits is completely sufficient. Because these upper limits would be allowed to rise over the course of the next decades, and only then would they have to be reduced, a lower CO_2 price would be reached. This would increase acceptance by all industrial and developing countries. Only a global market-economic trading with CO_2 certificates can ensure that limited financial resources will be used for effectively reducing emissions.

That the sun and its effects will usher in colder times is very likely. We cannot be certain whether or not the sun will even fall back into a Maunder-type minimum. And in the end we cannot influence whether a new Maunder-type minimum, partly compensated by CO_2-caused warming, would have less of an impact. We are simply forced to accept it. However, our ultimate duty is to supply nine billion people with enough affordable and readily available energy that is generated in a way that does not damage the environment or climate, and to produce sufficient potable water using water processing systems, and to ensure that enough areas are available to feed the population.

As is the case with every highly developed organism, there is no single solution; it comes down to the interaction of *all* sustainable energy sources, from renewable energy, including low-carbon technology and natural gas, including an inherently safe nuclear energy, or even nuclear fusion energy. Indeed the fusion energy of the sun is the source of our fossil and renewable energies. There's a lot out there indicating that the cold sun is giving us the time to unravel the secrets of the earth's power – fusion – and to use it without first exposing our planet to uncontrollable climate warming.

References

1. BP Europe SE (2011) *BP Energy Outlook 2030.* http://tinyurl.com/3fbopla, Bochum.
2. PBL Netherlands Environmental Assessment Agency (2012) Trends in Global CO_2 Emissions – 2012 Report, 6. http://edgar.jrc.ec.europa.eu/CO2REPORT2012.pdf.
3. *Neue Zürcher Zeitung* (2011) Klimapolitik verteilt das Weltvermögen neu. 14 November. http://www.nzz.ch/nachrichten/politik/schweiz/klimapolitik_verteilt_das_weltvermoegen_neu_1.8373227.html.
4. Hoskins, E. (2011) Worldwide CO_2 Emissions and the Futility of Any Action in the West. http://wattsupwiththat.com/2011/08/07/worldwide-co2-emissions-and-the-futility-of-any-action-in-the-west.
5. Stern, N. (2007) *The Economics of Climate Change: The Stern Review.* Cambridge University Press, Cambridge.
6. The Guardian (2008) Cost of tackling global climate change has doubled, warns Stern. 26 June. http://www.guardian.co.uk/environment/2008/jun/26/climatechange.scienceofclimatechange.
7. Google (2011) Weltentwicklungsindikatoren. http://tinyurl.com/42zk4bz.
8. *Frankfurter Allgemeine Zeitung* (2011) Der Klimaschutz kostet Billionen. 17 February. http://www.faz.net/artikel/C30770/plan-der-eu-kommission-der-klimaschutz-kostet-billionen-30327918.html.
9. United Nations (2011) World Economic and Social Survey 2011. http://www.un.org/en/development/desa/policy/wess/wess_current/2011wesspr_en.pdf.
10. United Nations (2009) World Economic and Social Survey 2009 – Promoting Development, Saving the Planet. http://www.un.org/en/development/desa/policy/wess/wess_archive/2009wess_overview_en.pdf.
11. Vahrenholt, F. and H. Gassner (2011) Erneuerbare Energien in Europa. *VGB Powertech* 1–2.
12. IPCC (2007) Climate Change 2007: Synthesis Report. http://www.ipcc.ch/publications_and_data/publications_and_data_reports.shtml.
13. IPCC (2000) *IPCC Special Report Emissions Scenarios.* Cambridge University Press, Cambridge.
14. International Energy Agency (2010) *World Energy Outlook 2010.* Organisation for Economic Cooperation and Development. http://www.iea.org/weo/docs/weo2010/weo2010_es_german.pdf.
15. IPCC (2007) *Climate Change 2007: The Physical Science Basis. Kapitel 10.2.1.1 The Special Report on Emission Scenarios and Constant-Concentration Commitment Scenarios.* Cambridge University Press, Cambridge and New York.
16. Shell (2009) Shell Energieszenarien bis 2050. http://www.shell.de/home/content/deu/aboutshell/our_strategy/scenarios_2050.
17. International Energy Agency (2011) *World Energy Outlook 2011.* Organization for Economic Cooperation and Development. http://www.iea.org/weo/docs/weo2011/es_german.pdf.
18. Shell (2011) Signals and Signposts – Shell Energy Scenarios to 2050. http://www-static.shell.com/static/aboutshell/downloads/aboutshell/signals_signposts.pdf.
19. NOAA-NGDC (2011) Tsunami Events Search. http://tinyurl.com/5szbhtx.
20. BDEW Bundesverband der Energie- und Wasserwirtschaft (2011) 1 February, Berlin, SP-V/Ba.

21. International Energy Agency (2010) *Energy Technology Perspectives 2010: Scenarios and Strategies to 2050*. Organization for Economic Cooperation and Development, Paris.
22. klimaretter.info (2011) Schiefergas klimaschädlicher als Kohle. http://www.klimaretter.info/forschung/hintergrund/8390-schiefergas-klimaschaedlicher-als-kohle.
23. Gegen-Gasbohren.de (2011) NRW-Umweltminister Remmel zum Thema Fracking. http://www.gegen-gasbohren.de/2011/08/09/nrw-umweltminister-remmel-zum-thema-fracking.
24. *Financial Times Deutschland* (2011) Zwangspause fürs Schiefergas. 1 August. http://www.ftd.de/politik/deutschland/:fracking-zwangspause-fuers-schiefergas/60086258.html.
25. *Le Monde diplomatique* (2011) Schiefergas. 8 July. http://www.monde-diplomatique.de/pm/2011/07/08.mondeText.artikel,a0026.idx,10 9540.
26. European Environment Agency (2010) Atmospheric Greenhouse Gas Concentrations (CSI 013) – Assessment. November. http://tinyurl.com/3u4p942.
27. Wissenschaftlicher Beirat Agrarpolitik beim Bundesministerium für Ernährung, L. u. V. (2007) *Nutzung von Biomasse zur Energiegewinnung - Empfehlungen an die Politik*.
28. European Environment Agency (2011) Opinion of the EEA Scientific Committee on Greenhouse Gas Accounting in Relation to Bioenergy. http://tinyurl.com/6eucukb.
29. BioÖkonomieRat (2010) Weltweit steigender Bedarf an Energie und Nahrungsmitteln erfordert Maßnahmen von Wissenschaft und Wirtschaft: BioÖkonomieRat weist auf verschärfte Nutzungskonkurrenz bei Biomasse auch in Deutschland hin Pressemitteilung 8 September. http://www.biooekonomierat.de/files/downloads/presse/BOER-Presseinfo02_10.pdf.
30. *Frankfurter Allgemeine Zeitung* (2010) Grüne Energie macht den Strom teuer. 11 July. http://www.faz.net/artikel/C30350/erneuerbare-energien-gruene-energie-macht-den-strom-teuer-30289308.html.
31. Weimann, J. (2009) *Die Klimapolitik-Katastrophe: Deutschland im Dunkel der Energiesparlampe*. Metropolis, Marburg.
32. Die Welt (2011) Stillstand und tiefrote Zahlen in der Solarbranche. 12 August. http://www.welt.de/wirtschaft/article13542122/Stillstand-und-tiefrote-Zahlen-in-der-Solarbranche.html.
33. Beurskens, L. W. M. and M. Hekkenberg (2011) *Renewable Energy Projections as Published in the National Renewable Energy Action Plans of the European Member States*. Energy Research Centre of the Netherlands and European Environment Agency. http://www.ecn.nl/docs/library/report/2010/e10069.pdf.
34. EurObserv'ER (2010) The State of Renewable Energies in Europe. http://www.eurobserv-er.org/pdf/barobilan10.pdf, 1–200.
35. Bundesministerium für Wirtschaft und Technologie (2010) *Energiekonzept für eine umweltschonende, zuverlässige und bezahlbare Energieversorgung*. http://www.bmu.de/files/pdfs/allgemein/application/pdf/energiekonzept_bundesregierung.pdf, Berlin.
36. Wikipedia (2011) Sabatier-Prozess. http://de.wikipedia.org/wiki/Sabatier-Prozess.
37. 50hertz (2011) Deutsche Übertragungsnetzbetreiber sehen erhebliche Auswirkungen des Kernkraftwerk-Moratoriums auf das elektrische System. *Pressemitteilung*, 22 May. http://www.50hertz-transmission.net/de/file/20110522_PM_Moratorium.pdf.
38. NTV (2011) Stromnetze sind 'am Limit'. 8 June. http://www.n-tv.de/politik/Stromnetze-sind-am-Limit-article3523506.html.

Glossary and abbreviations

2-degree target: The global temperature rise limit to 2° C, the target of international climate policy to prevent a catastrophic manmade disruption to the climate. The target was agreed at the 2010 Cancun conference.

Aa Index: Depicts the earth's geomagnetic activity, which is closely coupled to the activity of the sun's magnetic field.

Aerosol: Liquid or solid particles that are small enough to be suspended in the air. Because they reflect sunlight back into space and contribute to cloud formation, they often have a cooling effect on the earth's temperature.

AMO: Atlantic Multidecadal Oscillation, a fluctuation in the North Atlantic's current, which leads to a change in sea surface temperature. The AMO lasts 50–70 years and has cool and warm phases.

AR4: Fourth Assessment Report of the UN Intergovernmental Panel on Climate Change, 2007.

Cap & Trade: The producers of emissions are issued with certificates allowing them to emit a certain amount of greenhouse gas (e.g. CO_2) over a defined period. If the producer emits more than the allotted amount, additional certificates must be purchased. A decrease in the emission limits is achieved by reducing the number of certificates that are issued. The target is to cut emissions where it is most economical to do so. In the European Union, the EU Emissions Trading for CO_2 was enacted in 2005.

CCS: Carbon capture and sequestering, a technology to reduce the quantity of CO_2 emitted into the atmosphere and storing it.

CERN: Conseil Européen pour la Recherche Nucléaire, the European Organization for Nuclear Research, Geneva. With

3000 employees, it is the world's largest research centre for particle physics. It is funded by twenty European states.

CLOUD: Cosmics Leaving OUtdoor Droplets (the Cloud Experiment), An experiment in progress since 2006 at the CERN nuclear research facility. Directed by Jasper Kirkby, scientists are investigating the impact of cosmic rays on the formation of condensation nuclei (aerosols) in the atmosphere and thus cloud formation.

Cosmic rays: High energy particles (protons, electrons, ionized atoms) streaming from outer space which are mostly prevented from reaching the earth's atmosphere by the sun's magnetic field. The more active the sun, the fewer cosmic rays that penetrate the earth's atmosphere.

CRU: Climatic Research Unit, a facility at the University of East Anglia, Norwich. The CRU records land-based temperature measurements worldwide. They form the HadCRUT temperature series of the Hadley Centre, Exeter and have been providing data on the global temperature development since 1850.

Dalton minimum: A period of low solar activity occurring between 1790 and 1830 associated with a strong cooling of the earth's surface temperature.

Eddy cycle: A solar activity cycle with a mean length of 1000 years.

El Niño: A powerful warming of the upper water layer of the equatorial Pacific occurring irregularly every 2–7 years. High and low pressure zones switch location so that atmospheric and ocean currents in part reverse during an El Niño event. As a result, weather anomalies occur over much of the globe and result in pronounced warm peaks in the global temperature curve with magnitudes of 0.2–0.7° C.

ENSO: El Niño Southern Oscillation, an oceanic-atmospheric circulation system in the Pacific region involving the El Niño phenomenon and the Southern Oscillation (SO). The latter describes a pressure shift between the South Asian low pressure zone and the South Pacific high pressure zone.

FAR: First Assessment Report, the first climate assessment report issued by the IPCC in 1990.

Feedback: Self-amplification (positive feedback) or damping (negative feedback) of a process.

Fossil fuels: Coal, lignite, peat, natural gas and crude oil, all created from the biodegradation of plants and animals over geological time.

GISS: Global surface temperature series from the Goddard Institute for Space Studies of NASA based on measurement stations located globally.

Gleissberg cycle: A solar activity cycle with a mean length of 87 years.

GRACE: Gravity Recovery And Climate Experiment; satellite-supported measurement of the earth's gravitational field. Based on changes of the gravitational field over time, an attempt to determine changes in the polar ice mass is made.

HadCRUT: Global surface temperature data series of the Hadley Centre in Exeter (ocean data) and the Climatic Research Unit of the University of East Anglia (land-based temperatures).

Hale cycle: A solar activity cycle with a mean length of 22 years.

Hallstatt cycle: A solar activity cycle with a mean length of 2300 years.

Hockey stick: Temperature reconstruction made by Michael Mann and colleagues for the last 1000 years depicting a temperature development that was more or less constant before the Industrial Revolution, and then rose sharply. The curve served as the mainstay of the IPCC's Third Assessment Report (TAR) of 2001. Data and statistical analysis methods used for the hockey stick's construction were later shown to be flawed and thus the curve is no longer considered valid.

Holocene: The current interglacial time, i.e. the postglacial time. It began after the last ice age ended 12,000 years ago.

Ice age: Over the last two million years the earth has been steadily switching between ice ages (glacial phases lasting about 90,000 years) and warm phases (interglacials lasting about 10,000 years). The last ice age ended about 12,000 years ago. The earth is currently in an interglacial.

Interglacial: A warm period between the ice ages.

Internal oceanic oscillation: Rhythmic fluctuations in sea surface temperatures, atmospheric pressure relationships and water current of oceanic regions having imprecise lengths ranging from a few years to many decades. It is unknown if it involves pure self-oscillations of the climate system or if external factors such as solar activity have an impact. Examples are the PDO, AMO, NAO and ENSO.

Ionosphere: The earth's atmosphere 80–300 km in altitude where the sun's UV radiation generates large quantities of ions and free electrons.

IPCC: Intergovernmental Panel on Climate Change, an advisory group established by the United Nations.

Isotopes: Atoms of the same elements, but having different mass numbers.

La Niña: Cold counterpart of the El Niño. La Niña has impacts over the globe and results in cold peaks in the earth's temperature curve.

Little Ice Age: A natural cold phase lasting from the fifteenth to the nineteenth centuries, which coincided with strongly reduced solar activity.

Maunder minimum: A period of profoundly reduced solar activity from 1645 to 1715, which coincided with the peak of the Little Ice Age.

Medieval Warm Period: A natural warm phase that occurred from the ninth to the fourteenth centuries and had a temperature level similar to that of today. This warm period coincided with a time of increased solar activity.

Milankovitch cycles: Periodic cyclic changes in the earth's orbital parameters with periods ranging from 10,000 years to 100,000 years which have profound long-term impacts on the earth's climate.

Millennium cycles: Natural, long-period climate variability with typical cycle lengths of 1000–2500 years. In many cases the climate runs parallel to the changes in solar activity.

Moving average: The smoothing of greatly varying data sets and their curves by computing the average values of successive data points.

Muon: An elementary particle created by a reaction between cosmic rays and elements of the atmosphere at an altitude of approximately 10 km. Muons are suspected of playing a decisive role in low-level cloud cover and thus to be involved in controlling the climate.

NAO: The North Atlantic Oscillation. It describes the rhythmic alternation of the atmospheric pressure difference between the Iceland low and the Azores high. Cycle lengths ranging from just a few years to several decades occur. The NAO to a large extent affects the winds and the climate over Europe.

NASA: National Aeronautics and Space Administration of the United States.

Neutron monitor: Neutron measurement system for determining the intensity of cosmic rays.

NOAA: National Oceanic and Atmospheric Administration of the United States, responsible for studying weather and oceans.

Nuclide: An atomic nucleus characterized by its number of protons and neutrons. Different nuclides of the same chemical element (i.e. having the same number of protons but different number of neutrons) are designated as isotopes of this element.

PDO: Pacific Decadal Oscillation. This is the change in the surface temperature of the north Pacific region; it has a decisive impact on global average temperature. The PDO lasts 40–60 years, having warm and cold phases.

PETM: Paleocene–Eocene Temperature Maximum. An abrupt temperature increase occurring about 55 million years ago the cause of which is still unknown. It is interpreted by some as analogous to the increase in CO_2 today.

PIK: Potsdam Institute for Climate Impact Research.

Postglacial: The last 10,000 years of relative warmth since the end of the last ice age.

ppm: Parts per million.

Proton: An elementary particle with a positive charge.

Proxy: An indirect indicator of the climate, such as a temperature reconstruction from isotopes of corals, ice cores or dripstones.

Radiative forcing: A term introduced by the IPCC to describes the degree an external disturbance of the earth's radiative budget has on the climate system. This can include variations in the irradiative intensity of the sun, changes in the concentration of greenhouse gases or aerosols, or changes in surface reflection (albedo) of the earth's surface. Radiative forcing is expressed in watts per square metre (W/m^2).

Renewable energy: Energy from sources that are self-renewing or that are endlessly available, such as hydropower, solar energy, wind energy, geothermal, tidal currents or biomass use.

RSS: Satellite supported global temperature data set generated by Remote Sensing Systems, Santa Rosa, CA.

SAR: Second Assessment Report of the IPCC, released in 1995.

Schwabe cycle: Solar activity cycle with a mean length of 11 years.

Sensitivity: Climate warming resulting from the doubling of atmospheric CO_2 concentration.

SIM: Spectral Irradiance Monitor. A measuring instrument on NASA's SORCE satellite which records the flux of individual solar irradiation types coming towards the earth.

Stratosphere: The layer of the earth's atmosphere above the troposphere at an altitude of 15–50 km.

Suess/de Vries cycle: Solar activity cycle with a mean length of 210 years.

Sunspot: A dark, cooler spot on the sun's surface accompanied by brighter zones, so-called faculae. The number and size of sunspots is the simplest measure of solar activity. The greater the number of sunspots, the greater the level of solar activity.

TAR: Third Assessment Report of the IPCC released in 2001.

Troposphere: The lowest atmospheric layer of the earth, up to 15 km in altitude. Some 90 per cent of the air is found in the troposphere and almost all water vapour, which means most weather events take place here.

TSI: Total Solar Irradiance; the total irradiation from the sun reaching the upper edge of the earth's atmosphere over 1 square metre. It includes all wavelengths.

UAH: Satellite supported global temperature data set gathered by the University of Alabama, Huntsville.

UV: Ultraviolet radiation; electromagnetic radiation that is invisible to the naked eye and with a wavelength shorter than that of visible light.

Volatility: Variations in the generation of solar and wind energy, or in the price of electric power.

WBGU: Advisory Council on Global Climate Change for the German government. The WBGU is an independent scientific advisory council and recommends action and research concerning global environmental and development problems.

WWF: World Wildlife Fund.